INTRODUCTION TO SOLAR PRINCIPLES

Thomas E. Kissell
Terra Community College
Fremont, Ohio

Prentice Hall

Boston   Columbus   Indianapolis   New York   San Francisco   Upper Saddle River
Amsterdam   Cape Town   Dubai   London   Madrid   Milan   Munich   Paris   Montreal   Toronto
Delhi   Mexico City   Sao Paulo   Sydney   Hong Kong   Seoul   Singapore   Taipei   Tokyo

Editorial Director: Vernon R. Anthony
Acquisitions Editor: David Ploskonka
Development Editor: Dan Trudden
Editorial Assistant: Nancy Kesterson
Director of Marketing: David Gesell
Marketing Manager: Kara Clarke
Senior Marketing Coordinator: Alicia Wozniak
Senior Marketing Assistant: Les Roberts
Project Manager: Holly Shufeldt

Art Director: Jayne Conte
Cover Designer: Suzanne Behnke
Cover Image: iStockphoto
Full-Service Project Management Composition:
    Moganambigai Sundaramurthy, Integra
Printer/Binder: Edwards Brothers
Cover Printer: Lehigh-Phoenix Color
Text Font: Palatino

Credits and acknowledgments borrowed from other sources and reproduced, with permission, in this textbook appear on the appropriate page within the text. Unless otherwise stated, all artwork has been provided by the author.

Library of Congress Cataloging-in-Publication Data

Kissell, Thomas E.
    Introduction to solar principles/Thomas E. Kissell.
       p. cm.
    Includes bibliographical references and index.
    ISBN-13: 978-0-13-510385-2
    ISBN-10: 0-13-510385-1
    1. Photovoltaic cells—Textbooks.   2. Solar energy—Textbooks.   I. Title.
TK8322.K57 2012
621.47—dc22

                                                    2010046855

10 9 8 7 6 5 4 3 2 1

Prentice Hall
is an imprint of

www.pearsonhighered.com

ISBN 10:   0-13-510385-1
ISBN 13: 978-0-13-510385-2

Dedication

I would like to dedicate this book to my wife Kathy, my brothers Jerry, Bruce, and Rich, and my sister Pat, who helped shape my early life.

PREFACE

Introduction to Solar Energy covers all aspects of how small and large solar photovoltaic panels operate and how other types of solar energy systems are used for solar lighting, heating water, and heating residential and commercial buildings. It is written specifically for students who want to learn enough about solar energy to enter the job market as a solar energy technician in sales, installation, or repair. It also provides enough content information so students can understand concepts for installing and troubleshooting all types of solar energy equipment and solar photovoltaic panels in particular. This book provides enough detail to give technicians the knowledge they need to handle even the most complex installations and maintenance tasks. It is clearly written and starts out with simple concepts and continues through the most complex issues. Each chapter contains photos and diagrams that help explain basic and complex concepts. The text explains how basic solar energy panels convert energy from the sun into hot water that can be used to supplement the hot water heater in a residence or commercial location, or it can be used as part of the main heating system. The text also explains the large variety of materials that are used to make the solar photovoltaic panels that convert energy from the sun into DC electrical power. Additional information is provided about electrical and electronic circuits used in inverters to convert DC voltages to AC voltage at 60 Hz frequency and battery charge controllers that are used to control the DC electrical power used for charging batteries. Many photos and diagrams are included with each topic, and all math and data are provided in tables so the reader does not need calculus or other mathematics to understand the concepts. The chapters in the text can be used individually or in consecutive order to meet the needs of all solar energy curricula.

ORGANIZATION OF THE TEXT

The first chapter provides an overview of the current solar energy industry and the history of solar energy systems used for heating, lighting, and converting the sun's energy into DC electrical power with solar photovoltaic panels. Chapter 1 also explains how to finance solar energy projects and discusses return on investments for current solar projects. This chapter also covers the types of job opportunities in the solar field.

Chapter 2 explains all the industrial, commercial, and residential electrical energy demands of the United States and how energy from solar energy fits into the supply of green energy. This chapter helps students understand all the variables about solar energy and how solar photovoltaic energy is used as a stand-alone electrical system and how it is used for grid-tied systems.

Chapter 3 explains the different ways solar energy is used in today's technical applications to provide heat and light for residential and commercial buildings. This chapter also explains in detail how these types of solar energy systems are as important as solar photovoltaic panels. Large solar parabolic trough systems that provide industrial-scale electrical generation and Stirling engines are also explained, as well as all types of solar photovoltaic panels and materials that panels are made of. Information about connecting solar panels in series and parallel to provide arrays with specific voltage levels and current capabilities is provided. Different types of solar panel mounting and tracking systems are also explained.

Chapter 4 explains how sites on which to install solar energy equipment and solar photovoltaic panels are selected. This chapter also covers the largest solar photovoltaic panels installations and solar farms located across the United States, as well as some of the largest companies that work in the solar industry today.

Chapter 5 explains the basic scientific principles of how photovoltaic materials convert sunlight to DC electrical power. The chapter explains how basic P-type and N-type materials are created by combining different types of materials into solar photovoltaic material and then how this material is assembled into solar panels and solar arrays.

Chapter 6 explains how solar panels and solar material are mass-produced to make large numbers of solar photovoltaic panels. A detailed step-by-step explanation of how automation processes are used to manufacture solar panels is presented.

Chapter 7 provides detailed information about electrical concepts such as voltage, current resistance, and power as they relate to photovoltaic panels. This chapter explains electrical concepts such as series and parallel circuits, true power, apparent power, and power factor, which will help students better understand problems that can occur when electrical power from solar photovoltaic panels is connected to the grid and when it is used to provide electrical power for residential and industrial applications.

Chapter 8 explains how electronic inverters are used to convert the DC electrical power produced by photovoltaic panels to AC voltage at 60 Hz. This chapter also explains the operation of DC battery charge controllers that are used to ensure that the electrical power used to charge batteries is controlled so batteries do not over charge.

Chapter 9 covers the different types of wet cell and dry cell batteries that are used for storage for electrical power for solar energy systems. This chapter also explains in detail all the terms associated with the different types of batteries. Information is provided about advantages and disadvantages for the different types of batteries available for use with solar energy systems, as well as information about how to select the type of battery and the size of battery for specific solar photovoltaic panels.

Chapter 10 provides a complete explanation of the electrical grid in its present form, and it also provides information about the new "Smart grid" and how it will help make electrical transmission more efficient. This chapter provides in-depth details of the equipment and switch gear used to connect the electrical power from solar photovoltaic panels to the grid. Connections between solar photovoltaic panels on solar farms and substations and power distribution for underground and overhead service are explained.

Chapter 11 explains the steps for installing residential and commercial solar photovoltaic panels. This chapter also provides detailed information about all the steps involved in installing solar panels on the roof of a large industrial site and connecting all the electrical switches, inverters, and transformers for the system. Information about troubleshooting processes and how to specifically troubleshoot problems with solar panels is provided. Troubleshooting electrical and solar panels and identifying typical problems are explained in detail so students can easily master solving problems.

Chapter 12 covers the theory of basic electricity that is needed to understand all the electrical parts of the electrical systems for solar energy systems. This chapter is a comprehensive, detailed study of DC and AC electricity, including three-phase and single-phase power, transformers, the National Electric Code, AC and DC motors, and basic electronics used in rectifiers and inverters. This chapter is in-depth enough to provide students with all the electrical knowledge needed to install, troubleshoot, and repair electrical parts of the solar energy systems.

ACKNOWLEDGMENTS

I would like to thank my wife Kathy for all her help in keeping this project on track for her support during this time.

I want to thank Kevin Walker, John Carpenter, Bruce Meyer, and Jayne Bowersox, at Terra Community College, who helped me develop the individual topics, obtain pictures and evaluate the early content. I would also like to thank Ron Swenson of ElectroRoof® in Santa Cruz, California, and Bruce Kissell at Plantronics Inc. in Santa Cruz, California, for their help and the information they provided about the solar installation at Plantronics.

I would also like to thank Dan Trudden, Developmental Editor at Pearson Higher Education in Columbus, Ohio, for all his help during this project. He has gone above and beyond what I expected in guiding me through the project and ensuring that deadlines were met. I am especially indebted to his attention to detail and his help in getting many of the permissions and high-resolution images for his text.

I would also like to personally thank Ron Swenson of ElectroRoof® in Santa Cruz, California, who generously gave his time and energy to help me get the additional background information and pictures I needed to complete this book. I also want to thank him for the vast number of pictures he provided to make the book a success.

I would personally like to thank the following people and their companies for help in providing pictures and diagrams and the permissions to use them in this book:

Lin DeBeaulieu, Schneider Electric (Xantrex)
Rebecca Samuel, NXP Semiconductors
Jay Worley, Jacksonville Electric Authority
Jennifer Kramer at PSEG
Jackie Anderson, Florida Power and Light
Kee Kalwara, Schott Solar Albuquerque, NM
Matthew Kraft, Corporate Communications, Schott Solar Albuquerque, NM

I would also like to thank the following people who reviewed this text:

Yelleshpur N. Dathatri, Farmingdale State College
Thomas Hart, Tarrant County College
Troy Wanek, Red Rocks Community College

I would like to offer additional information to students and educators on a wide variety of topics that support solar energy technology. This information is available in the following textbooks I am currently writing and have recently written: *Introduction to Wind Energy; Industrial Electronics, Applications for Programmable Controllers, Instrumentation and Controls, Electrical Machines and Motor Controls; Electricity, Electronics and Control Systems for HVAC; Motor Control Technology for Industrial Maintenance; Electricity, Fluid Power and Mechanical Systems for Industrial Maintenance;* and *Electricity and Electronics for Industrial Maintenance.*

Thomas Kissell

CONTENTS

Introduction to Solar Energy

OBJECTIVES FOR THIS CHAPTER

When you have completed this chapter, you will be able to

- Explain how photovoltaic cells create electricity from energy from the sun.
- Identify the basic types of solar energy converters for heating, heating water, and producing electricity.
- Identify basic types of photovoltaic cells.
- List 10 jobs that will be needed in future solar energy industries, and the skills that are required to do these jobs.

TERMS FOR THIS CHAPTER

Equalization	Nickel-Cadmium (NiCad) Battery
Generator	Passive Solar Energy System
Grid-Tied System	Photovoltaic Cells
Heat Transfer Fluid	Return on Investment (ROI)
Hybrid Lighting Systems (Solar Lighting)	Solar Energy
Inverter	Solar Farm
Lead-Acid Battery	Stirling Solar Engine

1.0 OVERVIEW OF THIS CHAPTER

This chapter will introduce how solar energy is used today as an alternative energy source. These applications include residential and commercial heating; residential and commercial solar water heaters, hybrid lighting systems for large commercial and smaller residential buildings, photovoltaic (PV) cells that convert sunlight to direct current (DC) electricity, dish/engine systems that use concentrating mirrors to power a sterling engine that creates electricity, and parabolic trough systems that use mirrors to focus and concentrate sunlight on oil-filled tubes and use the heated oil with the aid of a heat exchanger to change water to steam to power traditional generating turbines. All of these types of solar energy systems will be introduced and explained.

Other topics that will be covered include how energy from the sun reaches the earth and how the earth and sun relate to each other. Details of the movement of the earth on its axis and around the sun and how this movement affects the amount of solar energy available day to day and season to season will be presented.

Other sections of this chapter will include information about solar initiatives, solar energy classifications, solar energy specifications, solar energy standards and certification. This chapter will also explain the need for the United States and other countries throughout the world to obtain energy from green technologies. Green technologies produce or provide energy from renewable sources that have lower emissions of pollutants. The future directions and current initiatives for solar power will also be discussed, as well as ways to make solar energy more financially competitive. Other financial implications and return on investment (ROI), pay back for all types of solar energy, as well as current tax considerations for solar energy will be discussed. The final section of this chapter will cover the types of skills needed for green energy and solar energy jobs, as well as the types of jobs in the solar energy sector.

1.1 MODERN SOLAR ENERGY SYSTEMS

Modern *solar energy* systems include solar heating, solar hot water heating, and *photovoltaic cells* that create electricity from light, and solar lighting systems. This chapter will introduce you to each of these areas of technology, and you will begin to see the science and technology that is involved in each, as well as the job opportunities in each of these areas.

1.1.1 Solar Heating Systems for Residential Applications

Solar heating equipment is now available to augment traditional heating systems, such as electric heat, gas heat, oil heat, or a heat pump. An active solar energy system has several methods of collecting heat from solar energy and applying it directly to heating system or storing it for future use. This includes active collector systems that are located on the roof or in the backyard that absorb solar radiation into a liquid. Liquid is used because it is easily distributed or moved from the outside where the heat is added to the inside of the house where the heat is removed. The collectors provide a means to concentrate solar radiation or solar energy into the liquid where it is absorbed. Systems such as these typically require a storage capability to ensure heat is available after the sun goes down or when collection is at a minimum. The storage system must be large enough to provide heat energy during the night or during days when the amount of solar energy is not at its peak.

There are basically two types of active solar heating systems, one that uses liquid and fluids that may consist of alcohol or glycol, which freeze at much lower temperatures than the home will be exposed to, or an air system that involves moving large volumes of air through the solar collection system and back into the home or storage area. Each of these systems will have a large collector system that is installed on the rooftop or any location in the backyard. Some of the problems with the liquid system include expending large amounts of energy to move the liquid with pumps, storing the liquid in insulated liquid containers, and preventing leaks of the liquid from the plumbing and piping system. The air heating system is basically a large collector area that is heated by the solar energy and a large volume of air that is moved over the collector system. The air can be moved directly into the home for heating or over a storage medium that will increase the temperature of the storage material so that the heat can be removed at a later time. The air heating system can be designed to heat any individual room or the entire house. Certain materials that are used in the home may increase the efficiency of the systems, such as a large mass of concrete or masonry products that tends to heat up more easily and hold more heat energy over longer periods of time than do carpeting and wood structures. The efficiency of the solar heating systems will also improve the overall control of temperature in a home, which will increase the satisfaction of the people living there. If the solar energy system is not designed correctly, it will allow the temperature in the home to vary between 10°F and 15°F around the set point, which becomes uncomfortable, and the system will not be considered usable. Figure 1-1 and Figure 1-2 show an active solar heating system in which liquid is heated in a solar collector that is mounted on the roof. The active solar energy heating system will be discussed in greater detail in other chapters of this text.

Another type of heating system is called a *passive solar energy system,* which collects solar energy and uses it as part of the residential heating with the minimum amount of mechanical equipment. The passive solar heating system is most efficient and effective when it is designed into the home during construction. This is not

FIGURE 1-1   The collectors of a rooftop solar energy system.

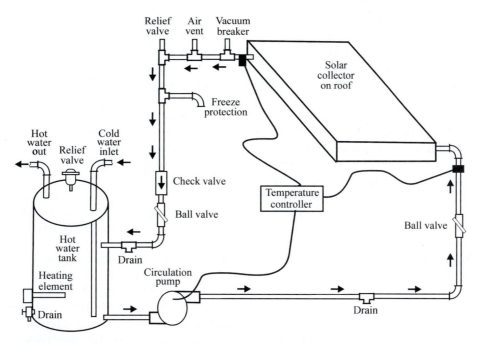

FIGURE 1-2 The plumbing diagram of the rooftop solar energy system.

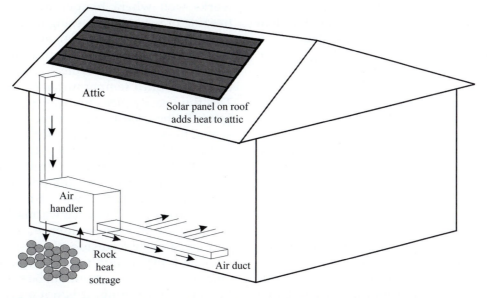

FIGURE 1-3 A diagram showing the parts of a type of solar heating system that uses rocks to store the heat.

always practical, since many homes were built 10 or 20 years ago and the passive solar heating system must be added on a later date. When passive solar heating systems are added on to an existing home, fans and blowers may be required to move a sufficient amount of air throughout all areas of the home, whereas a newly designed and constructed home can build natural air movement and heat movement into the home's design. The passive solar heating type system uses collectors on the roof, and the heat energy is moved to a storage area, which is usually made of rocks or dense material that can take on and release large

amounts of heat energy quickly. Figure 1-3 shows an example of this type of system. When the sun is shining, heat energy is harvested and moved to the storage area where the rocks or storage medium is heated. Since the sun is shining during this time and the outside temperatures are generally warmer, the system will only use a portion of the heat energy that is captured. The remainder of the heat energy is stored for use on cloudy days or at night.

If solar heat is used as a primary source of heating for residents, some building codes may require a secondary or backup heating system. The supplementary heating system will provide enough heat to maintain comfort or safety to keep plumbing from freezing during the coldest periods of time if the solar energy system fails to operate properly. This may include electric heat, wood-burning systems, gas heat, or other primary fuel heating systems.

It is necessary to check local covenants, zoning ordinances, and building codes that may restrict or even prohibit the installation of a solar system on your property. A restrictive home owners' association or local codes may not allow solar collector systems showing on the rooftop or in backyards. As the need for solar energy has become more important, some of these codes and restrictions have been rewritten to allow for specially designed systems that are more eye appealing or can be built into the outside structure of the home.

1.1.2 Solar Water Heating Systems

Solar water heating systems provide a means to warm the water that is used as hot water in the home to a desired temperature, or it may be designed so that the additional energy needed to bring the water up to the final temperature is minimal. This type of system may be called a solar domestic hot water system, and it can often be integrated into a solar heating system. The solar hot water system typically includes a large collector system that sits on the roof or in the backyard. Storage tanks, piping and plumbing control valves, and pumps are usually located inside the building. Some systems are called *passive systems* because they do not need as much hardware to move the water. The storage part of the system is very important since solar energy is available only for a portion of the day, and the hot water needs from the system may continue into the night. Some solar hot water systems are designed to augment the traditional hot water system. The goal of this type of hot water system is to offset or reduce the cost of heating water, and it does not necessarily need to work every day of the year or it may not be expected to provide 100% of the hot water needs. Other systems are designed with the expectation that they will provide the majority of the hot water needs, and the traditional hot water system will become the backup.

Several types of solar hot water collectors are used in residential applications and for small commercial applications. These include the flat plate collector system, which uses an insulated weatherproof box that contains a dark absorbent material under one or more transparent covers that may be glass or plastic. Some systems do not require a cover and the material used to absorb the heat is exposed directly to the environment. A second type of system is an integral collector storage system, and it

includes a collector system that is located on the roof. This system provides preheating of the water that goes through a traditional hot water heating system. This system is designed to augment or offset the need for heating water that may be as cold as 50°F as it comes from the city source or the ground. Since the water is preheated through the solar system, less traditional energy is required than with the traditional type of hot water heating system.

A third type of solar hot water heating system is an evacuated tube collector, and it consists of a tube that is inside another tube. The usable water travels through the inside tube, and the outside tube is filled with an energy-absorbing material or liquid that easily transfers heat to the water in the inside tube. The outside tube may also have fins or other collecting areas to increase the amount of solar energy that is transferred.

Solar hot water heating systems may also be classified as direct circulation systems or indirect circulation systems. The direct circulation system pumps the primary water for the residence that is used for hot water purposes directly through the solar collecting system and into a storage tank. This type of system works well where temperatures seldom get below freezing or get below freezing for only short periods of time. The indirect circulation system uses a *heat transfer fluid* that is usable in temperatures well below freezing. A heat transfer fluid is typically a glycol fluid that does not freeze until temperatures reach well below 32°F. This fluid easily picks up heat from the sun and transfers it to the fluid (usually water) inside the dwelling, so that the water is not subjected to the colder temperatures outdoors. This fluid is circulated through the collector system that is outside the house and may be exposed to freezing temperatures from time to time. This fluid, when warmed, is pumped into the home through a system that warms the hot water that is being used by the residence. If the temperature becomes too cold outside, the system stops pumping the heat transfer fluid until the temperature increases. Since the transfer medium can be exposure to temperatures below freezing, it is not a problem to stop the fluid from flowing when it is too cold. It is important to understand that if the water that is being used in the residence is pumped through a collector and allowed to set when the temperature becomes too cold, it may freeze and cause the plumbing and piping to split or damage other valves and controls for the system. Figure 1-4 shows a typical solar hot water system.

Passive solar energy water heating systems are practical in some warmer climates. These work best where solar energy is available over longer periods of time. The system is passive because pumps and other mechanical components are not used in the system.

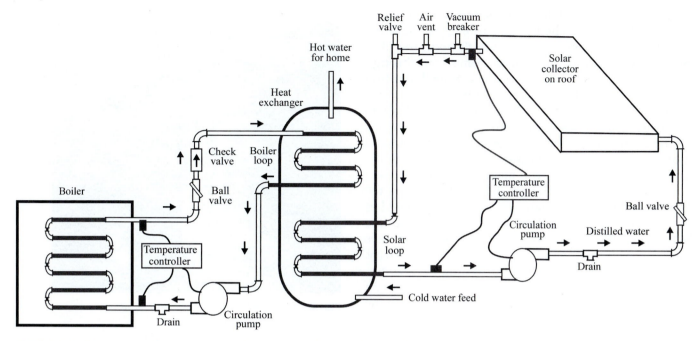

FIGURE 1-4　Diagram of solar hot water heating system with an auxiliary boiler.

Instead, it uses the law of physics that causes liquids to move when a temperature difference occurs. Typically warm water will rise in the system, and the cooler water will move to the lower locations. This causes the warm water to move upward to the collector system where it is able to add additional heat, and then gravity is used to move the water back down to where it is needed.

Solar energy water heating systems typically require a backup system to provide hot water at times when the sun is down or on cloudy days and solar energy is not available in sufficient amounts to heat the water. Today systems are available to work together so that solar energy is used when practical, and traditional hot water systems such as electric, gas, and other fuels are used to heat the water. Some green energy systems may use a small solar energy system to produce part of the energy that is used specifically for heating the home and heating hot water. The solar energy is used to augment the main energy that is used to heat the home and the water directly. In each of these cases, the solar energy must be collected and stored in batteries or in other forms such as water storage to be useful over longer periods of time when solar energy is not available.

1.1.3 Solar Lighting and Hybrid Solar Lighting

Solar lighting is also called hybrid solar lighting or a *hybrid lighting system*. It is a lighting system that distributes natural light within the interior of residential and commercial buildings directly and indirectly from the sun. This type of lighting system utilizes sunlight and moves it from the exterior of the building to the interior of the building through one or more lighting fixtures that utilize fiber optic cable. The solar lighting system is fully integrated into the existing electrical lighting system, and it delivers the benefits of natural lighting with the convenience and reliability of electrical light. When solar lighting or hybrid solar lighting is used in commercial applications, it has been shown that natural light increases retail sales, reduces energy consumption, and improves employee productivity and wellness.

For years glass windows have been used to allow sunlight to enter a building or residence. Recently it has been determined that the glass windows allow unwanted heat energy to enter the living space during the summer and heat energy to be lost from the living space in the winter when the outside temperature is colder. For this reason large windows may not be an efficient method to bring sunlight into a building. The hybrid solar lighting system consists of a roof-mounted solar platform that is a 45-foot-long plastic fiber optic bundle and a number of special "hybrid" luminaries. The technology concentrates natural sunlight into a small bundle of optical fibers, and the fiber optic cables installed through the structure of the commercial building or residence and over irregular contours such as ceiling joists and inside walls without disturbing the sunlight that is within the optical fibers. In this way sunlight is piped directly into a building or enclosure to one or more special lighting fixtures called hybrid luminaries that diffuse the light throughout the space.

Some of these hybrid luminaries can deliver up to 25,000 peak lumens. The hybrid luminaries blend the sunlight that is piped from outside through the fiber optics with natural light that comes in through windows and adds it to existing artificial lighting to provide controllable interior lighting. When the sunlight levels vary during the day due to changes in the position of the sun or from clouds and storms moving in, the light that is harvested is controlled and automatically increases or decreases the light with the natural lighting. This control makes the light pleasant and unnoticeable as it changes during the day. As the sun sets, with decreased sunlight, the system switches over to electrical lighting seamlessly. In some cases the electricity for this lighting comes from photovoltaic cells that have converted solar energy into DC electricity. When the sunrise occurs and sunlight becomes available again, the system begins harvesting sunlight and the controller begins to automatically increase the hybrid light and diminish the artificial electrical lighting proportionally. This type of system provides energy efficient lighting that is free from the sun. In locations where the sunlight is continuous and strong throughout the day, this type of system can provide the majority of the lighting. In other locations where electrical lighting must be used as well, the hybrid lighting significantly lowers the amount of electricity that is used for lighting, thus reducing the electric bill.

On new homes this type of lighting system can be built directly into the structure, and companionways can be built directly into the walls and ceilings so that the tubes can be easily installed and maintained. Since the structure has these openings built in, additional tubes and other solar energy systems can be installed at a later date. Another type of lighting system uses fiber optic cables that are very flexible and this system is also easily added to older homes.

Another type of solar lighting is indirect solar lighting. Figure 1-5 shows an example of this type of system mounted on the roof. An opening is cut into the roof and a special fitting that has a glass or plastic lens protrudes above the roofline. This opening is connected to a tube that has a shiny surface, and the other end of the tube is connected to a second lens that is mounted in the ceiling inside the house in the living space. The sunlight is collected through the lens on the roof and is directed through the tube that has a shiny surface into the lens that is in the ceiling; the light diffuses from the lens into the living space. Any time the sun is shining, it is directed into the living space as though it came through a window. The advantage of this type of lighting system is that the light can be mounted in the ceiling of the home or other places where it may not line up with a traditional skylight or window. Since light is available only when the sun is shining, the system must be augmented with a traditional electric light. In many cases the electric light operates from electricity that has

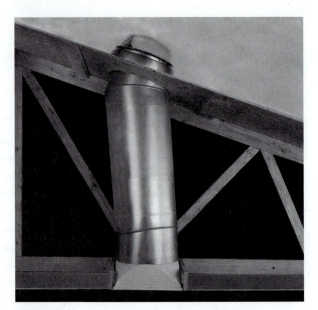

FIGURE 1-5 A solar lighting system.   (Courtesy Solartube International.)

been converted through PV cells and stored in batteries for use on cloudy days and at night. This type of lighting system can be installed in new homes, or it can be retrofitted into older homes.

1.2 PRODUCING ELECTRICITY FROM SOLAR ENERGY

The process of producing electricity from solar energy was discovered in 1839 by a French physicist Alexandre Edmond Becquerel. He discovered that conductance of electricity rises when a metal electrode and an electrolyte were illuminated. This discovery was studied and the technology has been improved upon to the point today at which solar energy panels called *photovoltaic (PV) cells* efficiently and effectively convert solar energy into DC electricity. Today a wide variety and types of PV cells are manufactured and installed worldwide. The manufacturing processes and designs of these photovoltaic cells are continually improving, which makes the technology more efficient and less expensive. This section explores the history of photovoltaic cells and how they have evolved from their earliest discovery to the most modern solar energy conversion systems available today. Figure 1-6 shows a typical solar photovoltaic collector mounted on a residential rooftop. This type of collector is designed to convert solar energy to DC electricity through photovoltaic cells. Figure 1-7 shows an industrial installation of rooftop photovoltaic cells. The roof of the building has a large number of PV cells that are used to collect sunlight and convert it to DC voltage. This type of application uses one or more *inverters*. An inverter is an electronic system that is used to convert the DC voltage

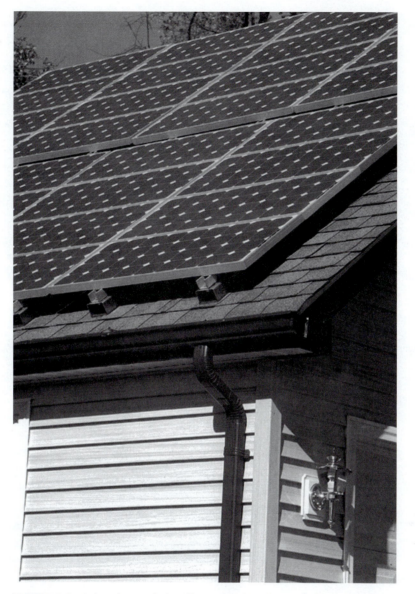

FIGURE 1-6 Solar photovoltaic cells mounted on the roof of a residence. (Courtesy U.S. Department of Energy.)

to either single- or three-phase alternating (AC) voltage at the appropriate frequency (60 Hz). Transformers are used to increase the voltage level to 480 volts (V) so that is usable in a factory. The three-phase voltage is also available to be sent to the grid, and additional transformers can increase the voltage to 12,470 V so that it is compatible with the grid. A net meter is used so that electricity can flow into the grid when the solar panels are producing maximum power, and power from the grid can seamlessly flow back through the meter into the building when the solar panels are not producing enough to meet the needs of the building. As much of the electricity as possible is used directly in the buildings, and any excess is sold to the grid. These types of systems are designed to place as many solar panels on the roof of the buildings as possible.

Another typical application of solar photovoltaic cells is a *solar farm*. A solar farm is a large group of solar panels that have been installed in a single location, and the voltage that is produced from the photovoltaic cells is sent to the grid through a local connection. Figure 1-8 shows a typical solar farm. You can see a large number of connected photovoltaic panels to harvest as much electrical energy as possible and convert it to DC voltage.

Additional information about residential, industrial, and solar farms provided in later chapters will give you the information needed to install, troubleshoot, and repair the systems. A more in-depth study of the PV cells will be provided in Chapters 4 and 5. The next section of this chapter will introduce you to the types of photovoltaic cells and the science and technology behind them, as well as their history.

FIGURE 1-7   Solar photovoltaic panels mounted on the roof of industrial buildings.   (Courtesy Ron Swenson, www.SolarSchools.com.)

FIGURE 1-8   Solar photovoltaic panels mounted on a solar farm. (Pgiam/iStockphoto.com)

1.3 TYPES OF PHOTOVOLTAIC CELLS

Photovoltaic technologies are used to convert solar energy to DC electricity. Today a number of photovoltaic technologies are in use. This section will introduce each of these types of technologies briefly so you'll begin to get an idea of the advantages and disadvantages of each. Each of these photovoltaic cells will be discussed in greater detail in Chapters 5, 6, and 7. The photovoltaic panels are mounted on the roof of residential, commercial, or industrial buildings where they convert the sunlight to DC electricity.

Currently there are four basic types of silicon photovoltaic cells that are manufactured and used in the solar energy field: the single-crystal silicon cell; the polycrystalline silicon cell (also known as multicrystal silicon cell); the ribbon silicon cell; and the amorphous silicon cell, abbreviated "Si" and also known as thin-film silicon cell.

Most of the commercial photovoltaic cells are manufactured from silicon, which is the same material that sand is made of. The silicon that is used in manufacturing photovoltaic cells is extremely pure. Other materials such as gallium arsenide are beginning to be used in manufacturing photovoltaic cells. The reasons that silicon is so widely used are that (1) it is a semiconductor material and (2) sand that is used as the primary manufacturing material is easily available.

1.3.1 Single-Crystal Photovoltaic Cells

During the manufacturing process for single-crystal PV cells, silicon is purified, melted, and then crystallized into blocks called "ingots." The ingots are sliced into thin wafers to make individual cells, and the cells are connected to make a solar panel. The body of the cell has a small amount of positive electrical charge and a thin layer at the top of the cell has a small negative charge. During the manufacturing process the cell is attached to a base material called the "backplane." The backplane is usually a layer of metal that reinforces the cell physically and provides an electrical contact at the bottom.

The top of the cell must be open to allow sunlight to enter. To accomplish this a thin grid of metal instead of a continuous layer is applied to the top. The grid must be thin enough to admit enough sunlight to cause the photovoltaic cell to convert solar energy to electricity, yet be wide enough to carry the electrical energy that is produced. The physical process that occurs to make the photovoltaic cells produce electricity is caused by light particles called photons that strike the silicon material, where they are absorbed into the cell. When a photon strikes an electron in the silica material, it dislodges an electron and causes it to move. The movement of electrons through the material is called current flow. When the electron is dislodged and moves, it leaves an empty space called a "hole." The movement of electrons in the silicon wafer occurs from the body of the wafer to the top layer where it comes into contact with the metal grid. Since an electrical path exists outside the silicon cell, between the grid and the backplane, a flow of electrons continually moves as long as light strikes the material. This flow of electrons becomes a useful current that the photovoltaic cells produce. As more and more light strikes the body of the material, more and more electrons come loose from their atoms and begin to move this free current through the cell. The sunlight produces a force that makes the electrons break loose from their atoms, and this force creates an electrical potential difference that is called voltage. Typically the voltage at the cell is approximately 1.5 V. The amount of available light affects the flow of electrons, which is the electrical current for the system. The electrical potential difference (voltage) that is created when light strikes the silicon atoms is increased when the amount of sunlight increases, and it is also affected by temperature.

1.3.2 Polycrystalline Silicon Cells

Polycrystalline cells were developed to reduce the cost of solar panels. The polycrystalline silica cells can be manufactured at a lower cost than the single-crystal cells, and they are manufactured in a similar manner. The difference is that lower-cost silicon is used that is not as pure. The operation of the polycrystalline silicon cell is similar to the single-crystal cell in that light strikes the silicon material and causes electrons to begin the flow. Since the polycrystalline silicon is less pure, it typically has a slightly lower efficiency. Overall a lower efficiency is balanced out by the lower manufacturing cost. The lower cost means that it can be installed more widely in applications for which the higher cost cells are not affordable.

1.3.3 Ribbon Silicon

The ribbon photovoltaic cells are made by growing a ribbon of silicon cells from the molten silicon. The ribbon is used instead of the silicon block or ingot. The individual cells operate much the same as the single-crystal and polycrystalline cells by creating a flow of electrons (current) when sunlight strikes the surface of the silicon.

1.3.4 Amorphous Thin-Film Silicon

The single-crystal cell, the polycrystalline cell, and the ribbon cell silicon have distinct crystal structures at the atomic level. Amorphous silicon is also called thin-film silicon and it is abbreviated a-Si, and it does not have this structure. Amorphous silicon units are made by depositing very thin layers of vaporized silicon onto a backplane of glass, plastic, or metal. This process is carried out while the material is in a vacuum; it is a relatively new process used to create solar photovoltaic cells less expensively.

1.3.5 Other Types of Photovoltaic Cells and Manufacturing Processes

Newer manufacturing processes have created a number of new types of photovoltaic cells. These new processes allow some new materials to be added to the silicon material and in some cases the new material is used instead of silicon. The main driving force for the new processes and products is to provide some advantages in cost and efficiency.

1.3.6 Cadmium Telluride Thin-Film PV

One of the newer material combinations used to produce a less expensive photovoltaic cell is cadmium telluride (CdTe). It is achieving manufacturing costs low enough to market solar panels at less than $1 per watt, which is close to the cost target that is considered low enough for solar energy to compete directly with coal-burning electricity on the grid.

Cadmium has several advantages over traditional silicon technology. The cells are manufactured from two types of cadmium molecules, cadmium sulfide and cadmium telluride. Since the two types of cadmium can simply be mixed together to achieve the required properties, which simplifies the manufacturing process compared to the multistep process of joining two different types of doped silicon in a silicon solar panel. Doping is a process in which two or more elements are combined at the atomic level to change their electrical properties.

Another advantage of cadmium telluride is that it absorbs sunlight at close to the ideal wavelength for converting sunlight to electrical power by capturing energy at shorter wavelengths than is possible with silicon panels. The final advantage of cadmium is that it is very abundant, and it is produced as a by-product of producing other industrial metals such as zinc.

There are several drawbacks to using cadmium telluride, however, including that cadmium telluride solar panels currently achieve an efficiency of approximately 10.6%, which is significantly lower than the typical efficiencies of silicon solar cells, which is near 16%. Since the efficiency is less than that of silicon cells, more cadmium telluride solar cells than silicon cells are needed to get the same amount of electrical energy. This makes it imperative that the cadmium telluride is sold at a lower cost so that the overall cost of a project is about the same.

Additional information about all of these types of photovoltaic cells and other types of PV cells is provided in greater detail in later chapters.

1.4 HISTORY OF PHOTOVOLTAIC CELLS

The history of photovoltaic cells spans more than 170 years. A list of some of the historical events is provided in Table 1-1.

TABLE 1-1 History of the Photovoltaic Cell

Year	Event
1839–1899	**Discovery of photovoltaic effect** (Discovery of the basic phenomena and properties of PV materials.) A physical phenomenon allowing light-electricity conversion, the photovoltaic effect, was discovered in **1839** by the French physicist Alexandre Edmond Becquerel. Experimenting with metal electrodes and electrolyte, he discovered that conductance rises with illumination.
1887	Heinrich Hertz discovered that ultraviolet light changes the voltage at which sparks between two metal electrodes would be initiated.
1900–1949	**Theoretical explanation of the photovoltaic effect and first solar cells**
1904	**Theoretical explanation of the photovoltaic effect**
1921	The author of the most comprehensive theoretical work about the photovoltaic effect was Albert Einstein, who described the phenomenon in 1904. For his theoretical explanation he was awarded a Nobel Prize in 1921. Einstein's theoretical explanation was practically proven by Robert Millikan's experiment in 1916.
1918	**The first silicon solar cells**
	Jan Czochralski, a Polish scientist, discovered a method for monocrystalline silicon production, which enabled monocrystalline solar cells production.
1941	The first silicon monocrystalline solar cell was constructed in 1941.
1932	**The photovoltaic effect in other materials**
	The photovoltaic effect in cadmium-selenide was observed. Nowadays, CdS belongs among important materials for solar cells production.
1950–1969	PVs used in intensive space research
1955	**The first satellites and solar-powered cars**
	The preparation of satellite energy supply by solar cells began. Western Electric put up for sale commercial licenses for solar cells production. Hoffman Electronics Semiconductor Division introduced a commercial photovoltaic product with 2% efficiency for $25 per cell with 14 mega watt (mW) peak power.
	The first sun-powered automobile was demonstrated in Chicago, Illinois, on August 31.
1957	Hoffman Electronics introduced a solar cell with 8% efficiency.
1958	Hoffman Electronics introduced a solar cell with 9% efficiency. The first radiation-proof silicon solar cell was produced for the purposes of space technology. On March 17, the first satellite powered by solar cells, Vanguard I, was launched. The system ran continuously for 8 years. Two other satellites, Explorer III and Vanguard II, were launched by Americans, and Sputnik III by Russians. The first telephone repeater powered by solar cells was built in Americus, Georgia.
1959	Hoffman Electronics introduced commercially available solar cells with 10% efficiency. Americans launched Explorer VI with photovoltaic field of 9,600 cells and Explorer VII.
1960	Hoffman Electronics introduced yet another solar cell with 14% efficiency.

1961	**The first photovoltaic conferences**

A United Nations conference on solar energy application in developing countries took place.
The Defense Studies Institute organized the first photovoltaic conference in Washington.

1962	The first commercial telecommunications satellite, Telstar, developed by Bell Laboratories, was launched. The photovoltaic system peak power for satellite power supply was 14 W. The second photovoltaic conference took place in Washington.

1963	**The first solar modules**

Sharp Corporation developed the first usable photovoltaic module from silicon solar cells.
The biggest photovoltaic system at the time, the 242 W module field was set up in Japan.

1965	The Japanese scientific program for Japanese satellite launch commenced.
1966	An astronomic observatory that was in orbit had 1 kW peak power photovoltaic module field.
1968	The OVI-13 satellite with two CdS panels was launched.
1972	Solar Power Corporation was established as a commercial business.
1973	A Solar Power Corporation sales office opened in Braintree, Massachusetts. The French implemented a CdS photovoltaic system enabling educational TV programs to broadcast in the province of Niger.
1973	Solarex Corporation was established. At Delaware University a photovoltaic-thermal hybrid system Solar One, one of the first photovoltaic systems for domestic application, was developed. Besides the photovoltaic system, the system incorporated a warmth keeper of phase-changeable materials. A silicon solar cell costing $30 per W was produced.
1974	The Japanese Sunshine project commenced.
1975	Solec International and Solar Technology International were established. The American government encouraged JPL Laboratories to do research in the field of photovoltaic systems for application on earth.
1976	Under NASA's protection LeRC commenced photovoltaic system installations for application on earth.

1985	**The first photovoltaic systems for the Third World rural areas**

The systems were meant for refrigerators, telecommunication equipment, medical equipment, lighting, and water pumping, and as a power supply, as well as for other applications. NASA LeRC introduced several demonstration projects. The first amorphous silicon solar cell was developed by RCA Laboratories.

1977	The world production of photovoltaic modules exceeded 500 kW. NASA LeRC commenced implementing photovoltaic systems in six meteorological stations in different locations within the United States. NASA LeRC introduced additional trial demonstration projects. Solar Energy Research Institute located in Golden, Colorado, launched its operation. On Native American reservation NASA LeRC set up a 3.5 kW system—the first system ever to satisfy the demands of an entire village. It was used for water pumping and power supply for15 households.
1979	ARCO Solar of Camarillo, California, built the biggest solar cells and photovoltaic systems production plant premises at that time. NASA LeRC built a 1.8 kW water pumping photovoltaic system in Burkina Faso. The system peak power was enlarged to 3.6 kW the same year. In Mt. Laguna, California, a trial 60 kW hybrid diesel-photovoltaic system was built for radar station power supply.

1980	**Large stand-alone photovoltaic systems**

ARCO Solar was the first to produce photovoltaic modules with peak power of more than 1 MW per year. A trial photovoltaic system installation was made in the center of the volcano observatory in Hawaii. A new company, BP, appeared in the market. ARCO Solar built a 105.6 kW system in Utah. The modules integrated in the system were produced by Motorola, ARCO Solar, and Spectrolab.

1981	NASA LeRC began to build systems for vaccine refrigerators' power supply on 30 locations around the globe (the project was closed in 1984). Solar Challenger, the first plane ever powered by solar energy, took off. A system with peak power of 90.4 kW with modules produced by Solar Power Corporation was built in Square Shopping Center in Lovington, New Mexico. A similar system was built for Beverly High School in Beverly, Massachusetts. A seawater desalination system with 10.8 kW peak power was built in Jeddah, Saudi Arabia, the same year. Helios Technology, the oldest European solar cells producer, was established.
1982	The world production of photovoltaic modules exceeded 9.3 MW. Solarex established Solarex Aerospace division. At the Vienna conference NASA LeRC introduced a trial case of terrestrial satellite reception station and public lighting electricity supply. Volkswagen began testing photovoltaic systems placed on vehicle roofs with 160 W peak power for vehicle start-up. Solarex production premises rooftops in Frederick, Maryland, were equipped with photovoltaic systems with 200 kW peak power. ARCO Solar built a 1 MW PV power plant with modules on more than 108 acres.

(Continued)

TABLE 1-1	(Continued)
Year	**Event**
1983	The world production of photovoltaic modules exceeded 21.3 MW peak power, with product worth of $250 million. Solar Trek vehicle with photovoltaic system of 1 kW drove 4,000 km in 20 days of the Australia Race. The maximum speed was 72 km/h, and the average speed was 24 km/h. The same year the vehicle surpassed the distance of 4,000 km between Long Beach, California, and Daytona Beach, Florida, in 18 days. Solarex Corporation bought an amorphous cells production technology from cells producer RCA and built its own trial power plant in Newtown, Pennsylvania. ARCO Solar built a 6 MW photovoltaic power plant as a subsystem of the public electricity grid for Pacific Gas and Electric Company application in California. The system satisfied the demand of 2,000 to 2,500 households. Solar Power Corporation built four stand alone photovoltaic systems for the needs of a village in Tunisia with total peak power of 31 kW per system. A 1.8 kW photovoltaic system was built to satisfy the needs of a local hospital in Guyana. The applications, such as vaccine refrigerators, indoor lighting, ordination lighting, and radio appliances were powered by the system. The system was planned and built by NASA Lewis Research Center and Solarex. A similar yet more powerful photovoltaic system of 4 kW was set up in Ecuador. A 1.8 kW photovoltaic system was set up in Zimbabwe for the same purpose. Solarex Corporation merged with Amoco Solar Company, owned by Standard Oil Company.
1984	A 1 MW photovoltaic power plant began to operate in Sacramento, California. ARCO Solar introduced the first amorphous modules. NASA LeRC placed 17 photovoltaic systems to satisfy the demands of the local schools, lighting, medical equipment, and water pumping in Gabon. BP Solar Systems with EGS donations built a 30 kW photovoltaic system connected to a public electric grid near Southampton, Great Britain. Solarex Corporation supplied the equipment for photovoltaic system for Georgetown University Intercultural Center demands with total peak power of 337 kW and 4,464 modules. BP Solar bought Monosolar's thin-film division, Nortek, Inc.
1985	Researches of University of New South Wales in Australia constructed a solar cell with more than 20% efficiency. BP built a power plant in Sydney, Australia, and shortly after another one near Madrid, Spain. A photovoltaic system was built in Sulawesi, Indonesia, for the purposes of a terrestrial satellite station.
1986	ARCO Solar introduced a G-4000, the first commercial thin-film photovoltaic module.
1987	At the Pentax World Solar Challenge race through Australia, a General Motors Sunracer vehicle won with an average speed of 71 km/h.
1990	Energy Conversion Devices Inc. (ECD) and Canon Inc. established a joint company, United Solar Systems Corporation, for solar cells production. Siemens bought ARCO Solar and established Siemens Solar Industries, which is now one of the biggest photovoltaic companies in the world. Solar Energy Research Institute (SERI) was renamed as the National Renewable Energy Laboratory (NREL).
1991	BP Solar Systems was renamed as BP Solar International (BPSI) and became an independent unit within British Petroleum.
1992	A photovoltaic system of 0.5 kW was placed in Antarctica for laboratory, lighting, personal computer, and microwave oven needs. A silicon solar cell with 20% efficiency was patented.
1994	NREL, an important institution in the field of renewable energy sources in the United States, launched its website on the Internet. The U.S. Department of Energy built several trial systems for agriculture, hospitals, lighting, water pumping, and so on in Brazil. ASE GmbH from Germany purchased Mobil Solar Energy Corporation technology and established ASE Americas, Inc.
1995	The first international fund for the promotion of photovoltaic system commercialization was established, which supported projects in India. The World Bank and the Indian Renewable Energy Sources Agency sponsored projects in cooperation with Siemens Solar.
1996	BP Solar purchased APS production premises in California and announced a commercial CIS solar cells production. Icar, the plane, powered by solar energy, with 3,000 solar cells in total surface of 21 m^2 flew over Germany.
1997	Greece agreed to sponsor the first 5 MW of a total planned 50 MW photovoltaic systems on Crete. Because of a misunderstanding among investors, the system was not realized. The activities, which will result in 36,400 50 W systems within the next three years, started in Indonesia.
1999	Solar Cells, Inc. (SCI), True North Partners, and LLC of Phoenix, Arizona, merged to become First Solar, LLC.
2000	**Photovoltaics and stock exchange in Europe**
	Mostly in Germany, some photovoltaic and renewable energy resources companies have shares listed at the stock exchange. Capital mergers in Germany led to large photovoltaic corporation establishments. During 2000 and 2001 production of Japanese producers increased significantly. Sharp and Kyocera each produce modules with peak power equivalent to the annual consumption in Germany, the most demanding European market. Sanyo is close as well.
2002–2010	Several large power plants were built in Germany. On April 29, 2003, at that time the world's largest photovoltaic plant was connected to the public grid in Hemau near Regensburg (Bavaria), Germany. The peak power of the "Solarpark Hemau" plant is 4 MW. Due to renewable energy law "EEG" many other large systems up to 5 MWp were built in Germany in 2004. Some of them are Solarparks Geiseltalsee, Leipzig, Bürstadt, and Göttelborn.

1.5 FACTORS THAT AFFECT THE AMOUNT OF SOLAR ENERGY REACHING THE EARTH

The amount of solar energy that reaches the earth changes daily, due to several distinct processes. First the earth spins on its axis and makes a complete rotation once every 24 hours. This process creates the sunrise, midday sun, sunset, and nighttime darkness. To study this effect it is best to select a single location such as Cleveland, Ohio. As the earth rotates on its axis, it begins to move Cleveland into the sunlight, which creates a condition we refer to as sunrise. Because the earth is rotating with respect to the sun, the sun appears to move across Cleveland from east to west. The sun moves completely to the western horizon at sunset. During sunrise and sunset the sun's light and energy rays do not reach the location directly; rather they are received at an angle, which produces less energy than when they strike the same area from directly overhead at midday. When the earth moves so that Cleveland is directly under the sun during the day, the energy from the sun's rays will be at its strongest level. If photovoltaic cells are used to convert sunlight into electrical energy, they will produce the maximum amount of energy during the time when the sun is directly overhead. During sunrise and sunset the sun's rays will not strike the photovoltaic cells directly and the amount of electrical energy they will be able to produce will be less than when the sun strikes them directly at midday.

Other conditions will also cause the amount of solar energy that reaches Cleveland to vary during the year. In addition to rotating on its axis every 24 hours, the earth also revolves around the sun in an orbit that takes approximately 365 days. Figure 1-9 shows the path the earth takes as it orbits around the sun. From this figure you can see that the path is not an exact circle; rather it is an elliptical path, which means the earth will be farther from the sun at certain times of the year and closer to the sun at other times. Another condition that affects the amount of sun each location on earth receives each day is the tilt of the earth on its axis. The earth tilts on its axis at approximately 23°, but it actually changes its tilt slightly from 21° to 25° as it

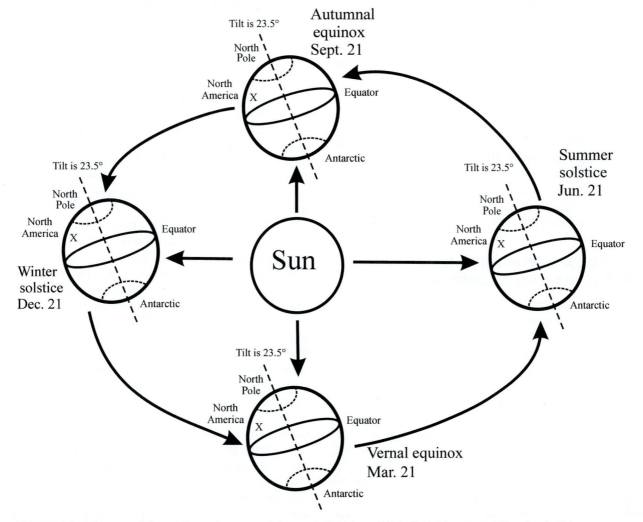

FIGURE 1-9 Diagram of the earth moving around the sun in 365 days. "X" indicates location of Cleveland, Ohio.

moves around the sun. The tilt of the earth constantly stays at the 21° to 25° angle as the earth travels around the sun. Since the earth travels around the sun in one year, cities such as Cleveland in the Northern Hemisphere tilt toward the sun during the summer months and away from the sun in the winter. During the winter the tilt of the earth remains the same, but because the earth is at the other end of its elliptical route around the sun, the Northern Hemisphere is now pointed away from the sun, which causes the Northern Hemisphere to have shorter days and receive less solar energy.

In the Northern Hemisphere the earth is tilting away from the sun during the winter months. December 21 is called the winter solstice, and it is the day with the shortest number of hours of exposure to the sun in the Northern Hemisphere. On June 21 the summer solstice occurs and the Northern Hemisphere has the most number of hours of sunlight. On March 21 the earth is at its vernal equinox and the amount of daylight and darkness is approximately the same. On September 21 the earth is at its autumnal equinox and again the amount of daylight and darkness is approximately the same.

Since Cleveland, Ohio, is in the Northern Hemisphere, and it is located approximately 3,000 miles north of the equator, it gets more sunlight during the summer when the Northern Hemisphere is tilted toward the sun, and it gets fewer hours of sunlight during the winter when the Northern Hemisphere is pointing away from the sun. The equator is an imaginary line that runs around the center of the earth at a location that is approximately halfway between the North and South Poles.

This information is useful to understand that the amount of solar energy is always changing, month to month and season to season. It is also useful to help identify the best places for solar energy to be installed and where the largest solar energy farms can convert the most amount of solar energy.

1.6 HOW WEATHER AFFECTS HOW MUCH SOLAR ENERGY CAN BE CONVERTED

Another factor that affects how much solar energy can be converted is the weather. Weather conditions are not controlled completely by the seasons or by the way the earth orbits the sun; rather other issues influence the weather. For example a weather front may cover a large portion of the United States, and it will cause a heavy dense cloud cover for several days. When the cloud cover becomes too dense, a large amount of solar energy is blocked from reaching solar panels. This type of condition may also occur for a shorter period of time during a thunderstorm or snowstorm. Some areas of the country that are near the Great Lakes may be more prone to overcast skies, heavier cloud cover, and fog. Fog is caused by a temperature differential between the air and ground, but the effect on solar energy equipment receiving sunlight is similar to that caused by a heavy cloud cover. Another condition that may affect solar cells is their being covered by snow.

In some northern climates and periodically in some southern climates, snow showers can provide a light covering of snow on solar panels and stop them from receiving sunlight. Typically this is not a long-term problem, because most solar equipment is a dark color, which will absorb any sunlight and create a warming effect that would usually melt the snow away. A similar problem to snow is the lack of sufficient rain that periodically washes off the outside cover of the panels. One of the problems with any solar panel that lies nearly parallel or flat, is that any dust or dirt that drops out of the air will fall onto the face of the panels. Over time this leaves a layer of dirt and dust that begins to diminish the amount of sunlight that reaches the panel itself. If an area receives a coating of dust, and it does not rain for a period of time, maintenance personnel will have to wash the panels. A rain shower every 4 to 6 weeks is sufficient to keep the panels clean.

1.7 OTHER TERMS FOR SOLAR ENERGY SYSTEMS

Energy from the sun is called solar energy. The energy consists primarily of light energy that travels to the earth as light waves or photons. When these light waves are absorbed by material, the temperature of the material rises. Solar energy applications include heating a space, such as a room in a house, by sunlight entering through a window or being absorbed through a dark-colored roof on a building or house. Another solar energy application is heating water for domestic (household) or commercial use. Solar energy can also be converted to DC electricity through photovoltaic cells.

Other solar energy applications include solar intensifiers or collectors that consist of a group of mirror surface that intensify the heat energy of the solar rays into a specific location, which creates a very high temperature. This high temperature is focused on a pipe that contains a liquid such as oil or molten salts that is heated and piped through a heat exchanger that has water in the second loop. The temperature of the oil or molten salt is high enough to cause water in the second loop of the heat exchanger to boil and change state into steam. The steam is used to turn a turbine that drives a *generator* and creates electrical power. A generator is a machine that converts rotational motion into electrical energy.

There are several terms that need to be defined so that you can understand solar energy more easily. *Energy* is simply defined as the ability to do work. It is important to understand that energy cannot be created or destroyed, and any form of energy can be transformed or changed into any other form of energy, but the amount of total energy always remains the same. This principle or concept is called the "conservation of energy" and it was discovered early in the 19th century.

One form of energy is heat, which is defined as a form of energy that is transferred by a difference in temperature. Heat transfer is the movement of heat from a warmer area or object to a cooler area or object. Another way of explaining heat is that it is a form of energy that can flow from one energy system to another energy system by virtue of temperature differences.

Temperature is defined as the degree of hotness or coldness of a body or environment. It can also be defined as the measure of the average kinetic energy of the particles in a sample of matter, and it can be expressed in terms of units or degrees designated on a standard scale. The two basic scales of measuring temperature are the Fahrenheit scale and the Celsius scale. The Fahrenheit scale is based on the freezing point of water, 32°, and the boiling point of water, 212°. On the Celsius scale water freezes at 0° and boils at 100°.

The sun radiates energy primarily in the form of light. Light waves, or photons, carry the energy away uniformly in all directions. The sun also transfers a miniscule proportion of its energy in other ways. It emits streams and bursts of kinetic energy and electric potential energy when it ejects high-speed charged particles into space, commonly called the solar wind and solar flares. The sun also transfers kinetic and thermal energy to its orbiting planets through gravitationally induced tidal friction.

1.8 STORING SOLAR ENERGY

Solar energy that is used for residential and commercial applications typically has a storage medium available to hold electrical energy for times when sunlight is diminished due to clouds or at night when the sun goes down. Batteries are the main products to store electrical energy long and short term. The batteries can be lead-acid, lithium-ion, nickel-cadmium (NiCad), or nickel-metal-hybrid.

The battery needs to be able to accept a charge quickly and allow it to be discharged deeply. Battery chargers are needed to maintain the charge on the battery and protect it from completely discharging or overcharging.

The DC voltage produced by solar cells is not usable by the large number of appliances or equipment in residential, commercial, or industrial applications that use AC electricity. DC voltage produced by solar cells can be used directly for heating elements, lighting, or DC motors.

If the voltage stays in a DC form, batteries will be used to store it. One of the electronic controls for this type of system is a battery charger and controller that ensures the batteries do not become overcharged or go completely dead. If AC voltage is required to power electrical motors and appliances, or if the power is sent to the grid, an electronic power inverter is required to convert DC voltage to AC voltage. A special switch is also required to connect and disconnect the solar system from the grid. Sometimes the battery charging equipment and the inverter are designed into the same piece of equipment.

Not all solar energy systems require batteries; but if batteries are used, they are an integral part of a solar energy system, as they store power that can be used when the system is not able to provide power during the night or on cloudy days. The voltage level for the batteries can be 12, 24, or 48 V. The two basic types of batteries that are typically used in solar energy systems are the *nickel-cadmium (NiCad)* battery and *lead-acid battery*. Nickel-cadmium batteries are more expensive than lead-acid batteries, but they last longer, require less maintenance, and may be discharged more fully than lead-acid batteries. The reason NiCad batteries can be discharged more deeply than lead-acid batteries is that they maintain nearly full rated voltage until they are nearly completely discharged; then the voltage level falls off quickly. One drawback of the NiCad battery is that it is expensive to dispose of because cadmium is a hazardous material. One of the problems with a lead-acid battery is that it begins to lose its voltage level quickly after it discharges down to only 40% or 50% of the total charge. When the voltage level drops off below its full charge value, the battery voltage is no longer useful to power loads, so the lead-acid battery needs to be kept at nearly full charge to maintain its highest voltage level. If the lead-acid battery is not sealed, the water level will need to be checked often and distilled water must be added if the level gets low. The reason the lead-acid water level drops is that the plates in a lead-acid battery get hot enough to boil off the liquid. When this occurs, the liquid boils off the battery as a vapor, which results in the level dropping. If the acid liquid level drops too low, the top of the lead plates will be exposed, which will cause severe damage and the inability of the battery to produce voltage. Another problem that is created when the vapor from the acid is boiled off is that it will corrode any metal it comes into contact with, so it is important to keep the top of the battery and its terminals clean.

Batteries used in a solar energy system are typically deep-cycle batteries. They discharge more of their energy over a longer period of time and have a longer life than a shallow-cycle battery, such as a car battery. Shallow-cycle batteries discharge a great deal of energy in a short period of time and recharge over a longer period of time. When lead-acid batteries are used with solar energy systems, they are designed as deep-cycle batteries. They will continue to recharge as long as it is daylight and the PV cells are producing power.

Lead-acid batteries may need to be equalized every 60 to 90 days. *Equalization* involves a controlled overcharging of the battery to ensure that each is fully charged. During the equalization process the battery must be monitored closely to make sure that no individual cell has a much higher charge than any of the others. The battery area needs to be well ventilated, as the process will release gas from the battery acid that is explosive and dangerous to breathe. The level of the acid in the battery needs to be checked and the battery refilled with distilled water after equalization if the water level is low.

When the solar energy system produces electrical power, it must be sent through a charge controller before it goes to the battery. The charge controller protects the battery from overcharging by stopping the voltage from the PV cells from flowing to the battery when the battery is fully charged. The charge controller will also provide a way to monitor how much of a charge the battery has. Another job of the charge controller is to prevent the battery from being drained down so far that it cannot be recharged.

Initial costs for a solar battery depend on the type of battery. A mid-sized battery will cost anywhere from $200 to $1,000. An industrial battery will cost several thousand dollars. Smaller batteries, such as marine or golf cart batteries, might cost less than $100. Freight and handling costs need to be added to get a true final price.

Replacement and disposal of the batteries represent an ongoing cost in the solar energy system. Solar batteries must be replaced after several years. Although industrial lead-acid batteries may last as long as 20 years, intermediate-sized batteries will last approximately 7 to 12 years. The smallest batteries, such as those used by a golf cart, may only last 3 to 5 years. NiCad batteries last longer in years but not in power cycles. All batteries must be disposed of properly. This can be quite expensive if the battery is NiCad. Figure 1-10 shows a typical battery storage system for a solar energy system. Battery charging systems will be covered in greater detail in Chapter 9.

FIGURE 1-10 A typical battery storage facility for solar energy system. (Courtesy NREL and the U.S. Department of Energy.)

1.9 GRID-TIED SOLAR PHOTOVOLTAIC SYSTEM

Photovoltaic cells can be grouped together in large-scale power production, and the electrical power can be stored or connected directly to the electrical grid. This type of system is called a *grid-tied system*. The grid-tied system can be part of a residential system, a commercial system, or a larger solar installation on the roof of an industrial site. Since the solar photovoltaic system has no moving parts anywhere in the entire system, grid-tied photovoltaic systems are the most reliable renewable energy technology systems available. PV manufacturers today are making solar panels at such quality that they are able to warranty their panels for 25 years, and the expected lifespan of PV panels is now closer to 50 years.

The advances in electronic power converters (inverters) over the past 10 years has helped make DC voltage conversion to AC voltage and integration to the electrical grid much easier. Today the inverter enables the electrical grid to replace batteries which are typically the most expensive cost in a system. Since the lead-acid battery storage systems are considered to be toxic, their disposal must be controlled. Now, instead of connecting solar panels to a battery bank, the solar electric system is connected directly to the power grid and the inverter allows the solar electrical system to provide the voltage needs of the building or residence and to feed surplus electrical power to the grid through a net meter. The net meter allows the residence or industry to seamlessly draw electricity from the grid when there is not enough sunshine to meet the demand; during the summer months, when the system is likely to be producing more than is needed, the surplus feeds the grid through the meter and the user gets a credit on its utility.

1.10 PARABOLIC TROUGH SYSTEMS THAT GENERATE ELECTRICITY

Several types of parabolic trough technologies are used today. One type of parabolic trough system uses mirrors to focus solar energy to a receiver pipe that is filled with a fluid, such as synthetic oil or molten salt. The concentrated sunlight causes the temperature of the fluid to increase above the boiling point of water, and this fluid moves through a heat exchanger. The heat exchanger has two loops that keep the synthetic oil or molten salt separate from the fluid that turns to steam. The steam is directed onto a turbine wheel that turns a traditional electrical generator. The turbine and generator for this system are the same type of generating system used in a coal-fired or nuclear power station. The trough system is generally located in parts of southwest United States where the sun shines more constantly and outdoor temperatures are high enough to heat the oil or molten salt. In some locations the conditions for heating are so good that the amount of heated synthetic oil or molten salt produced is more than

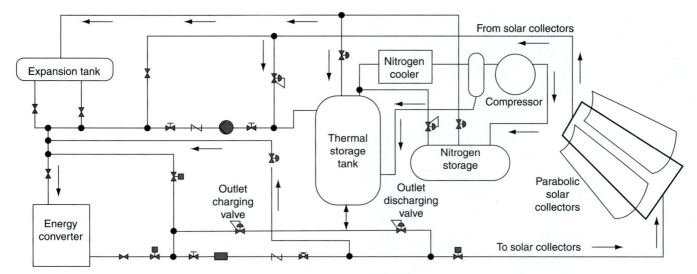

FIGURE 1-11 A diagram showing the plumbing components of a solar parabolic trough energy-producing system.

can be used during the day, so it can be stored, which allows the heat exchanger to produce steam even after the sun is no longer shining at the end of each day. When the fluid begins to cool below the temperature that steam can be created in the heat exchanger, another energy source such as natural gas can be used to provide heat. Then steam can be produced at night or on cloudy days, and the electrical generator can be turned by the steam striking the turbine on a continual basis.

With increased research, the parabolic trough system should become competitive in operating costs and cost per kilowatt-hour with traditional coal-fired and nuclear power plants. This is possible because the parabolic trough system is capable of generating sufficient steam to operate larger megawatt plant generators. Figure 1-11 shows the plumbing and components for the parabolic trough type system. Figure 1-12 is a photograph of this type system showing the relative size of the mirrors and the piping.

1.11 DISH/ENGINE SYSTEMS FOR GENERATING ELECTRICITY

Another type of solar energy system that generates electricity is the dish/engine system, and it uses a dish-shaped array of mirrors to concentrate sunlight into a receiver that is located in the dish. The receiver consists of a sealed system of piping. It has a tank that is filled with gas that is under pressure. When the gas is heated, the energy in the gas is transmitted to a **Stirling solar engine**, and the pistons in the Stirling engine cause a crankshaft to rotate and turn a generator. Figure 1-13 shows a picture of this type of system, and you can see that the dish is rather large (approximately 35 to 40 feet in diameter).

The basic principle of operation of a Stirling engine is that it has pistons and a crankshaft with a fixed amount of a gas sealed inside the engine. When the gas inside the

engine is heated, the pistons begin moving, which turns the crankshaft. As long as energy is received, the gas inside the engine goes through a cycle of becoming pressurized from the heat and absorbs additional heat from the solar energy collector and then uses this energy to move the pistons, which in turn moves the crankshaft and generator. This type of system has about a 30% efficiency of converting solar energy into electricity. When the sun begins to produce thermal energy in the early morning, the gas in the Stirling engine begins to take on energy and expand, which causes the pistons to move and the engine begins to run. When it comes up to speed and has enough horsepower, the electrical generator is energized and the system begins to produce electricity. The amount of electricity generated depends on the amount of solar energy that is converted into heat and transferred to the gas in the engine. Since the gas and the engine are sealed, the gas continually expands and causes the pistons to operate; then the gas condenses and is ready to be heated again. At the end of the day when the sun starts to go down, the amount of energy begins to draw down, the generator is brought off line, and the engine eventually stops since there is not enough heat energy in the gas. As soon as the sun begins to show again in the morning, the collector will begin heating the gas and the engine will start to cycle again for another day. The Stirling system starts and stops automatically each day and produces electricity whenever the solar energy is sufficient to turn the generator.

Because the Stirling engine has a dish for collection, it must have a complex tracking system that moves it through a range of positions to ensure it is always collecting the maximum amount of solar energy. The initial cost of the hardware of the Stirling system is rather high, so the cost of producing a kilowatt-hour of energy from the generator is approximately $2 per kilowatt-hour. This is rather expensive when compared to the cost of nuclear power at approximately 15 cents a kilowatt-hour and to coal, which is approximately 10 to 12 cents per kilowatt-hour. With

FIGURE 1-12 A parabolic trough solar energy producing system. (Courtesy U.S. Deptartment of Energy.)

FIGURE 1-13 The collectors for a Stirling engine solar generating system. (Courtesy U.S. Department of Energy.)

continued research and additional installations, the cost per kilowatt-hour will continually be reduced. You will learn more about the Stirling engines in later chapters in this text.

1.12 FEDERAL AND STATE DEPARTMENTS THAT PROVIDE RESEARCH FOR SOLAR ENERGY

Solar energy research has expanded tremendously over the past few years. The research is controlled by the federal Department of Energy and the National Renewable Energy Laboratory (NREL). This section will discuss the relationship among companies that produce solar energy equipment, university researchers, and the federal agencies that provide research dollars to make the studies possible.

1.12.1 Department of Energy

The Department of Energy (DOE) is a cabinet-level department in the U.S. government. The head of the DOE works directly with the president to help set policy for renewable energy technology. The DOE creates policies and directs research funding for solar energy and other energy development.

1.12.2 National Renewable Energy Laboratory

The National Renewable Energy Laboratory (NREL) provides standards and data for the solar energy industry. This certification is called "type certification" and is used to ensure that any solar energy equipment is designed, manufactured, documented to follow specific standards and other technical requirements. This certification also includes additional evaluations completed during installation, operation, and maintenance to ensure all of these functions are in compliance with the design standards. The type certification is applied to all turbines of the same type and it relates to how they conform to design evaluation, type testing, and manufacturing.

1.12.3 Sandia National Laboratory

Since 1949, Sandia National Laboratory has developed science-based technologies that support our national security. Sandia is a government-owned/contractor operated (GOCO) facility. Sandia Corporation, a Lockheed Martin company, manages Sandia for the U.S. Department of Energy's National Nuclear Security Administration.

Today, the 300+ million Americans depend on Sandia's technology solutions to solve national and global threats to peace and freedom through science and technology, people, infrastructure, and partnerships. One of Sandia's missions is solar energy research and development. Sandia National Labs provides research funds and knowledge for developing solar and wind energy. Sandia's solar energy research lab is located in Albuquerque, New Mexico.

1.12.4 Oak Ridge National Laboratory

The Oak Ridge National Laboratory (ORNL) is located in Roane County, Tennessee, about 7 miles from the center of the city of Oak Ridge and about 25 miles from Knoxville. Oak Ridge National Laboratory is the Department of Energy's largest science and energy lab. Managed since April 2000 by a partnership of the University of Tennessee and Battelle, ORNL was established in 1943 as part of the secret Manhattan Project to pioneer a method for producing and separating plutonium. During the 1950s and 1960s, ORNL became an international center for the study of nuclear energy and related research in the physical and life sciences. With the creation of the DOE in the 1970s, ORNL's mission broadened to include a variety of energy technologies and strategies. Today the laboratory supports the nation with research on solar energy.

1.12.5 National Center for Photovoltaics

The National Center for Photovoltaics (NCPV) is part of NREL. It focuses on innovations in PV technology that drive industry growth in photovoltaic manufacturing nationwide. By working together, government, academic, and private entities address technical challenges with stronger solutions and benefits to PV. The mission for the center is to help the U.S. photovoltaic industry maintain a competitive position. The U.S. Department of Energy created NCPV, and it is now part of the its Solar Energy Technologies Program.

1.12.6 National Electrical Code and Underwriter Laboratory

The National Electrical Code (NEC) pertains to electrical fire and safety standards, and it was written by the National Fire Protection Association (NFPA). The NFPA was established in 1896 as an international nonprofit association. The NEC is part of NFPA 70 that pertain to electrical safety and is published and updated every three years. The latest publication was in 2008. Solar energy electrical systems and wiring must conform to the NEC in the United States. This includes wiring in panels, switches, and controls. The NEC is also followed to ensure correct grounding of all electrical components and circuits to ensure personnel safety and to prevent electrical shock hazards.

The NEC works with other agencies such as Underwriter Laboratory (UL) to ensure electrical safety. For example Article 280 of the NEC and UL 1449 apply to lightening protection and surge to help prevent damage or losses due to lighting strikes. UL also provides standards (UL 1741) for electrical inverters. The electrical inverter is used in some solar energy applications to convert the voltage output from the solar energy equipment to exactly 60 Hz. You will learn more about these standards and codes in later chapters.

1.12.7 ANSI American National Standards Institute

The American National Standards Institute (ANSI) is a private organization that was founded in 1918. Its membership consists of private and nonprofit organizations. Their standards are considered open standards because all stakeholders helped develop them. The goal of any organization that establishes standards is to get its standards accepted by ANSI. ANSI helps ensure the integrity of the standards developers. ANSI identifies organizations that use its structure to create and maintain standards, such as the Solar Energy Industries Association (SEIA), which creates performance standards for solar energy equipment.

1.13 POLITICAL IMPLICATIONS

The political implications for solar energy include the tax credits and incentives that encourage solar energy installations and laws that will limit the expansion of some carbon-based fuels and some other energy sources. Federal, state, and local governments will continue to enact laws and regulations that will move toward a more aggressive integration of solar energy with current coal, nuclear, and other methods of generating electricity.

Regulations for clean air will limit the increased use of coal and other carbon fuels, while increasing the need for more electricity produced by solar energy. Federal, state, and local governments can also improve the cost factor by creating an artificial pricing structure for electricity produced from solar energy that will allow it to compete with electricity produced from coal, nuclear, or other forms of carbon fuels. These same governments can also create structures that make it more profitable for individuals to produce electricity from solar energy and sell it back to the electrical utility for higher rates.

Federal, state, and local governments can provide tax breaks, grants, and secure loans for installations of solar energy and the ongoing maintenance and repair of system equipment once it is installed. The government also is providing large amounts of money for updating and expanding a smart grid. This research and development of the smart grid and the infusion of millions of dollars of new investments will allow the smart grid to be expanded and make it possible for more solar energy systems to connect to the grid and sell back excess generated electricity.

1.14 FINANCIAL IMPLICATIONS AND RETURN ON INVESTMENT

Several factors need to be considered to determine the *return on investment* (ROI) for solar energy. ROI is the amount of money or other benefits that become available after money is invested in a project. For example if a solar project costs $100,000 to install and start up, the money that the investors receive from the energy the solar turbine produces over a specific period of time is the return on investment. The return on investment must be large enough to make it feasible for investors to invest in solar energy projects. Individuals or companies that invest in a project will calculate the amount of return and how many years it will take to repay their investment. The factors used in determining the ROI include the ability to harvest solar energy and turn it into electricity and how often the sunshine is enough to do so. The more efficient a solar energy system is and the longer the sun shines at maximum value, the better its return on investment. Basically the larger solar energy equipment is, the more efficient the system may be, leading to a higher ROI than would come from a smaller unit. The next issue that is important to ROI is the initial cost of the solar energy system and the installation costs. When governments make overt efforts to recruit solar projects and installations and lower the costs of initial planning, ensuring that installation costs remain low, the ROI is vastly improved. When the government offers to pay grants and guarantee loans, it also lowers the initial cost immediately. When loans include lower interest rates, the ROI again increases.

Certain installations will cause the output of the solar energy system to be increased over other installations. For example, solar farms in the desert will ensure that a significant amount of solar energy is available for harvest, and the solar energy will be collected more continuously and more strongly to ensure that the solar energy system is producing at near peak levels. When the amount of sun and the duration of the sunlight increase, so does the ROI.

Another factor that improves the ROI is the installation of larger numbers of solar energy systems in a cluster called a solar farm. When multiple solar energy collectors are installed near one another, the installation costs are reduced slightly because of improvements to the site infrastructure, such as electrical power substations, grid connections, Supervisory Control and Data Acquisition (SCADA) systems, and other technology that can be shared.

1.15 OTHER WAYS TO GET INCOME FROM SOLAR ENERGY

Another way companies, farmers, and other landowners can make money from solar energy systems is through contracts with energy companies that install solar energy on their property and pay them a fee for this. The sites are selected because they produce the best solar energy. Since the solar energy companies do not own the land, they must either purchase land outright or lease it and pay the owner in some way. Payment methods include fixed payments over a specific period of time that may include an established increase at specific dates of the contract. The advantage of this type of contract is that it provides a stable income for the landowner and provides a consistent known cost to the energy producer.

Another way to receive payment from a solar energy company is to design a contract that pays a royalty or percentage of the produced income. This type of contract needs to have a means by which both parties can get access to all of the records of production and other costs. A problem with this type of contract is that the income may vary from month to month due to the amount of solar energy that is produced each month. A similar plan uses a fixed payment amount plus a percentage from a royalty. Yet another type of plan involves an equity partnership, which involves the landowner selling all or a portion of the land to the solar energy company. If the ownership is percentage based, then the income is split along the same percentage.

1.16 GREEN ENERGY PAYBACK CALCULATIONS

One of the most important concerns with deciding to install a solar energy system or a series of solar panels on a solar farm is determining when they will produce enough income dollars to pay off the investment and begin to earn income for the owner. The calculation that helps determine the point at which the solar panels produce enough energy to pay back the investment is called the pay back calculation. Basically the pay back calculation involves two elements: all the costs on one side of the equation and all the income factors on the other side. Table 1-2 lists all the cost factors and all the items that should be included for producing income.

For example the cost may include site preparation; permits and supporting documents for environmental impacts and inspections; shipping the solar energy equipment to the site, which may include shipping from an overseas location; installing brackets; installing the panels on the brackets; start-up costs and making start-up adjustments; connecting to the electrical system and to the transformers, substation, and grid; and installing the SCADA system, which consists of software that continually monitors the solar energy output. The other costs include annual inspections and periodic maintenance, as well as a budget for minor and major maintenance that will occur during the life of the electrical and mechanical equipment. In larger systems or on solar farms you may need to factor in salaries of onsite maintenance personnel.

On the income side of the calculation you would need to estimate the amount of electrical energy that solar energy will produce per day, week, and month so that you can determine how much electrical load you will be replacing with the solar energy output if the power is being consumed onsite as a stand-alone system. To estimate the output of the solar energy equipment, you would need to use the size of the panels and the estimated amount of sunlight for the location where you are installing the equipment (there are formulas for this on several websites). Next you would calculate the total cost of the replacement power the solar energy equipment is producing and use the utility billing cost factor to determine the total cost you would have paid the utility for this power, including any penalty charges such as exceeding a demand factor or other penalties for excessive usage for peak periods. The electrical rates will be different for residential users as compared to commercial or industrial users. Once you have the income and cost calculated, you can determine the length of time it will take to pay back the investment for a solar energy system that will supply power directly to a home, commercial establishment, or industry.

If the system is a grid-tied system and the power is being sold directly to the grid, you would need to determine two things: the output of the solar energy system over a period of time, and since the power is being purchased

TABLE 1-2 Cost and Income Factors for Solar Projects

Cost Factors	Income-Producing Factors
Site preparation	How much electrical power will be produced daily, weekly, or monthly?
Permits	How much electrical power is offset or will be replaced by the solar energy system?
Environmental studies	Price of the electrical power that the solar energy equipment will be replacing.
Inspections	Cost of power that is purchased or bought back by the electric utility.
Shipping from manufacture to site	Efficiency of the solar energy system and any storage capability.
Cost of the solar energy equipment	Location of the solar equipment and how many hours of peak sun it will have per year.
Installing the panels	Grants or tax credits.
Bringing the solar system online	Accelerated depreciation for tax purposes.
Electrical gear and connecting to the electrical system	Renewable energy credits.
Transformers and substations	
Installing SCADA systems	
Annual inspections	
Periodic maintenance	
Estimated minor maintenance	
Estimated major maintenance	
Salaries of any personnel	

back by a utility, the "buy back" price that is determined by the utility that is purchasing the energy. You need to be aware that the price the utility pays for electrical power it is buying back may be substantially less than the price of the power it is selling. For this reason some larger solar farms act as a small utility company and sell the power directly to other end users, instead of selling back to the utility.

1.17 ONLINE PAYBACK CALCULATORS

Several online calculator programs or computer programs will help you determine the exact values for each of these factors. Some calculations include more or fewer factors and may weight some factors more than others. Some companies that manufacture and install solar energy can provide payback calculations that are very accurate and can

define the values more accurately since they have a long track record of installing and operating solar equipment.

Payback calculators are found at a variety of websites. The information in Table 1-3 can also be used to determine the payback. After the efficiency of the solar energy system is established, the amount of power the system will be capable of producing is determined. For this step you will need to determine how long the sun is available each day and what the average electrical output of the solar energy system is during that day for the location where it will be installed. This will help determine the amount of electrical power the solar energy system will create for each day. The information used for this part of the calculation must be as accurate as possible to provide an accurate payback. This information is also specific to the site location. These data may need to

TABLE 1-3 Data Used to Calculate Solar Energy Payback Period

Data	Notes
The mean power output as a function of mean solar energy input	From the manufacturer's data
Estimates of the cost per kilowatt-hour as a function of mean solar energy equipment cost	From a calculation of output solar energy equipment and where it is installed
Estimates of the payback period as a function of mean electrical energy produced and referenced to local electricity costs	Need to know the cost of electricity that is being replaced, or if it is sold to the grid, the cost of the electricity buyback rates
Power output profile showing the percentage of time that the solar energy system produces different levels of power including the zero-power output case	From calculation of how long the solar equipment will operate each day

be averaged over a day, a week, or month and adjusted for month-to-month variations depending on the site. Some of these data can be taken from the data provided by the SCADA system from another solar energy site that has been running at this location previously. This is useful when additional solar energy systems are added to a location or to a solar energy farm.

Once the efficiency of the solar energy equipment is determined, the amount of time it will be operating and the total amount of energy it will produce in a year can be calculated. The next step in the payback calculation is to determine the cost of the electrical energy the solar energy system is replacing, or the buyback cost of the electricity if its electrical power is produced primarily for a utility on the grid. These amounts will vary from location to location and state to state depending on electrical power rates, legislation, and the current regulations for producing electrical power from solar energy. Once these two values are known, and the total cost have been defined, the payback period can be determined.

1.18 TAX CONSIDERATIONS FOR SOLAR POWER

Financial incentives are available in the form of tax relief programs that include income tax incentives, corporate tax incentives, property tax incentives, and sales tax incentives. The tax incentives include both federal and state tax incentives as well as loans and other incentives to offset the cost of installation for solar energy. These incentives are designed to help accelerate installation of new systems. Some of the incentives are designed for home owners and personal use systems, and others are designed for corporations that are installing and operating larger systems to help add electrical power to the utility supply. Table 1-4 shows some of these programs. This table

is provided by the database of State Incentives for Renewables & Efficiency (DSIRE), a program that was established in 1995 as an ongoing project of the North Carolina Solar Center and the Interstate Renewable Energy Council (IREC). This is just a sampling of federal programs, and you need to check these types of websites for the latest programs as they become available and are funded under various budget proposals from the U.S. government. You should also be aware that many of these programs have limitations and deadlines. You can go online to the DSIRE website to check the latest incentives.

At this time the programs from the federal government include corporate tax deductions, corporate tax depreciation, corporate tax exemptions, loans, corporate tax credits, federal grant programs, federal loan programs, industrial recruitment support, personal tax exemption, personal tax credit, and production incentives.

At the federal level the current tax incentive for residential systems is for 30% of the total installed cost of the system, not to exceed $4,000. For commercial systems the credit is equal to 30% of expenditures, with no maximum credit for small solar systems placed in service after December 31, 2008.

1.19 ADDITIONAL U.S. INVESTMENT TAX CREDIT

In 2005 the U.S. Congress passed the Federal Energy Policy Act. This act included a federal tax credit for residential property initially applied to solar electric systems, solar water heating systems, and fuel cells. In January 2008 H.R. 1424 extended the tax credit to small solar energy systems. This program will extend the tax credit through December 31, 2016. This tax credit must be used against the alternative minimum tax, so some low-income home owners may not be eligible to use it.

TABLE 1-4 Federal Incentives/Policies for Renewables and Efficiency	
Financial Incentives	**Name of Program**
Corporate deduction	• Energy efficient commercial building tax deduction
Corporate depreciation	• Modified accelerated cost-recovery system (MACRS) + bonus depreciation (2008–2009)
Corporate exemption	• Residential energy conservation subsidy exclusion (corporate)
Corporate tax credit	• Business energy investment tax credit (ITC) • Renewable electricity production tax credit (PTC)
Federal grant program	• U.S. Department of Treasury—renewable energy grants • USDA—Rural Energy for America Program (REAP) grants
Federal loan program	• Clean renewable energy bonds (CREBS) • U.S. Department of Energy—loan guarantee program • USDA—Rural Energy for America Program (REAP) grants
Industrial recruitment support	• Qualifying advanced energy project investment tax credit
Personal exemption	• Residential energy conservation subsidy exclusion (personal)
Personal tax credit	• Residential energy efficiency tax credit
Production incentive	• Renewable energy production incentives (REPI)

A taxpayer who uses the residence as a primary home located in the United States may claim a credit of 30% of qualified expenditures for a system that provides power to that dwelling unit. Equipment expenditures can be deducted when the installation is completed, power is being produced, and the home is occupied. Expenditures can include costs for onsite preparation, labor, assembly, or installation of the solar panels and for piping or wiring used to connect the system to the home electrical system. If the federal tax credit exceeds tax liability, the excess amount may be carried forward to the succeeding taxable year. The maximum allowable credit will vary by technology and equipment requirements. Owners of small solar energy systems with 100 kilowatts (kW) of capacity and less can receive a credit for 30% of the total installed cost of the system, not to exceed $4,000. The credit will be available for equipment installed through December 31, 2016. For solar panels used for homes the credit is limited to the lesser of $4,000 or $1,000 per kW of capacity.

Another program is the investment tax credit (ITC), which is a federal-level program that is available to help consumers purchase small solar energy systems for homes, farms, or businesses. The ITC was originally created as part of the Emergency Economic Stabilization Act of 2008, and it was modified through the American Recovery and Reinvestment Act of 2009. The ITC is available for equipment installed from October 3, 2008, through December 31, 2016.

1.20 EXAMPLES OF STATE PROGRAMS

Many states have implemented tax programs to help stimulate new installations of solar energy equipment and systems. These tax programs are in addition to the federal tax programs and can be used with federal tax deductions. This means that home owners, business owners, and commercial energy producers can put together state and federal tax incentives to help with the financing of installing a solar energy system.

The Ohio Department of Development's Ohio Energy Office has a grant program to implement renewable energy projects limited to solar electric and solar thermal systems for commercial, industrial, institutional, and governmental entities in Ohio. The maximum incentive is 40% of eligible system costs, with a maximum incentive of $200,000 of eligible system costs. Ohio also has a residential grant that provides up to $25,000 for residential systems through the Residential Solar Energy Incentive (NOFA 09-02). What is unique about the program in Ohio is that this grant opportunity is an Advanced Energy Fund program funded by a rider on the electric bills of customers of investor-owned electric distribution utilities in the state. The utility companies include American Electric Power (AEP), Duke Energy (formerly CINergy), Dayton Power and Light, and First Energy.

1.21 CHECKING OUT THE LATEST OFFER

It is important to check with local, state, and federal programs prior to planning the installation of solar energy equipment or solar energy systems. These programs change almost monthly as new legislations is approved at the local, state, and federal levels. You can check the most up-to date programs through federal, state, and local government websites. You can also contact the appropriate energy offices or the offices that control incentives for solar energy systems.

Some utility companies also provide the latest information about tax incentive and loan programs. Some solar energy equipment manufactures and companies that install solar energy equipment are also aware of all the tax incentive and loan programs available at any time.

1.22 SKILLS NEEDED FOR GREEN TECHNOLOGY JOBS

The solar energy industry will require a large number of electrical technicians and mechanical technicians to assemble, install, troubleshoot, and maintain all sizes and types of solar energy systems. The industry will also require a large number of architects and designers as well as planners who will help determine the exact location to place individual or larger solar energy projects and solar energy farms. Most of these jobs will require the individual doing the work to be certified by the North American Board of Certified Energy Practitioners (NABCEP). Courses are available that will prepare you to take the test to receive this certification.

When you identify all the parts that are required to operate a solar energy system, you find another larger group of technicians who will be needed to manufacture the metal for the different types of brackets, the electronic companies to make the programmable logic controllers (PLCs), control boards, inverters, and switch gears for the electrical part of the system. Machinists will be needed to manufacture all the gears in the gearboxes and transmissions in tracking systems, as well as the machining needed for all the hydraulic pumps, valves, and cylinders that are used in some solar tracking systems. Machinists will also supply the mounting brackets and other hardware for solar panels.

Another group of jobs will be available for site preparation, which involves everything from road building to preparation of footings used to anchor solar panels. This involves excavating the holes for the concrete and mounting any conduits into the concrete for the electrical wires to run underground to the solar panels.

Other jobs include installing solar panels on the roofs of homes and businesses. On smaller units this will consist of operating medium-size cranes to large-scale cranes to lift panels onto the roof of a home or business. Another group of technicians will be needed to extend the high-voltage transmission lines between the solar farms and the nearest grid tie point.

In some applications, a complete large-scale electrical power distribution system will be needed where large numbers of solar panels or solar farms are located. The power distribution system will consist of large banks of transformers, as well as the power distribution towers for transferring high-voltage lines over long distances. Another growth in jobs will be the companies that manufacture the transformers, towers, and high-voltage cables, as well as the companies that locate and install the towers and transformers in the field.

1.22.1 Skills Needed to Install and Maintain Solar Energy and Solar Heating Systems

When solar energy systems operate for long hours, they need routine or periodic maintenance. Technicians doing this work must be able to climb on rooftops to work on the switch gear, battery charging systems, solar tracking systems, and more complex mechanical systems used to provide electrical power by a generator that is turned by steam passing over turbines. All the components must be inspected periodically. Sometimes vibration analysis will be required, and adjustments to shafts and couplings must be made on these larger generation systems.

In some cases the technicians will need mechanical and rigging skills to install or change out solar panels on a room. Some of this work may be completed with larger lifting and crane equipment that are used to place solar panels or solar heating equipment on the roof. In other applications, the parts may be removed and replace individually and all of the work is completed at the roof tops. Technicians will be needed for overhaul and periodic maintenance projects for solar energy systems and these technicians may be assigned to a specific site, or they may travel from site to site on a contract basis since the work at any one site may be limited to a few weeks per year.

1.22.2 Jobs in Solar Energy Industries

In the next 10 years the solar energy industry will need a large number of skilled technicians and other professionals to meet expansion needs. Some projections put the total number of new jobs at 1 million. The jobs in the solar energy technology industries are referred to as "green-collar jobs." These jobs include construction managers, leaders of solar energy installation crews, maintenance technicians, engineers, senior mechanical and electrical engineers, and environmental engineers.

1.22.3 Planning and Sales Jobs for Solar Energy

Other types of jobs in the field of solar energy include planning and sales jobs. The planners need environmental backgrounds and training, and they must understand the requirements of solar energy availability and other details about site preparation, including connecting to the grid. Sales specialists are needed to meet with clients who express an interest.

1.22.4 Educational Requirements for Solar Energy Jobs

Some of the solar energy jobs will require only a high school or a vocational school degree. These types of jobs will be the lowest-level jobs. Certificates from community and technical college programs in electrical, mechanical, automation, computer-aided design or computer aided drafting (CAD) will be needed, as well as architecture or project management degrees. Some of the jobs will require a bachelor's degree from technology programs or engineering and engineering technology degrees. Some of the jobs will also require a four-year business degree or a master's degree. Some employers will want technicians to have the proper training and get a solar certification once they have been on the job for some time.

1.22.5 Teachers and Technical Trainers

As the field of solar technology requires more and more trained technicians and trained personnel, the need for teachers and trainers in this area will expand also. Companies that manufacture and maintain solar energy systems will need a number of technical trainers to provide proper instruction and training for the maintenance and operation of these large systems. Community and technical colleges as well as universities will need a number of skilled individuals who have the ability to teach these technical programs.

1.22.6 Field Service Technicians

The field service technician is an employee of the company that manufactures and installs solar energy systems. These technicians are on call 24 hours a day, 7 days a week (24/7) so that they can quickly respond to problems at solar energy facilities. The field service technician will have electrical skills to work on the electrical panel and the electrical conversion equipment and mechanical skills to work on the mechanical drive train for tracking systems for solar panels. Since some solar systems are used to produce high-temperature or high-pressure steam that is used to turn generator turbines, some additional jobs will include additional mechanical skills. The specialty skills for these technicians may also include troubleshooting and repair of hydraulic systems and their components, PLCs, and electronic systems used in inverters and other monitoring systems for electrical generating turbines that are similar to coal-fired and nuclear electrical generating sites.

When a problem occurs in a solar energy system and it stops producing electrical power, the field service technician is called to troubleshoot the entire system and quickly identify not only the symptoms of a problem but also its root cause. Field service technicians are also used to complete routine periodic maintenance on solar energy systems.

1.22.7 Project Manager and Architectural Jobs

When a solar energy project is being designed, specified, and installed, personnel are required to design and implement the system. These tasks include researching specifications, fulfilling permit requirements, and managing the entire installation process. Some of the skills are similar to managing a construction project, but a thorough understanding of the requirements of solar energy systems and building structural design is needed since much of the solar energy system is mounted on rooftops or must be integrated into building structures when the solar energy is used for heating or lighting systems. The skills become more complex when a number of solar energy packages are installed at the same location as part of a larger solar energy farm.

The project manager must understand all the requirements for preparing the site, installing solar energy systems into new and existing buildings, and making the final connection between solar energy electrical production equipment and the electrical grid. The project manager will rely on specialists in each of these areas, but the project manager's skills will keep the project on schedule and within budget. Architectural skills for these types of projects will be similar to those for other large construction projects, but additional knowledge about solar energy systems will be required.

1.22.8 Solar Energy Panel Technicians

In the next 10 years the numbers of solar energy systems that are manufactured will nearly double. A number of new companies will begin to manufacture and assemble solar panels throughout the United States and overseas. The jobs in this area will include assembly of the component parts to make an operating solar panels and other solar heating system equipment. Manufacturing and assembly technicians will need to be able to read and interpret blueprints and follow procedures to assemble the mechanical and electrical components in the solar energy heating components and photovoltaic panels. This may involve assembling all the components at the manufacturing site or at the location where they are installed. After these components are completely assembled, they will be put on a test stand and operated under test conditions, so that final adjustments can be made. The final assembly is checked for alignment of electrical connections. If the photovoltaic system uses tracking hardware, it must be set up and tested, and control of the electrical production system verified. If an inverter is used, it is connected to the batteries or the output of the PV panels and the battery charging and control systems and inverter must be tested to ensure that their output produces voltage at a set frequency of 60 Hz.

Another job in this process is a quality-control technician who ensures that procedures are followed and specifications are met when the products are manufactured. The quality technician ensures that every solar energy product that leaves the assembly facility is ready to be installed and will operate without problems.

1.22.9 Solar Heating Technician

As the demand for solar energy heating products increases in both the United States and abroad, a large number of companies are beginning to produce solar heating products and other components used in the control systems for them. Modern solar heating systems have complex control systems that must be installed and calibrated to ensure customer satisfaction and energy conservation to provide the best return on investment for the solar heating products. The solar heating products can be used to heat water for the hot water needs of a home or commercial establishment, or they can be used to heat water for heating the home like a boiler system. Technicians who install and service solar energy heating systems will need to be trained in the theory, operation, and control of traditional heating, ventilating, and air-conditioning (HVAC) systems, since the solar system must integrate with the systems that will be needed for backup or subsidiary heating or cooling. The integration to traditional systems is required because solar energy systems cannot stand alone because of limitations of solar energy availability.

1.22.10 Electricians and Electrical Transmission Technicians

When a number of large solar photovoltaic panels are connected to produce large volumes of electricity, the electrical controls and switch gear to connect directly to the grid or to charge battery systems must be set up and maintained to ensure that they are operating safely and efficiently. These systems require electrical technicians to install and maintain them so that they meet the requirements of the electrical codes and are properly connected to the grid. This process is similar to connecting any electrical generation system, such as a coal-fired, nuclear power, hydroelectric, or gas-fired generation system to the grid. If a large amount of electricity is being transmitted from the solar farm to the existing grid, additional power poles and transmission towers may be required. This system may also require a number of new transformers and substations to provide the power at the proper voltage level. Electricians and electrical transmission technicians will be required to ensure that the power transmission system meets the requirements of the existing transmission utility.

Some of the new installations to transmit a high-voltage power from the solar energy site to the existing grid may be installed underground. Underground cable installation involves burying large conduits (plastic piping) underground and then pulling electrical cables through them. Modern installation processes allow the conduits to be placed underground with a drilling/ boring machine. This machine uses a drill bit that is mounted on the end of the drilling pipe and is mechanically controlled to ensure precise positioning as the drill pushes through the earth at depths of 3 to 10 feet. A very sensitive detection system follows the drilling head as it progresses so that it can be located within inches of its intended path. The drilling pipe is added in 8- to 10-foot sections until the entire length of the run is completed. After the drilling head reaches its final destination in a hole that is dug to the required depth, the plastic pipe that is used for the conduit is connected to the end of the drill. The plastic conduit is stored on large rolls, which allow it to be unrolled as the drill pulls the drill pipe sections back to the drilling machine. When the process is completed, the two ends of the conduit remain 3 to 6 feet out of the holes at each end. Technicians use a pull wire that is installed into the conduit and attached to the cables to pull them through the underground conduit. The electrical cables are connected to the solar electrical cabinet on one end and to the service disconnect equipment where it is installed to the grid at the other end.

1.22.11 Riggers and Solar Panel Installers

The installation of solar panels on rooftops or on large solar farms requires rigging skills. These include attaching cables and lifting slings to the solar panels so that they can be lifted safely to their final position. The lifting cables and slings are connected to a crane that is used to lift them into position. The crane must be large enough to lift the weight of the solar panels and tall enough to ensure that they can be safely positioned to the center or far reaches of a rooftop. Rigging also includes the skills needed to precisely move the solar energy components when they are located above the ground.

1.22.12 Where to Look for Solar Energy Jobs

When you are looking for a solar energy job, the best places to look are websites that advertise these jobs. You can also go to the websites of the companies that manufacture and/or install solar energy equipment and check for the jobs that they are posting. Some of the companies that want to hire personnel for their solar energy jobs will host job fairs or other types of activities that directly recruit highly skilled personnel.

Questions

1. Name four types of solar energy systems.
2. Explain how a photovoltaic cell converts solar energy into voltage and current.
3. Identify the main parts of a solar hot water system and how it works.
4. Name four agencies that are doing research for solar energy.
5. Identify the factors you would need to consider to determine the return on investment for a solar energy system.

Multiple Choice

1. Which of the following best describes the difference between a passive and an active solar heating system?_____
   a. The active system uses more energy than the passive system because it has pumps and fans.
   b. The passive system uses more energy than the passive system because it has pumps and fans.
   c. There is no difference between the active and passive types of solar heating systems.
2. The solar energy from the sun is _____
   a. Different at the same location from season to season because the sun rotates around the earth.
   b. Different at the same location from season to season because the earth rotates around the sun.
   c. Is always the same at the same location from season to season because the sun is in a stationary position in relation to the earth.
3. If electricity from a solar energy system is connected to the grid _____
   a. It must be 24 volts DC because the voltage on the grid is DC.
   b. It must be sent through an inverter because the frequency needs to be 60 Hz to match the voltage on the grid.
   c. Electricity from solar energy can never be connected to the grid.
4. What does NREL stand for?_____
   a. National Resource and Electricity Laboratory
   b. National Renewable Energy Laboratory
   c. National Renewable Electricity Laboratory
5. Which of the following jobs will be involved in solar energy?_____
   a. Planning and sales
   b. Field service technicians
   c. Project managers and architects
   d. Electricians and electrical transmission technicians
   e. All of the above

2

Electrical and Energy Demands for the United States and the World

OBJECTIVES FOR THIS CHAPTER

When you have completed this chapter, you will be able to

- Explain the need for an uninterruptible and continuous power source for residential and industrial uses.
- Identify the total daily electrical demand for residential, commercial, and industrial uses of electricity.
- Explain the difference between regular electrical demand and the peak electrical demand.
- Explain how the electrical grid connects all parts of the residential, commercial, and industrial electrical systems throughout the United States.

TERMS FOR THIS CHAPTER

Brownout	Peak Electrical Demand
Electrical Demand	Uninterruptible Power Supply
Kilowatt	Volts AC (VAC)
Megawatt	Volts DC (VDC)

2.0 OVERVIEW OF THIS CHAPTER

Electrical power from solar energy is becoming a viable source for providing electricity from alternative energy. This chapter will provide an overview of solar energy from small residential and commercial systems to larger solar farms in use today. This chapter will also cover how electricity is currently used in the United States in residential, commercial, and industrial applications and the need for an uninterruptible and continuous electrical power supply. Information will be provided about the limits of transmission for electricity through the grid and data will also be provided to show the total electrical usage for industry and residential applications in the United States on a daily and annual basis. Other important information such as the total commercial and industrial demand and total peak demand will be explained.

This chapter will cover the types of solar energy systems used today that are stand-alone and provide power directly to a residence, or are small commercial application, or are tied to the grid where they supply power to a utility. Some applications are tied to the grid, but they also provide power to be used at the site. The availability of electricity from solar energy systems will also be discussed in regard to the amount of intermittent electrical power and the amount of power that is available 24 hours per day. The information in this chapter will help you understand how large the demand for electrical power is and how electrical power from solar energy will help add to the electrical power that is produced, but it will never be the exclusive power source for the entire country. Electrical power from solar panels can become a sole source of electrical power for some small homes with the aid of batteries to store energy for times when the panels will not produce, such as nighttime. The ideal application for sole source energy is for remote sites or locations where electricity from the grid may not be available.

2.1 THE NEED FOR AN UNINTERRUPTED AND CONTINUOUS POWER SUPPLY

The problems with electricity generated by the solar energy include that it is not consistent in most areas and it may not be available at night. The sun may shine in excess of 12 hours per day in the summer in the Northern Hemisphere, but it may provide sunshine for less than 9 hours a day in the winter. The sun does not provide the same amount of solar energy each day of the week because of storms and weather fronts that may cause cloudy days. Such conditions cause the electrical power from the solar energy system to be intermittent and not steady enough to provide the electricity needed to operate industry and commercial establishments on a continual basis.

Today's technology requires electricity at homes for telephones, computers, lighting, heating, and air conditioning, as well as refrigerating food. All of these items are needed for health and safety. Our police and security rely on electricity for communications and operations. Food stores, convenience stores, and homes require refrigeration to keep foods frozen or cooled to temperatures so they do not spoil. Many small businesses and homes require electricity to operate computers. Stores use electricity to power computers and telephone lines to complete credit card and other financial transactions. Banks and other financial institutions need electricity to keep their networks operating.

Many industries require large amounts of electricity on a continual basis to produce products, process foods and chemicals, and carry on other industrial processes. Many of these processes, such as making steel, chemicals, and pharmaceuticals, are continuous and cannot be shut down. For this reason, the electrical supply must be continuous and adequate to supply all of these operations.

The fact that electricity that is generated from solar energy must be augmented with traditional sources has been a problem since solar energy equipment was first introduced. Some applications simply tie the solar energy equipment into the existing grid to offset the use of hydrocarbon fuels such as coal or fuel oil that are used to

generate electricity. This problem is not unique to solar energy systems; it has also been a problem for wind energy generation, which becomes minimal at night in some locations or on days that the wind does not blow at a sufficient speed to generate electricity. The problem of intermittent electrical production also occurs with electricity that is generated from hydroelectric dams, which produce little or no electrical power in dry seasons or when there is not a sufficient amount of runoff from rains.

2.2 WAYS TO WORK AROUND VARYING AMOUNTS OF ELECTRICITY GENERATED BY SOLAR ENERGY

If solar energy is used to produce electricity for a remote home site, cabin, or farm, a bank of batteries can be used to store electricity that is generated when the sun is shining, so that electricity is available when the sun goes down. The batteries must be continually checked to make sure they do not overcharge or drain below their minimum rating. If this system uses a bank of batteries, the batteries have a cut-out relay to drop out the battery bank before the loads discharge them completely. Some smaller solar energy systems that are not tied to the grid must rely on batteries.

Other applications use a small solar generation panel to help keep the batteries at minimum charge so that a small amount of field current is always available. In the worst case if there are a number of days during which the level of sunshine is below the minimum needed to supply the residential need, the batteries can be used to supply the voltage for the loads during these times of low solar energy output. If the batteries discharge below the minimum rating, the control switch will disconnect them until the solar energy system can begin to charge them again.

If a home or small commercial site uses batteries to store excess power, and it needs AC voltage, it will need to use an inverter to change the DC voltage back to AC when it is used from the storage batteries.

Newer batteries that are stronger and can withstand deep discharge are being developed. Some research has been completed using large capacitors for storage also. These new technologies are intended to extend the range of storage capabilities for these applications.

2.3 TRANSMISSION LIMITATIONS FOR ELECTRICITY

The electrical power produced by solar photovoltaic panels is low-voltage direct current (DC), which cannot be transmitted very far unless it is converted to alternating current (AC) and stepped up through transformers to a higher voltage. Most residential users of electricity use single-phase 115 volt AC (VAC), which comes from two wires: one from line voltage called Line 1 and the other from a neutral point on a transformer. In most residential applications a second line called Line 2 is also provided, and it can be used in conjunction with Line 1 to provide 230 VAC for loads such as the air conditioning system, electric range, or clothes dryer. Each line (Line 1 and Line 2) provides 115 VAC to neutral. The 230 VAC provided by Line 1 and Line 2 are typically supplied from the power company through a step down transformer and are connected to the residence through an electrical meter and onto an entrance panel that houses the main circuit breaker for all the electrical loads in the house. This voltage is supplied through two wires that each have a voltage level of 115 V to neutral. If the DC electrical power from a solar panel is to be used in a residence as a replacement for the 230 VAC, it will need to go through an electronic inverter to change the DC to AC voltage, and it will need to pass through step up transformers to match the 230 VAC that are coming from Line 1 and Line 2. The transformer will also provide the neutral point so that each line can provide 115 volts to neutral for any 115-volt loads in the home. If the residence only had 115-VAC loads and did not have any 230 VAC loads, then it would only need one line and neutral.

If the solar energy DC power is produced on a solar farm and it is intended to be connected to the electrical transmission grid, it will need to be converted to three-phase AC voltage through an inverter and stepped up to the higher transmission voltages through step up transformers. If the power from the solar panels is used directly onsite, such as rooftop panels used on an industrial building, the DC power from the solar panel must be converted to three-phase AC voltage by an inverter and the step up transformers must be used to get the AC voltage to the 480 VAC, which is the same level as the voltage that is supplied by the grid through step down transformers to the industrial application. The AC voltage in this case would be supplied by Line 1, Line 2, and Line 3. For more information about the electrical power used in residential and industrial applications, inverters, and transformers, see Chapters 7 and 12.

Typically AC voltage is produced at a coal-fired or nuclear power generating site, and the voltage is passed through a set of step up transformers to get the power ready for its trip to its destination. Typical voltage levels for transmission depend on how far the voltage needs to be transmitted. For the longest distances, the voltage needs to be stepped up to 765 or 500 k volts (kV). For middle distances it will be transmitted at 345 or 230 kV, and for the shorter transmission distances the voltage is stepped up to 138 or 69 kV for smaller projects.

Once voltage reaches its destination, at sub-transmission stations called *substations* that are at the edge of cities or near large industrial users, the voltage is stepped down to 69 or 26 kV for distribution around these areas. If the voltage is routed around a residential area or light commercial area, the voltage will be stepped down to 13 or 4 kV so that it can end up at the end user at 480 or 240/208 V for industrial users and 240 or 120 V for commercial or residential customers. Figure 2-1 is an example of these voltages showing typical transmission patterns.

The limitation of voltage transmission will determine where solar energy generating systems are located and what amount of voltage is supplied to the transformer. Some of the solar energy systems will need larger step up transformers to get their voltage levels high enough to transmit voltage over the longest distances possible, up to 400 miles. Research is ongoing to design and implement larger solar energy farms up to multi-**megawatts** and put them together with multiple installations for solar energy farms so that they can use the same step up transformers and transmission lines. A megawatt is 1 million watts of electricity.

2.4 THE ELECTRICAL DEMAND FOR THE UNITED STATES

In order to understand how solar generated electricity will help provide electricity for an individual application or for the grid, you need to have some idea of how large the demand for industrial, commercial, and residential electricity in the United States is. The *electrical demand* is composed of several sectors: industrial, commercial and residential. First the total demand for

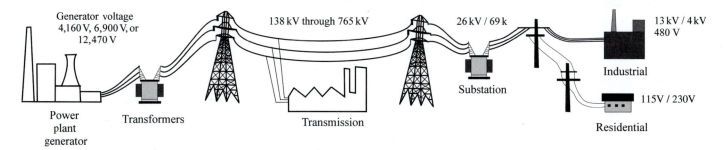

FIGURE 2-1 Diagram of typical power distribution system with the amount of voltage at each point indicated.

electricity can be measured by the total amount of electricity that is needed to supply all consumers over a period of an hour, day, week, month, or year. One of the reasons to understand the electrical demand is that it takes several hours to burn or convert the energy source and turn the generator prior to the consumption of the electricity. For example at a coal-fired plant, the electrical utility needs to anticipate the usage as far as an hour ahead so it can grind the coal and transfer it to the boiler that creates the heat to make the steam to turn the turbine. Once the electricity is produced, it is transmitted through the grid to its final destination where it is consumed.

The electrical power that is generated by the solar power equipment is created only when sunlight is available. During heating and air conditioning seasons, larger demands for electrical power may occur around the clock, so in many cases solar energy must be stored or supplemented to provide electrical power 24/7. Since the larger demand for industries occurs during the daylight hours, solar energy makes a good addition to electricity produced from fossil fuels or nuclear power.

Another way to calculate the total amount of power that is needed each day is to break down users by sections, such as industrial users, which are the larger industrial complexes in an area, or commercial users, which include the larger grocery stores and small and large shopping areas and malls that need large amounts of electrical power for lighting, air conditioning, and refrigeration. The commercial users may also include smaller businesses. The last sector is the residential demand and in some areas this demand may be equal in volume to the industrial or commercial demand.

Yet another way to measure the demand is to determine the *peak demand* during each day. For example larger industries may reach their peak demand during the early part of the day when machinery is first started up, and this demand may become lower on the second or third shift when some companies do not operate all of their machinery. This type of demand may also be larger on weekdays and lower on Saturday and Sunday.

Demand may also be measured by the time of year, when demand in southern and southwestern states may be much higher during long hot periods when air conditioning is used for extended hours. This same area may not use as much electricity when the weather cools and the air conditioners can be turned off.

What this means to you as someone learning about electricity that is created by solar generation is that the demand for using electricity in the United States fluctuates, and the ability of the solar energy systems to produce electricity will fluctuate with the conditions that determine the amount of sunshine an area receives. When these two issues (the supply and the demand) are combined, it makes the problem of supplying electricity from solar energy systems much more complex.

Table 2-1 shows the demand for electricity expressed as electricity consumption in billions of *kilowatt*-hours (kWh) per day for 2008 and 2009 and the estimate for 2010. A kilowatt is one thousand watts. This table is from the Energy Information Administration (EIA) and shows that in 2008 the residential sector consumed 3.51 billion kWh per day; it is estimated that the residential sector will consume 3.82 billion kWh in 2010. In 2008 the commercial sector consumed 3.64 billion kWh per day; the 2010 estimate is 3.70 billion kWh per day. In 2008 the industrial sector consumed 2.49 billion kWh per day, and it is estimated that this sector will consume 2.56 billion kWh per day in 2010.

In 2008 the transportation sector, which includes electric rail and electric modes of transportation, consumed 0.02 billion kWh per day, and this sector is estimated to consume the same amount of energy in 2010. The direct use column is defined as the amount of electricity the electrical generation plants use to create the electricity. This includes electrical power to pump cooling water and for other applications such as conveyor belts to move coal in coal fired plants. The direct use in

TABLE 2-1 U.S. Electrical Consumption by Sector			
Electricity Consumption (billion kWh) per day	**2008**	**2009**	**2010 est.**
Residential sector	3.51	3.79	3.82
Commercial sector	3.64	3.71	3.70
Industrial sector	2.49	2.74	2.56
Transportation sector	0.02	0.02	0.02
Direct use (power used to generate electricity)	0.43	0.40	0.41
Total electrical Consumption	10.06	10.67	10.49
Total electricity Consumption Retail Sales	9.66	10.26	10.10

Source: Energy Information Administration/Short-Term Energy Outlook—March 2009.

2008 is 0.43 billion kWh, and the estimate for 2010 is 0.41 billion kWh per day.

The total daily electrical energy consumption for the United States in 2008 was 10.06 billion kWh per day, and in 2009 this value increased to 10.67 billion kWh per day. The total electrical energy consumption for the United States is predicted to be 10.10 billion kWh per day in 2010. The amount consumed in 2010 has decreased slightly due to the economic recession in the United States. The bottom row of this graph shows the total amount of electricity sold or billed in the United States. In 2008, 9.66 billion kWh per day were sold, and the estimate for 2010 is 10.10 billion kWh per day.

2.5 TOTAL INSTALLED ELECTRICAL GENERATION CAPACITY

Additional information about the amount of electricity that is used in the United States includes data on the amount of capacity to generate electricity. Table 2-2 from the EIA includes all of the facilities that can generate electricity and shows their capacity for 1980 and then for 2000 to 2006. These figures include electricity generated from coal-fired, gas-fired, oil-fired, nuclear power, hydroelectric, wind, and solar sources. The capacity in 1980 was 579 million kW, and in 2006 that capacity has increased to 1,139 million kW. The trend line indicates that the demand is expected to increase every year. One of the problems for the energy producers is that the demand and capacity must both grow at the same rate to keep pace with future needs.

2.5.1 U.S. Electricity Generation by Fuel and Sector

The next area to understand about the amount of electricity generated in the United States is to identify the sources or sectors that generate electricity. Table 2-3 shows sources of electrical generation and the amount each one produces in billion kWh per day. You can compare the amount of electricity generated by solar energy to the total electricity produced by other methods. When the list is put into order, in 2006 the top-producing sectors were coal (5,397 billion kWh), nuclear power (2.157 billion kWh), and natural gas (2,012 billion kWh). The remaining nonrenewable-producing sectors include other gases, petroleum, residual fuel oil, distillate fuel oil, petroleum coke, other petroleum, and pumped storage hydroelectric. and their combined production is less than 0.3 billion kWh.

Electrical production from renewable sources is shown separately on this table. In 2006 conventional hydroelectric produced 0.784 billion kWh, wind produced 0.073 billion kWh, geothermal produced 0.04 billion kWh, wood waste produced 0.028 billion kWh, and solar produced 0.001 billion kWh.

TABLE 2-2	Amount of electrical generation capacity for the United States from all sources (in billion kWh)							
1980	**2000**	**2001**	**2002**	**2003**	**2004**	**2005**	**2006**
0.579	0.942	0.981	1.040	1.089	1.110	1.128	1.139

Source: Energy Information Administration.

In 2008 conventional hydroelectric produced 0.715 billion kWh, wind produced 0.127 billion kWh, geothermal produced 0.041 billion kWh, wood waste produced 0.030 billion kWh, and solar produced 0.002 billion kWh. From 2006 to 2008 electricity produced from wind energy increased from 0.073 billion kWh to 0.127 billion kWh, and the projection for 2010 is 0.208 billion kWh, an increase of nearly three times.

It is important to understand that the production of electricity from solar energy is increasing at a very rapid rate each year, but the amount produced from solar energy in 2006 was 0.001 billion kWh, and it was only a small percentage of the total electricity generated, which was 11.136 billion kWh. The projected output for 2010 is 0.004 billion kWh, which amounts to less than 1% of the total of 11.324 billion kWh.

2.6 TOTAL DAILY DEMAND FOR ELECTRICITY IN THE UNITED STATES

The daily demand for electricity in the United States fluctuates from month to month and week to week depending on the time of year and the location across the country. The total demand on a daily basis also changes by the hour throughout the day. For example the demand may be largest during the day and diminish significantly during the late hours of the evening when people are sleeping. In the South, Southeast and Southwest air conditioning is one of the largest residential uses of electricity, so the residential demand for electricity is stronger in the summer than in the winter. The other large part of the daily electrical demand is the industrial demand from large industries. This demand is more consistent on a daily basis, but it does change on Saturday and Sunday when these plants do not operate at full capacity. The industrial demand also changes slightly in the evening and at night when the industrial usage drops.

Commercial demand comes mostly from stores and malls, and this demand changes over the hours of each day and is different on different days of the week. For example the commercial stores may open at 9 a.m. and close at 9 p.m., and the demand during those hours will be at its maximum. The other hours of the day, their demand is much less. The biggest problem with the fluctuating demand is that the amount of electricity

TABLE 2-3 U.S. electricity generation by fuel and sector (in billion kWh per day)

Electric Power Sector	2006	2007	2008	2009 (projected)	2010 (projected)
Coal	5.397	5.475	5.392	5.273	5.255
Natural gas	2.012	2.232	2.169	2.195	2.236
Other gases	0.012	0.011	0.012	0.011	0.012
Petroleum	0.164	0.168	0.116	0.139	0.164
Residual fuel oil	0.093	0.104	0.061	0.062	0.066
Distillate fuel oil	0.018	0.022	0.018	0.018	0.019
Petroleum coke	0.049	0.038	0.034	0.056	0.077
Other petroleum	0.003	0.004	0.003	0.002	0.002
Nuclear	2.157	2.209	2.2	2.209	2.218
Pumped storage hydroelectric	−0.018	−0.019	−0.017	−0.016	−0.016
Other fuels	0.019	0.019	0.02	0.022	0.023
Renewable:					
Conventional hydroelectric	0.784	0.674	0.715	0.684	0.717
Geothermal	0.04	0.04	0.041	0.043	0.043
Solar	0.001	0.002	0.002	0.003	0.004
Wind	0.073	0.094	0.127	0.156	0.208
Wood and wood waste	0.028	0.029	0.03	0.03	0.03
Other renewables	0.038	0.039	0.039	0.042	0.044
Subtotal electric power sector	10.707	10.974	10.845	10.79	10.937
Commercial Sector					
Coal	0.004	0.004	0.004	0.003	0.004
Natural gas	0.012	0.012	0.012	0.012	0.012
Petroleum	0.001	0.001	0	0.001	0.001
Other fuels	0.002	0.002	0.002	0.002	0.002
Renewables	0.004	0.004	0.004	0.004	0.004
Subtotal commercial sector	0.023	0.023	0.023	0.022	0.023
Industrial Sector					
Coal	0.053	0.046	0.047	0.045	0.047
Natural gas	0.213	0.213	0.2	0.189	0.194
Other gases	0.027	0.026	0.026	0.025	0.026
Petroleum	0.012	0.012	0.007	0.009	0.01
Other fuels	0.014	0.013	0.007	0.007	0.007
Renewables:					
Conventional hydroelectric	0.008	0.004	0.005	0.005	0.005
Wood and wood waste	0.078	0.077	0.075	0.072	0.074
Other renewables	0.002	0.002	0.002	0.002	0.002
Subtotal industrial sector	0.406	0.392	0.371	0.354	0.365
Total All Sectors	11.136	11.388	11.239	11.166	11.324

Source: Energy Information Administration.

generated must change hourly, daily, and weekly to meet this changing demand. The companies that control the grid can help shift electrical power from where it is generated to where it is consumed to make sure all levels of demand are provided for. The electrical output of solar PV panels may not match the peak demand periods, but they will help add to the amount of total electrical power that is available. Currently utilities use peaking generators to provide additional electrical power that is needed during peak times.

2.6.1 Total Peak Demand

The total *peak electrical demand* is different from the average electrical demand. The peak electrical demand for any given hour will fluctuate from a high demand for several hours and then be reduced by as much as 30% the remainder of the day. The peak demand is the highest amount of electricity that is needed at any given time. The reason the peak electrical demand is so important is that the electrical system must be designed large enough to meet the highest demand that occurs regardless of the time of day, day of the week or time of the year.

The peak electrical demand in the United States changes across the country by time zones. For example the peak demand may hit an hour or two hours earlier on the East Coast and central time zone before it occurs on the West Coast. This is called the rolling peak and it occurs when large cities and industrial users come online the first thing in the morning, and this cycle also occurs when the cities and industries shut down on the East Coast an hour or two before the cities on the West Coast. One of the larger demands occurs when nightfall comes and all of the lights in a city come on and then go off in the early morning. This is also adjusted from summer to winter when nightfall comes much later in the evening for cities in the northern half of the United States in the summer and much earlier in the winter. This means that the power utility must continually plan for changes in the time the peak will occur from day to day and month to month.

2.6.2 Peaking Generators

Over the past 10 years, electric utilities have installed a number of systems that can be brought online very quickly. These systems are powered by natural gas or oil and are called peaking systems. These systems can be fired up quickly and add large amounts of electricity to the grid in a short period of time during peak load demand periods.

These systems are also typically clean burning, which means that their impact on the environment is minimal. They are also installed close to the location where the extra demand is coming from, such as areas with a concentration of large industrial sites or large cities that may need extra power for air conditioning on hot summer days. The peaking generators may not run on other days, but since they are fired with natural gas, they can quickly be brought back up in case the need arises. Another time the peaking generators are used is when a large coal-fired or nuclear powered system needs to be shut down for a period of time. These systems can easily take care of the extra power that is

needed, and they can also be used to make up fluctuations in solar energy production.

2.6.3 Commercial and Industrial Demand

Commercial and industrial demand is usually large enough in any electrical system to use up to 60% of the total amount of electricity produced. This demand accounts for a large part of the total consumption of any system, but it tends to be predictable. When economic times are good, the commercial and industrial loads will tend to grow slightly, and these same loads will become smaller when an economic downturn occurs, such as the worldwide recession of 2009. This reduction in demand occurs because industrial orders become smaller or may be eliminated for short periods of time when the demand for the products is reduced. Peak electrical demand for industrial consumers is another important factor that must be considered when discussing electrical energy needs. The peak electrical demand is the largest amount of electrical energy that is needed during any hour at any time over a 24-hour period any day of the week or month. The peak electrical demand will be different in different parts of each state, as it may be much larger in industrial areas and occur mainly on Monday through Friday, whereas the demand in cities where the load is predominately commercial (malls and stores) may be larger on Friday and Saturday because the electrical loads for lighting and air conditioning will be higher on weekends.

2.6.4 Brownout

When the demand for electricity begins to outpace the supply on any given day, the voltage on the entire system will begin to become lower or droop. This condition of lower voltage is called a *brownout,* and it can begin to cause damage to appliances and other loads that remain connected to the electrical system. There is no defined level for a brownout, but it generally occurs when voltage drops or falls 8% to 12% below the typical supplied voltage. This may occur when demand is higher, such as on hot summer days, or if a major fault occurs in the power system, such as an automobile accident that takes out a power pole, or if a high wind takes out one or more power lines when trees blow into them.

2.6.5 Shedding Demand

In some cases, brownout may occur for several days during high demand periods or when a major storm such as an ice storm or a storm with high winds in the summer has damaged part of the electrical transmission system. If the brownout is going to last a longer period of time, the electrical utility must bring on all of its peaking generators or begin to shed the load. In some areas of California, the utility companies have used rolling brownouts to help keep the voltage up to standard levels. With the rolling brownout, the utility company plans ahead and notifies companies and cities that their power will be shut off completely for a short period of time so that sufficient power will be available for the remaining customers. It will also ask consumers to turn off as many electrical appliances as possible during these times. The power is shut off to the areas for a specific amount of time, and then it is turned back on and is shut off in a different part of the service area for the same amount of time. In this way the utility company is able to stay online even though it does not have enough generated power to supply all of its customers at the same time.

Another technique that is used to help balance the demand and production of electricity is to create contracts with larger users to provide them with a lower utility rate in return for their agreeing to shut down completely during the higher demand periods. This technique, which is sometimes called interruptible service, allows larger users to create a strategy to be able to shut down with a few hours notice if the overall voltage levels in an area are dropping lower or if a brownout occurs for any reason. The companies receive a large discount for all of the power they use, in return for shutting down completely on short notice.

2.7 BACKUP POWER

Some applications, such as hospitals, crucial city services, and safety services, may need to have a backup power supply that supports the power from the grid. For example hospitals may have a backup diesel generator that can provide up to 75% of the full load. This power will provide essential lighting and backup power for essential services such as surgeries that could be interrupted by a power outage. Some large industries have a backup for essential processes such as pouring molten iron or food processes that cannot be interrupted once they are started. You need to be aware of these systems when you are connecting a solar energy system to them, as the switch gear that moves between the systems must be able to isolate each system for maintenance or other activities.

An *uninterruptible power supply* is backed up with one or more batteries and an inverter that converts DC voltage to AC voltage. The uninterruptible power supply takes in AC voltage and converts it to DC voltage, which goes into the bank of batteries and then is converted back to AC electricity that has a frequency of 60 Hz. Some uninterruptible power supplies allow the AC voltage to pass through to the load and use only a small amount to keep the batteries charged at all times, and an electronic switch quickly switches the power over to the backup power in the event of a power

loss. This ensures the load never experiences the power outage and always has a source of AC power. The size of the battery in the backup system will determine how long the system can run on backup power. In many cases an uninterruptible power supply is used to provide immediate backup, and a longer-term backup energy source such as a diesel generator can be brought online shortly after the power failure and can continue to supply electrical power as long as the fuel is provided. Uninterruptible power supplies are used for computer equipment, telephone equipment, alarm systems, and other critical power systems.

2.8 RESIDENTIAL DEMAND

The residential demand in the United States consists of several large loads and several continuous loads. The larger loads in a home include the air conditioner, electric heating or heat pump, electric washing machine and clothes dryer, electric range, and microwave oven. Continuous loads may include lighting and the refrigerator and freezer, and these loads tend to run for longer periods throughout the day and throughout the year. The air conditioning and electric heating loads are seasonal. The electric washing machine and clothes dryer along with the electric range tend to be used periodically and on different days of the week. These appliances group together to create a residential demand. When thousands of homes are placed together around a city or suburban area, the electrical demand can be as large as or larger than the industrial or commercial demand.

The residential demand also tends to move across the United States with the time zones when all of the consumers in an area tend to turn on their lighting loads at the same time and then turn off the majority of the loads at late night when people go to bed. This load resumes again when everyone starts to wake up in the morning and turn on the loads again. Residential demand also grows when major building projects expand housing in a specific area or part of a city.

2.9 GRID-TIED SYSTEMS

"Grid" is the term that is applied to all the electrical power distribution system in the United States and North America. If a solar energy system is designed to produce power for the grid, it must be connected to an existing section of the grid and its power must meet the specifications for that section of the grid. The grid is used to distribute electricity across the United States, and it is actually made up of a number of interconnected wires that are used to move electricity from where it is generated to where it is consumed. When electricity is generated, it may be at values of less than 5,000 V. This voltage must be stepped up with transformers to higher voltages of

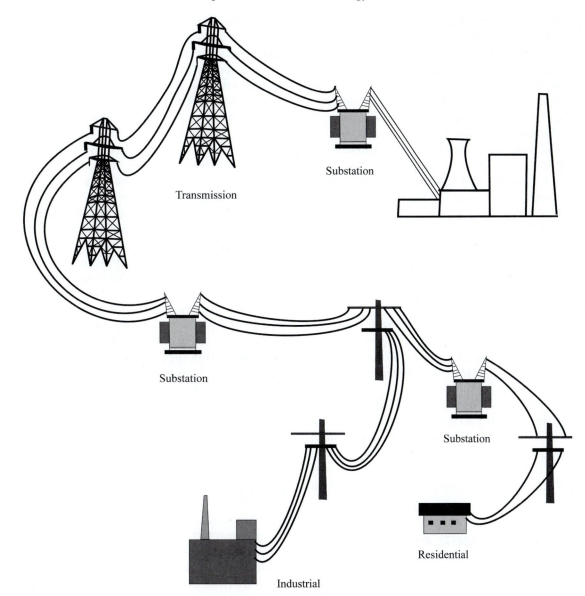

FIGURE 2-2   Diagram of electrical power distribution from generation station to residential customer.

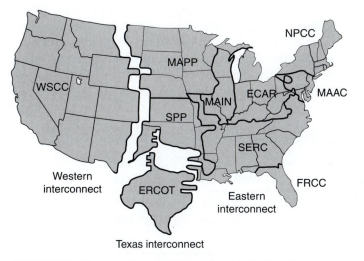

FIGURE 2-3   The electrical grid system for the United States
Source: U.S. Department of Energy.

345 k volts for long distances or 500 k volts for short distances, and it may be increased to even higher voltages up to 765 k volts for transmission over the longest distances. When the electricity reaches the end user, it is transformed back down at a substation near the city or at an industrial site to a level at which it is consumed. Figure 2-2 shows the distribution system from the generating site through transmission lines to a residential customer.

The grid is illustrated in Figure 2-3, which shows its location in each state. The grid is divided into three major sections, the Eastern Interconnect grid, the Western Interconnect grid, and the Texas Interconnect grid. The Eastern Interconnect grid sections include the Northeast Power Coordinating Council (NPCC), which includes Maine, Vermont, New Hampshire, New York, Massachusetts, and Rhode Island; the Mid-Atlantic Area

Council (MAAC), which includes Maryland, Delaware, New Jersey, and parts of Pennsylvania; the East Central Area Reliability Coordination Agreement (ECAR), which includes Michigan, Indiana, Ohio, West Virginia, and Kentucky; and the Southeastern Electric Reliability Council (SERC), which includes the southern states; the Florida Reliability Coordinating Council (FRCC), which includes all of Florida; the Mid-America Interconnected Network (MAIN), which includes Wisconsin and Illinois; the Mid-Continent Area Power Pool (MAPP), which includes the Dakotas, Nebraska, Minnesota, and Iowa; and the Southwest Power Pool (SPP), which includes, Kansas, and Oklahoma.

The Electric Reliability Council of Texas (ERCOT) includes most of Texas, and the Western Systems Coordinating Council (WSCC) includes all of the western states from New Mexico, Colorado, Wyoming, and Montana westward. ERCOT is now called the Texas Regional Entity (TRE).

Table 2-4 shows that supply, demand, and the amount of capacity margin for the Eastern Interconnect grid is 33.5%, for TRE it is 39%, and for the Western Interconnect grid it is 32.3%; the capacity margin for the entire United States is 30.4%

The grid provides electricity for each of the areas shown in the map. If you are providing electricity to the grid from a solar energy system, you must identify the part of the grid you will be connecting to and understand the specifications for that part of the grid. You will learn more about the grid in Chapter 10.

2.10 STAND-ALONE REMOTE POWER SOURCE

There are several types of stand-alone remote power supplies, including gas- or diesel-powered generators, wind turbines, and solar panels. The stand-alone units are typically used on remote farms or cabins that are located miles from electrical power that is available from the grid. A number of smaller solar energy systems that are designed specifically for this type of system have one or more batteries that can store the power for times when solar energy is not produced. The power system for the remote site can be DC or AC powered. If AC power is used, then an inverter is needed to convert DC voltage to AC. The battery bank is used to store the energy that is created for periods of time when the power source is not available. The solar energy system generates the voltage for the battery system and the batteries can store the electricity for long periods of time.

Some units combine a solar power system to augment the wind turbine. The solar system can provide battery voltage when the sun is shining and can back up the wind turbine even if the wind does not blow at sufficient speeds for several days at a time. The solar system charges the batteries when the wind is not moving the turbine. The wind turbine may be able to provide electrical power at night when the sun is not shining or on days that are stormy or rainy.

Another type of stand-alone solar energy system includes remote power for highway signs that are too far from an AC power source. Other applications include small solar panels to charge batteries that provide power for instrumentation systems for small oil wells in fields away from AC power sources. Figure 2-4 shows an example of this type of solar panel. Small solar panels also provide power for weather instruments that are located along highways or other remote places. Small solar panels are also used to recharge batteries on recreational vehicles and boats that may not have access to AC electricity at all times.

2.11 ENVIRONMENTAL ASSESSMENT FOR SOLAR POWER

Another issue that you will need to become aware of is the environmental and ecological assessment for solar power systems. The environmental issues include the environmental damage that is done when solar energy equipment is installed on large solar farms. In some cases the access roads and infrastructure must be improved prior to the installation process. The environmental impact also includes any changes to the area where the solar energy equipment is installed. Another environmental issue with solar panels is that they may be made from cadmium or other substances that are considered toxic and need to be recycled or disposed of properly when they reach the end of their useful life.

TABLE 2-4	U.S. Net Internal Demand, Actual or Planned Capacity Resources, and Capacity Margins 2009/2010			
	Eastern Grid	**TRE**	**Western Grid**	**U.S. Total**
Demand (megawatts)	476,057	46,068	113,504	635,629
Resource supply (megawatts)	684,015	80,424	170,745	935,184
Capacity margins (percentage)	33.5%	39%	32.3%	30.4%

Source: EIA (Energy Information Administration), Department of Energy.

FIGURE 2-4 Solar panel provides power for instruments for small rural oil well. (Courtesy T. E. Kissel.)

Questions

1. Explain some of the limitations of using a solar energy system as the sole source of electrical energy.
2. Explain why it is important to understand the total electrical residential, commercial, and industrial demand for electrical power when you study solar energy.
3. Explain the need for an uninterruptible and continuous power source for residential and industrial uses.
4. Identify the total daily electrical demand for residential, commercial, and industrial uses of electricity.
5. Explain the difference between regular electrical demand and peak electrical demand.
6. Explain how the electrical grid connects all parts of the residential, commercial, and industrial electrical systems throughout the United States.

Multiple Choice

1. In 2001 the total supply of electricity from all three sectors of the grid was _____
   a. 669 megawatts
   b. 793 megawatts
   c. 15.6 megawatts
2. A brownout occurs _____
   a. When the entire electrical system is completely lost.
   b. When more electricity is produced than can be utilized.
   c. When the voltage on the electrical grid falls 8% to 12% lower than its rated level.
3. A peaking generator is a generator that _____
   a. Can be brought online quickly to add large amounts of electrical power to the grid on a short notice.
   b. Produces electricity from hydroelectric sources.
   c. Runs continuously when demand for electricity is low.
4. The demand for electrical consumption for the United States for 2009 was _____
   a. 3.79 billion kWh per day for the residential sector.
   b. 3.71 billion kWh per day for commercial sector.
   c. 2.74 billion kWh per day for the industrial sector.
   d. All the above.
5. Which is the correct order of U.S. electrical generation by fuel from the largest to the smallest for 2008?
   a. Wind, solar, coal, nuclear, and natural gas
   b. Nuclear, natural gas, wind, solar, and coal
   c. Coal, nuclear, natural gas, wind, and solar
   d. Coal, natural gas, nuclear, wind, and solar

3

Types of Solar Energy Systems

OBJECTIVES FOR THIS CHAPTER

When you have completed this chapter, you will be able to

- Identify ways that solar energy is used to provide heating and lighting and how it produces electrical energy.
- Explain the differences among a photovoltaic cell, a photovoltaic panel, a photovoltaic module, and a photovoltaic array.
- Explain the operation of the parabolic trough electrical generating system.
- Identify five types of solar PV materials and/or processes used to manufacture PV material.
- Identify advantages and disadvantages of five types of solar PV materials and/or processes used to manufacture PV material.
- Explain the differences between active and passive solar energy heating system.

TERMS FOR THIS CHAPTER

Active Solar Heating
Current
Diffuse Light
Direct Sunlight
Electricity
Nanotechnology
Passive Solar Heating
Photovoltaic Array

Photovoltaic Cell
Photovoltaic Module
Photovoltaic Panel
Resistance
Solar Tube Lighting
Sub-Array
Voltage

3.0 OVERVIEW OF THIS CHAPTER

This chapter will explain the types of solar energy systems you will find installed around the world and in the United States. These systems include solar lighting, solar active and passive heating systems, solar hot water heating systems, solar photovoltaic (PV) systems, and parabolic arrays that concentrate the sun's rays onto tubing that is filled with oil. When the oil is heated to the degree that it turns water to steam, the steam is used to turn a large turbine wheel that causes a large generator to turn, producing electrical energy much like the generators at a coal-fired or nuclear-fired plant. In some applications the oil is heated by the sun to a temperature that is high enough to make water turn to steam. The oil is pumped through a heat exchanger that has water in one loop that receives heat from the hot oil at a temperature that is high enough to cause the water to turn to steam and the steam is used to turn a turbine that turns a generator. The solar PV system and the parabolic arrays are both used to create electrical energy, and the other systems are used primarily to heat the home and hot water and to provide solar lighting. The section on solar PV will explain all the types of PV systems that are currently used and will compare them to give you an idea of their efficiencies, advantages, and disadvantages. This chapter will review the various types of solar energy systems and provide a detailed background on each. It will also introduce all the different types of solar PV panels that are currently in use. Chapter 5 will provide a detailed explanation of the theory of operation of photovoltaic panels, as well as details that will allow you to understand the process by which light is converted into electrical energy. The objective of this chapter is to provide you with vocabulary and details describing different systems that are used to provide solar heating or systems that convert light to electricity.

3.1 MODERN SOLAR ENERGY SYSTEMS

Modern solar energy systems include using solar energy for lighting, to heat water for use as part of the hot water system, to heat water or air to be used as part of the central heating system for the home or commercial establishment, and with PV cells to convert solar energy to electrical power. All of these uses are primarily for residential and small commercial users. Larger solar photovoltaic systems are used for large commercial and industrial enterprises and for generating electricity for the grid. Another way to generate electricity from solar energy is to use concentrated solar energy panels, or parabolic trough systems, which concentrate levels of solar sunlight onto piping that has oil flowing through it. When the concentrated light is focused on the pipe, it causes the oil to heat to a high temperature and it then flows through a heat exchanger and causes water to turn to steam. The steam is used to turn a turbine that is connected directly to an AC generator that is large enough to produce electrical power that is similar to that produced by a coal-fired or nuclear power plant. This chapter will explain each of these types of solar energy systems. The technology used in each system will be introduced and explained in detail, including identifying the major parts of the system and their functions. There are two basic types of solar energy systems being used in small residential and medium-size commercial applications. These include the passive solar system and the active solar energy conversion system.

3.1.1 Active Solar Energy Systems

Active solar energy systems can be classified as solar energy systems that use electrical or mechanical energy to convert the sun's energy to heated water or air or to create electrical power. *Active solar heating* uses large-scale solar energy conversion systems that require large quantities of water or air to move the amounts of heat energy needed to make the heating system operate effectively. If the system contains solar *photovoltaic panels,* these may require tracking systems to be more efficient. All these activities require additional energy to be used. It is important to remember that an energy conversion system must use a certain percentage of its energy to make it efficient and cost-effective. This chapter will explain the operation of active solar energy systems that are used to heat water or air for the purpose of providing heat for a building or residence or to warm the water for a hot water system. It will also explain the types of photovoltaic systems that are used and how they are connected to create electrical power that is needed.

3.1.2 Passive Solar Energy Systems

Passive solar energy systems are designed to use the least amount of additional energy to cause them to operate. Basically passive solar design maximizes the use of the sun's energy for warmth in the winter, or to provide cooling in the summer, and provide natural light throughout the year, while creating comfortable living spaces and reducing energy costs. Many of the concepts of passive solar energy conversion must be designed into buildings and homes when they are first built, whereas others may be retrofitted after the building has been in place for some time. Some of the passive systems work better where the amount of sunshine and its intensity are stronger, such as the southwestern western, and southern United States. The concept and ideas of passive solar systems include design elements such as where to place windows and other components to receive the maximum amount of light and/or heat to the interior of the building. Other passive elements include larger amounts of insulation in the building that reduce the heat load in the summer and improve heating processes in the winter. Other passive features include additional attic vents and the use of construction products that reflect excess heat in the summer yet receive as much heat as possible in the winter.

3.2 SOLAR LIGHTING

One of the uses of solar energy is to provide lighting for residential and commercial buildings. The use of solar lighting allows traditional electrical lighting to be turned off during the daytime, which will save a large amount of money, especially in commercial locations such as shops and malls. Figure 3-1 shows a residence that uses a large amount of solar lighting. It has a large number of windows on its south side. The south side of a home in the Northern Hemisphere receives more sunlight in fall, winter, and

FIGURE 3-1   Passive solar lighting and heating.   (Courtesy U.S. Department of Energy)

FIGURE 3-2   Solar tube lighting.   (Courtesy Solatube International)

spring months. This ensures that direct sunlight can reach the windows for the majority of days during the year. *Direct sunlight* is light from the sun that enters the windows directly without needing to be reflected off another surface. The location of the windows in the residence is designed to let the sunlight enter the living space where it is needed. For a commercial application, the windows would be strategically located so that the light that enters through them is directed where it is needed.

One of the problems with using windows to provide lighting is that the heat from the solar energy that provides the lighting may also cause the interior space to heat up in the summer. To prevent this or limit it as much as possible, the windows are treated so that they allow light to enter, but heat is reflected and not let in. Another way to do this is to use awnings or overhang roofing that blocks some of the direct sunlight during the summer when the sun is high in the sky and allows more of the sunlight and heat energy to enter during the winter when additional heat is needed and the sun sets lower in the sky; the light will then flow into the windows under the awning. These lighting systems must be designed into the residence or commercial building as part of an overall solar energy system to ensure that the solar energy for the additional lighting does not cause more costs to be incurred for air conditioning.

3.2.1 Solar Tube Lighting

Another way to provide solar lighting where it is needed in an interior space without windows is to use a product called *solar tube lighting*. Figure 3-2 shows a solar tube lighting system mounted through the ceiling and roof. The portion of the system that is above the roof consists of a clear plastic lens that can receive light from 360°.

The light is directed through a piece of ductwork that has a shiny surface that will reflect the light continually through the tube. The tube connects the dome that is mounted on the roof to the lens that is mounted in the ceiling of the room where the light is needed. Figure 3-3

FIGURE 3-3   Solar tube lighting fixture in a kitchen.   (Courtesy Solatube International)

FIGURE 3-4 (left) Solar tube lighting installed in an open warehouse. (right) Solar tube lighting in a grocery store. The solar tube lighting fixtures direct sunlight from the rooftop directly into the building, much like a traditional electrical light.   (Courtesy Solatube International)

shows the light lens mounted in the ceiling of the kitchen. You can see in this image the amount of light that is provided by the system, and that the light looks similar to that produced by an electrical lighting fixture. An aperture is provided on the inside lens of some of the tube lights that can be partially or completely closed to limit the amount of light, similar to turning a traditional electrical light off with a switch. It is important to understand that since the light is solar sunlight, there is no need to turn it off as you would an electrical lighting fixture to save money. In some rooms of the house, such as bedrooms, the aperture may be required in order to dim the light if someone needs to sleep during the day.

Solar tube lighting can also be used extensively in commercial applications such as stores and malls or in grocery stores where lighting must be strategically positioned to highlight products. Figure 3-4 shows an example of this type of lighting system. The image on the left shows solar tube lighting in a warehouse, and the image on the right shows the tube lighting fixtures mounted in the finished ceiling in a grocery store, much like a traditional electrical lighting fixture. The solar light being directed into the store is strategically located to highlight products. Traditional electrical lighting fixtures are built directly into the solar tube lighting, and this lighting will automatically become energized when the amount of light from the solar tube becomes minimal. In some applications the backup electrical lighting is designed to run from stored electrical energy that has been created by solar PV cells. An electronic inverter is connected to the battery storage system to convert the DC electricity to usable AC electricity that is required by the lighting.

The main advantage to solar tube lighting is that it provides light that is basically free of charge any time the

sun is shining. Even on days when clouds block part of the sunlight, the amount of sunlight that is brought into the interior of the room or building by these lights is generally sufficient for its lighting needs. Once they are installed, they basically require minimal maintenance. The outside dome may need to be washed periodically if it does not rain enough to clean it to allow sufficient sunlight to pass through. If this type of lighting is used where snow falls, it is important to have backup electrical lighting in case the snow is deep enough to cover the lens. Typically the design of the lens minimizes snow build up on top of it that blocks the light from flowing through it.

3.2.2 Solar Lighting That Utilizes Fiber Optics

Another type of solar lighting system utilizes optical fibers that have one end protruding through the roof where they can receive light, and the other end located in the space where lighting is needed. The fibers provide sufficient lighting, just like a traditional electrical lighting fixture. Since this system uses long fibers to direct the light into the space where it is needed, the lighting system does not need to be extended through the roof directly above where the light is needed in the interior space. In most cases the lighting can be located where it cannot be seen and detract from the exterior façade or the looks of the building. Since the fibers can be virtually any length, they can be mounted in the space above the ceiling where they are out of sight and can be connected to the lighting fixture. Once the fixtures are purchased and mounted, the cost of lighting is virtually eliminated, since there is little maintenance and the fibers can last a long time.

3.2.3 DC Lighting Using Solar PV and Batteries

Another type of lighting that is available uses fixtures that are powered by 12 or 24 DC. The power for this system can be provided by a solar PV module that takes sunlight and converts it to electrical energy and stores it in a 12- or 24-V battery. A lead-acid battery is generally used to store the DC voltage. When the sun is shining, the solar PV cell creates electrical energy and stores it in the battery. The lights are connected directly to the battery; when they are turned on, they will utilize the stored DC voltage from the battery. This type of lighting can use traditional fixtures and switches to control the amount of light in the room. Typically this system is backed up by 12 V power supply that is connected to the grid, so that lighting would be available any time the battery power is not available or is completely used up. The advantage of this type of lighting is that the majority of the electrical power is provided by the solar PV panel, which reduces the overall cost of the electric bill. If the lighting system requires AC voltage, an inverter can be installed between the battery and the lights so that AC electrical power is provided to the lighting circuits.

3.2.4 Solar-Powered Street Lighting

Street lighting is required for safety and security to ensure that areas for pedestrians and automotive traffic have sufficient lighting at night. One of the newest innovations in this type of lighting is to mount a solar PV panel directly on the light pole where it can receive the maximum amount of sunlight. The electrical energy that is produced by the PV panel is used to charge a battery that is also located on the power pole. A battery-charging controller ensures that the battery is continually charged and does not become overcharged. This controller also acts as a switch to the grid-tied electrical system that is used to power the lighting in case the battery becomes discharged or cannot be charged due to the lack of sunlight. The solar panel is large enough to provide sufficient electricity to charge the battery, which will power the lighting when it is required. The solar PV panel is generally sized large enough to provide enough electricity to charge the battery completely on the shortest days of the year during the winter months. During the summer months when the sun is available for more hours of the day, the battery charge controller may need to disconnect the PV panel from the battery after it becomes fully charged, even though additional daylight hours are still available. Since the hours of darkness are shorter during the summer months, the amount of power needed from the battery will be less because the light is not energized for as many hours. Figure 3-5 shows a solar panel connected directly to a streetlight. The panel must be orientated on the lighting pole so that it faces in a southerly direction where it can gather the most solar energy and convert it to electricity. Since most street lighting is

FIGURE 3-5   Solar PV panels on lighting fixtures.

required for security and safety reasons, the streetlight must also be connected to the grid so that electrical power is always available even over extended periods of cloudy days when the system may not be able to charge the battery completely.

3.3 SOLAR HEATING USED TO PROVIDE HOT WATER

The water that is used for showering, washing clothes and dishes, and other uses for human consumption is called "potable water." A hot water tank is usually used to provide all the hot water needs of a household. The water in the tank is typically heated by an electric element in the bottom of the tank or by natural gas, propane, or fuel oil. The water that goes through the hot water tank is typically potable water. This means the water can be consumed as drinking water, so it must be free of microorganisms and possibly be treated with chemicals such as chlorine. The water that is used as potable water cannot be allowed to be contaminated or to sit for long periods of time where it will pick up anything to pollute it. If the solar hot water heating systems cannot keep the water clean and purified, a heat exchanger must be used. The heat exchanger system has two loops. One has the clean water in it, and the second contains a glycol solution that is pumped through the solar heating equipment. Since the glycol solution moves through the solar collector and is heated, it does not matter if it gets contaminated, because it never comes into contact with the clear pure water. The basic components in this type of system include the solar collector, pump, piping, drain back tank heat exchanger, hot water tank, and controller.

Figure 3-6 is a diagram of this type of system. The solar collector is mounted on the rooftop where it can

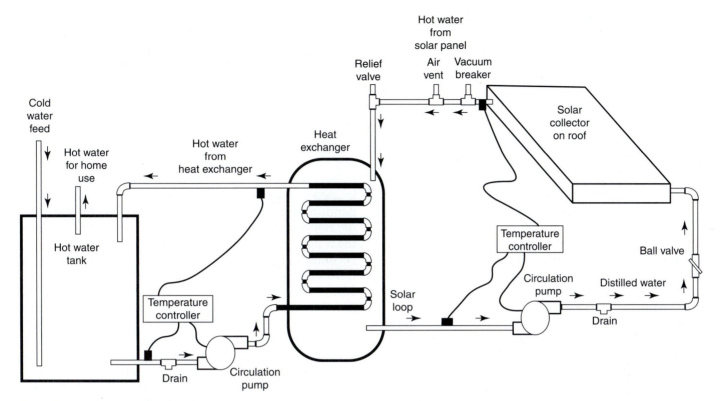

FIGURE 3-6 Solar water heating system used to provide hot water for a home.

receive the heat from the sun's rays. This system is a two-loop system. Each loop has a separate pump and separate piping. The outside loop consists of the solar collector, the pump, and the barrel of the heat exchanger. The solar collector is mounted outside on the roof, so the fluid that moves through it is usually a propylene glycol solution, and it is able to withstand colder temperatures without freezing. The pump moves the glycol solution through the collector, back into the house, and through the heat exchanger barrel. The heat exchanger is designed as a large barrel with copper tubing coil that is mounted inside it and has potable water running through it in a sealed system. The glycol solution fills the heat exchanger barrel and transfers heat into the potable water that is running through the inner copper tubing coil. As long as the temperature of the glycol solution is warmer than that of the potable water in the inner coil, heat will transfer from the glycol solution into the potable water, thus providing hot water for use in the house. If the temperature outside begins to drop, and the collector is no longer heated by the rays of the sun, the temperature of the glycol solution will quickly drop below the temperature of the warm water that is in the inner coil. If the system is allowed to continue running, the glycol solution will actually begin cooling the potable water in the inner coil. At this point the temperature sensors in the system will identify the temperature of the glycol as being too low to provide additional

heating of the water and will turn off the pump. The glycol in the collector will drain back to the heat exchanger. In some systems the glycol solution will simply no longer pump through the collector and is allowed to remain in the collector rather than being drained back inside the house.

The second loop in the system contains the potable water, and it has a pump that moves the potable water through the coils in the heat exchanger and on into the hot water tank that is located inside the house. The pump runs continually to ensure that the water is picking up heat from the glycol solution in the barrel as it moves through the sealed copper tube. The potable water will continue to pick up heat from the glycol solution as long as the collector is getting heat from the sun's rays and keeps the glycol solution warmer than the potable water. The pump will continue to move the potable water through the heat exchanger until the hot water tank inside the house reaches its set point temperature.

The fluid that moves through the outdoor solar collector is a propylene glycol solution usually mixed at a rate of 50% water and 50% propylene glycol. This makes the solution able to tolerate temperatures down to −29°F without freezing. Table 3-1 shows the freezing points of different mixtures by percentage of water and propylene glycol. It is important to protect the system when the temperature outside drops below freezing. Some systems drain back all the liquid to the inside tank when the

TABLE 3-1 Freezing point of propylene glycol mixture with water percentage by mass.								
Propylene Glycol Solution (% by *mass*)		0	10	20	30	40	50	60
Temperature	°F	32	26	18	7	−8	−29	−55
	°C	0	−3	−8	−14	−22	−34	−48

temperature gets below freezing outside. Sensors are strategically located to measure the outdoor temperature and drain the system if it gets too cold and diverts the glycol solution to the indoor system by switching valves. Since the glycol solution is able to withstand temperatures below freezing, some systems are designed to allow the solution to remain in the collector, and the pump is shut off so it does not move the solution through the heat exchanger, which would cause the potable water to begin to lose heat.

The sensors and controls must be set at the balance point for the system. The balance point is the temperature at which the glycol solution that is heated in the collector cannot provide enough temperature differential to heat hot water that is used as potable water. This balance point continually changes, as the temperature of the water in the indoor hot water tank rises and lowers as water is used. If no hot water is being used and the collector is providing glycol at a maximum temperature, the temperature difference between the glycol solution in the potable water in the heat exchanger will be large enough for the glycol solution to cause the potable water's temperature to rise. If no hot water is being extracted from the hot water tank, at some point the potable water will reach its desired maximum temperature. Then the water can no longer take on heat until someone uses hot water inside the house, which is then replaced in the system with fresh water that is at its coldest temperature. This causes the hot water tank temperature to decrease, and the temperature difference will be large enough for the potable water to increase in temperature from the glycol through the heat exchanger. This process continues throughout the day as long as the solar collector is taking heat from the sun's rays and adding it to the glycol solution and the temperature of the potable water inside is low enough to need heating. The system controls the heat transfer process when either the potable water temperature reaches its desired temperature level or sunlight conditions change and the solar collector is no longer able to add temperature to the glycol solution. This may be caused by a cloudy day or a storm that passes by, or when the sun begins to go down in the evening and no longer provides sufficient energy to the collector.

In some cases the solar water heating system is designed large enough to provide enough hot water to heat the home. In these applications the sun shines enough hours of the day and days of the month to provide sufficient heat to the water system to be used to heat the home. It is also important that the home is located in a region of the country where the outside temperature does not get too cold, so that the temperature differential from the outside walls to the inside walls of the home is not so large that the heat load for the home becomes bigger than the amount of heat energy that is transferred to the water. This means that if the temperature outside the home gets too cold, it will reduce the amount of heat the collector can gather from the sun's rays, and it will make the amount of heat energy needed in the home larger than the solar collector can gather from outside. At this point the room temperature inside the home will begin to drop and become uncomfortable. This point is somewhere around 68° to 70° F. If the system is large enough, it will be able to provide all the heat required during the day and to store additional heat for use at night or on cloudy days when the solar collector is not able to gather enough heat energy. In some cases a backup heating system that is powered by electrical heaters or a forced-air furnace that uses natural gas, propane gas, or heating oil is needed to ensure that the home never reaches an uncomfortable temperature. The advantage of this type of heating system is that the solar energy provides a sufficient amount of the heating energy to lower the utility bill of the electric company or gas company that is providing fuel for the alternative system. Typically the regions of the country that provide the most opportunities to harvest solar energy are also regions that are slightly warmer in the winter and do not require large quantities of heat energy to heat the home.

3.3.1 Normal Operation of a Solar Water Heating System

The normal operation of a solar water heating system occurs anytime the glycol solution in the solar collector is warmer than the potable water that runs through a coil of the heat exchanger. During this time the volume of glycol solution that can be heated and moved through the heat exchanger will determine the amount of solar energy that is harvested. The second part of the system includes the volume of potable water that is moved through the heat exchanger and the temperature difference between the glycol solution and the potable water. If the system is running and the glycol solution is receiving the sun's rays during the middle of the day, it will be harvesting the maximum amount of solar energy and converting it to heat energy in the glycol solution. The second part of the energy conversion depends on the temperature of the potable water and how different it is from the glycol solution. If the potable water is rather cool and the glycol is quite warm, the amount of heat transfer will be maximized. If the system has been operating for several hours and the potable water is near its maximum temperature, the temperature difference between the potable water and the glycol solution will be smaller, as will the transfer of energy.

The amount of heat energy the potable water will take on depends on how much potable water is being used and how much cooler water is added to the system, as well as how much sunlight is available to the collector. Another factor that may impact the efficiency of the system is the cleanliness of the heat exchanger tubes both inside and out. If the tubes become coated with residue from the water, the transfer of heat will be slowed down and the system will become less efficient.

3.3.2 Heating the Home with Water Heated by Solar Energy

Another way that solar energy is used for residences or commercial establishments is to use a loop to heat water or glycol solution and pass it through a heat exchanger in the heating system. Water or a glycol solution is pumped through a solar collector and is stored in a large tank where it can be later pumped through a water-heating coil that is located in a typical furnace. Figure 3-7 is an example of this type of system. The components and the system include the solar collector, heat exchanger, three pumps, a large water storage tank, the water-heating coil in the furnace,

and the furnace with a blower and ductwork. The solar collector is mounted on the roof in a similar fashion to the one used for heating hot water. If the collector is exposed to low temperatures, a glycol solution is pumped through it. If the outside temperatures does not go below freezing, distilled water may be pumped through the solar collector. A heat exchanger is connected with plumbing to the solar collector, and the heat exchanger is mounted inside the building. The glycol fluid flows through the barrel of the heat exchanger and is pumped back into the solar collector where it is heated.

The second loop of the heat exchanger is a set of coils that are mounted inside the barrel of the heat exchanger. This loop has water pumped through it by the second pump, and the water is pumped to the heating coil that is mounted in the furnace. This loop picks up heat in the water from the glycol solution that flows through the barrel of the heat exchanger. The pump for this loop runs continuously when heat energy is absorbed, and the water for the loop is stored in a large storage tank. This means that the water in the storage tank continually picks up heat from the glycol solution in the heat exchanger and most of the heat energy is stored in the water for later use by the furnace.

The third loop in the system consists of a third pump that is wired to the thermostat in the home or business. When interior rooms cool down and require heat, the thermostat turns on and calls for heat. This energizes the pump and furnace fan. The pump in this third loop pumps water from the storage tank through the water-heating coil that is mounted in the furnace. The furnace fan blows air from the interior of the house or business over the heating coil so that it picks up heat. The air is moved throughout the house through ductwork and registers, where it heats

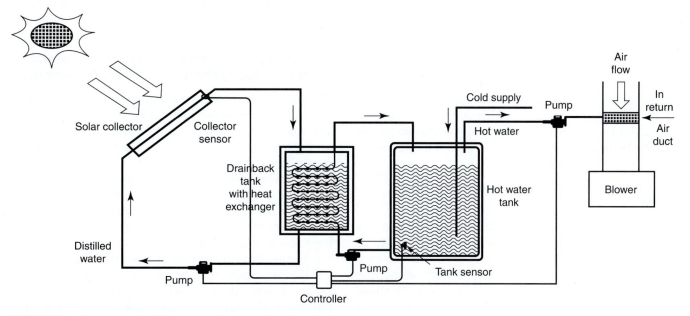

FIGURE 3-7 Solar space heating system that uses forced-air furnace.

the interior with the warm air. As the temperature indoors warms up and reaches the temperature set point, the thermostat is satisfied and turns off the water pump of the third loop and the furnace fan. When the third pump is turned off, the water in the third loop is no longer circulated through the heating coil in the furnace.

The solar collector for this system has glycol solution continually moving through it, which picks up heat from the solar energy. The pump for the first loop runs when heat is absorbed and moves the glycol through the collector and the heat exchanger. The second loop pump moves fluid causing the heat from the glycol to be transferred to the water that moves through the coil in the heat exchanger. This water comes from the large storage facility, so that it is continually providing heat to the storage water. The size of the water storage facility is designed to be large enough to provide heat throughout the night or for times when the solar energy system is not collecting heat. When the solar energy is no longer available, at night or on cloudy days, the hot water storage system should have enough hot water stored to pump through the furnace coil and enable the furnace to heat the home.

The furnace will have a backup heating system that will be natural gas, propane, or fuel oil that will provide all the heat necessary to heat the building at the coldest outdoor temperatures for as long as needed when the solar heating system cannot provide enough heat. This ensures that the home or business will always have heat regardless of external conditions such as large storms, blizzards, or other extended periods when solar energy may not be available. This system also provides the additional heat necessary if the outdoor temperature is extremely cold. The colder the outdoor temperature, the larger the temperature differential between the indoor and outdoor areas, and this causes the indoor space to need more heat. Typically the hot water coil and the furnace can provide the amount of heat the interior space needs down to about a 30° to 40° F temperature differential. This means that the solar system when working efficiently can provide the heat necessary for temperatures down to about 30° F outside. If it gets much colder than 30° F, the temperature differential becomes too large and the interior space needs additional heat energy, more than the hot water coil in the furnace can provide. A second stage thermostat is often used, and it automatically turns on the auxiliary heating any time the temperature differential inside the home is more than 2° or 3° F from the set point.

3.3.3 Controls for the Solar Space Heating System

The controls for the solar space heating system include temperature controls and pump controls for the glycol solution that moves through the solar collector. The controls ensure that the pump runs continuously so that the water from the storage unit is continually pumped through the heat exchanger as long as the glycol solution is being heated. In case a pump fails or other similar problems prevent the water from the storage unit from flowing through the heat exchanger, overheating controls and sensors must be used to protect the glycol fluid that is heated in the solar collector so it does not become too hot.

If distilled water is used instead of a glycol solution in the first loop, additional controls must be provided to ensure that the the outdoor part of the system drains down properly if the temperature gets below freezing. This usually includes sensors and valves that open and allow distilled water to drain back to the heat exchanger that is located inside the building so that it is not exposed to the freezing temperatures. The furnace will typically have controls to protect it from overheating when the auxiliary heat is used and to ensure that water from the storage tank is pumped through the coil when the system is energized to heat the interior. Additional controls may be used to provide damper controls on the ductwork to provide zone heating inside the building or home. Heating zone controls allow the temperature in one room of the interior to be controlled at a different temperature than another room. For example the temperature in a bedroom may be set to be cooler during the day when no one is using the room, which would save additional heat energy since that space does not need to be heated as much. The zone control would have a second setting that is based on a timer or is activated when people enter the room, and it would automatically bring on the heating system and heat that room to a warmer temperature.

This type of system could also be used with radiator coils that are placed directly in each room that is to be heated, and the controls would ensure that water flows through the radiator coils when that room requires heating. This is typical of a hot water boiler system that is used traditionally for heating. The zone controls for the system allow a thermostat to be located in each room that needs to be heated, and this allows each room to be set to a different temperature. This type of system would need additional pumps to move the warm water through each radiator individually.

3.3.4 Advantages and Disadvantages of the Solar Heating System for a Warm-Air Furnace

One of the advantages of using the solar heating system for heating is reducing the heating bill up to 30% or 40%. The solar heating system heats water that is used to heat air to provide heat for use inside the home or dwelling. In most cases the amount of heat that is provided from the solar heating system will not be enough to heat the space when the outdoor temperature is at its coldest or at night when the heat from the solar heating system is limited. In these cases the heating system must utilize an additional heat source such as a gas furnace. Any time the solar heating system can be used instead of the gas furnace, it has the advantage of offsetting the use of burning hydrocarbon-based fuels, which may cause additional air pollution. If a high-efficiency gas furnace is used, the amount of air pollution is less because the fuel is burned completely. If the backup heat is provided by an electrical heating coil, the electrical energy can come from an additional solar PV panel that produces DC electricity that could be used by the electric heating coil in the furnace. Otherwise the electricity needed by the furnace for heating would come from a grid-tied system that provides electricity to the rest of the home. In the fall and spring months, the solar heating system typically can provide all the heat required for the home or business since outdoor temperatures do not get too cold, and the amount of solar energy available tends to be larger due to the increased number of daylight hours in those months.

The disadvantages of this type of system include the initial cost of the installation and the pieces of the solar heating system, including the solar collector, heat exchanger, storage tank and pump, and plumbing. Another disadvantage of the system is that it will consume additional electricity to run the three pumps that move the glycol and water through the system. This should be offset by the savings from not using as much auxiliary fuel to heat the home. Since the system has additional piping and plumbing, there is also a chance of leaks that may cause damage to the structure, and traditionally the pumps will require additional maintenance.

3.3.5 Overheat Protection and Other Safety Controls

Solar energy systems are designed to operate automatically through the use of controllers. However, emergency procedures should be provided to protect the system equipment from overheating or from damage due to water leaks. The safety controls require a number of sensors to continually monitor temperatures and water flow and other sensors to monitor pump operation. Each safety system must have a preset condition that provides either an alarm to warn of the problem or alternate controls and panels that automatically make changes to the system to protect it from damage. For example if the pump that circulates the glycol solution fails to move enough solution to the collector, it may severely overheat, which can cause high pressures in the system that may damage the piping. In these cases the pressure must be relieved and the system shut down to a safe condition. This may include opening valves that allow the system to drain so that no additional solution is allowed to remain in the solar collector. If the system develops a leak, a series of valves may be installed to isolate various parts of the system so that the solutions (glycol or potable water) are not allowed to leak out completely.

Another important safety circuit is a temperature-monitoring system on the glycol fluid. If the outside temperature drops well below freezing, and glycol is allowed to continue to circulate through the collector system when the sun is not shining, its temperature will continue to decrease. If the glycol solution continues to loose temperature to where it is well below the freezing point of water, and it is allowed to circulate through the heat exchanger

barrel, it may cause the potable water that flows through the tubing of the heat exchanger to freeze. When water freezes it expands to a point where it would burst the copper tubing it is flowing through. Temperature controls are provided on the glycol part of the system and the potable water side of the system to ensure that this condition does not occur. If the temperature is sensed to be below the freezing point of water, the glycol pumps are shut down and the collector is drained down. These valves are usually designed to open when the power fails to ensure that the system can safely drain down.

Once the outside temperature begins to warm up, and the sun shines again, the collector will warm up quickly, and the fluid can begin to flow through it safely again. The system can come back up to temperature and begin to produce heat again as soon as the outside conditions change. The sensors and controls can quickly determine when the system can begin producing heat again after a shutdown period, which may be overnight or for a longer period of time in the coldest part of the winter if the system is installed in locations in the far northern regions.

3.3.6 Switching Over from Winter to Summer

During the winter months, the heat load for the heating system will be at its largest because that is when the temperature differential from the inside of the house to the outside is usually greatest. During the summer months the interior of the house does not need additional heat, so the solar water heating system is usually only used to provide hot water for washing dishes and clothes and bathing. This means the demand on the system will be much smaller because extra amounts of heat are not needed to heat the home's interior. Some systems are designed to pump a smaller amount of glycol solution through the collector, which basically reduces the size of the system and its ability to convert energy to heat the hot water. Other systems have sophisticated controls that allow glycol solution to bypass the collector so the system does not overheat, once the potable water is up to its set point temperature. Some systems have a larger indoor storage capacity that allows extra amounts of water to be heated and stored.

At some point in time either by a calendar date or by measuring the temperature differences, the system may need to be switched from a winter mode to a summer mode. This may be done automatically by the controls if they are sophisticated enough; in other cases it must be done manually.

3.3.7 Layup for Longer Periods of Time

If the system is not going to be used for long periods of time such as a week or longer, this glycol solution that circulates through the solar collector is generally drained back into the heat exchanger or a storage container so that is not exposed to the outside temperatures. The pumps may be turned off so that they do not waste energy circulating fluid that is not going to be heated or used.

Some systems have a built-in condition for changing the valves to the position required for laying up the system and allowing it to drain. If other adjustments to valves are required during the period when the system is not being used, it can be programmed into the control system so that the valves are in the proper position to automatically drain it down so it is ready when it is needed again. In some smaller systems, this must all be done manually, and the process must be reversed when it is ready to use again.

3.4 USING A SOLAR WATER HEATING SYSTEM TO HEAT A SWIMMING POOL

Another application for using a solar water heating system is to heat the water in a swimming pool. The water in a swimming pool can be pumped directly through the solar collector, since the system is typically not used when temperatures drop below the freezing point of water. If this is the case, the efficiency of the collector system is much better than if a heat exchanger were required.

A solar blanket is a clear plastic sheet that lays on top of the pool to prevent heat from escaping and to allow sunlight to filter through it and heat the water; it may be used in addition to a solar collector system. During the day the sunshine filters through the solar blanket and heats the water; during the evening hours when the air temperature drops below the water temperature, the plastic covering prevents the cooler air from coming into contact with the water in the pool and cooling it down. The combination of the pool water being circulated through the solar collector and the solar blanket will allow the water temperature to rise without the use of other methods of heating, such as a gas heater or heat pump. In some cases the solar water heating system is not needed during the summer months, because enough solar energy is absorbed into the pool water through direct sunlight. The solar heating system is usually used to extend the months when the swimming pool can be used by heating the water earlier in the spring and providing additional heat into the fall months. In some regions, such as the South and Southwest, the solar heating systems allow a swimming pool to be used year-round.

3.5 PASSIVE HOT WATER HEATING SYSTEMS

A passive hot water heating system uses gravity and temperature differences between hot and cold water to make water move through the collector. The advantage of the passive system is that it does not need to use large quantities of electricity to run pumps to move the water. Instead gravity and the temperature difference between hot and cold glycol solutions make the solution move through the solar collector outdoors and the heat exchanger indoors. The volume of glycol fluid and its pressure for this type of system are minimal, since they depend on passive systems such as gravity or temperature differential. When the glycol solution is heated, it will expand and cause the fluid to flow through the system. One way to ensure that the system is able to handle the hot water requirements is to make it larger than it typically would need to be if pumps were used.

Another type of system is a semi-passive system that uses solar PV cells to generate DC electricity in addition to the solar collector heating the glycol solution. The DC power is used to power the pumps in the system, so electricity from the grid is not required to cause the glycol fluid to flow and operate the system. An example of a typical passive solar water heating system is provided in the next section.

3.5.1 Small Passive Solar Water Heating System to Provide Residential Hot Water

Smaller solar collectors are available to be mounted on a rooftop, and this type of system can be added to a traditional hot water tank to provide an economical way to heat water for residential purposes. The diagram for this system is shown in Figure 3-8. The solar collector sits on the rooftop, and it is part of a sealed system that has glycol solution in it. The second part of the glycol system is a heat exchanger that is located inside the home by the traditional hot water tank. The glycol part of the system flows through a coil inside the heat exchanger. The potable water flows inside the barrel of the heat exchanger and flows around the coil where the glycol flows through. The system does not need a pump for the glycol solution or the potable hot water system; instead the glycol is moved through its piping system by a difference of temperature. When the glycol solution is heated, it expands slightly and moves through the solar collector as it picks up heat from the sun. As the heated glycol flows from the collector down to and through the heat exchanger, it looses heat. When it has moved completely through the heat exchanger, it looses enough temperature to create a temperature difference between the water leaving the heat exchanger and the water that is being heated inside the solar collector. Since the glycol system plumbing is completely filled with solution, this temperature difference is sufficient to cause the glycol solution to flow from the heat exchanger back up to the roof and through the solar collector again. An expansion tank is provided at the lower part of the circuit to ensure that the glycol solution has enough room when it gets to its warmest temperature and is completely expanded. When the sun begins to loose its ability to heat the glycol solution in the solar heater, the solution begins to cool down and condense, which slows the flow down to a point where it eventually stops. Since the glycol flow stops when it is cooler, it does not need expensive valves to divert it or stop it from cooling the potable water in the heat exchanger.

The potable water flows naturally through the heat exchanger and into the hot water heater when hot water is

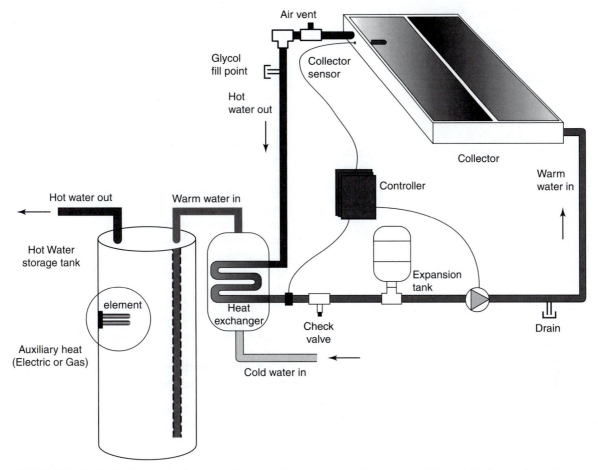

FIGURE 3-8   Active indirect solar heating system used to heat water that is part of the potable hot water system for a residence.

used anywhere in the house. If the solar heating collector does not add enough heat to the water in the hot water tank, the traditional heating element (gas or electric) will bring the water up to the required temperature. The potable water entering the system is at its coldest point when it enters the heat exchanger, and its temperature will be increased by the warmer glycol solution that flows through the coil in the heat exchanger. If additional hot water is needed at night or when the solar heating collector is not able to add heat to the hot water system, the gas or electric heating element will heat the water just as it normally would if the solar collector were not part of the system. This type of system is not meant to replace the hot water tank; rather it is intended to help supplement the hot water system and use the energy from the sun whenever it is available.

3.6 SOLAR HEATING USING AIR

Another way to heat the home using solar energy is to heat air, instead of heating water. This type of system moves a large volume of air through the solar collector system anytime heat is needed in the home. The air from the home that is inside each room is moved by fans through the solar collector system, which converts the sun's rays into heat energy. Since the collector system is

exposed to the outside temperatures, it is important that it is isolated when the sun is not available to provide heat to the collector. In these cases a backup system such as electric coils or a gas-fired or oil-fired furnace is used to provide the heat when the sun's energy is too low or at night when sunlight is not available. These types of systems are not designed to provide all the heat; rather they are designed to provide a significant amount of heat energy when the sun is shining.

Some systems that heat air provide a storage medium such as a large bin of rocks or concrete that takes on heat energy slowly during the day as excess heat is converted from the solar collector. The warm air continually heats the rock or concrete, which stores the heat energy in its mass. Then at night the air is diverted from moving through the exterior solar collector and is routed over the rock or concrete where it picks up heat throughout the evening. The transfer of heat energy will continue as long as the temperature in the rock storage medium is higher than the temperature in the house. This type of system must have a large area available for the rocks or concrete to provide sufficient storage space to last through the evening or through cloudy days. It is also important for this type of system to have sufficient filtering to keep dirt and mold from entering the living space if it begins to build up in the storage area. This type of system also must move large quantities of air in order for the heating system to operate correctly.

3.6.1 Passive Solar Heating with Air

Another way to heat a home is called *passive solar heating*. This type of heating system does not rely on fans to move the solar heated air through the home; rather it uses the temperature difference between the air that is heated by the sun as it moves through the solar collector and the temperature of the cooler air that is in the storage area. Since the air in the storage area is cooler than the air that is heated in the solar collector, the warmer air will move toward the cooler region, causing a draft or air movement over the storage medium. As long as the solar collector is receiving sunlight, it is typically much warmer than the space around the storage medium. The advantage of this type of passive system is that additional electrical energy is not used for fans to move the air.

Figure 3-9 shows a diagram of a residence being heated in the winter from the passive solar system and also how the system storage medium can be used to aid in cooling in the summer. In the summer months the outdoor temperature at night is typically cooler than 70°. This cool air can be moved over the storage medium, which causes the rocks or concrete to cool down to whatever the temperature is outdoors. The storage medium can also be large containers of water. During the day when it is warmer outside, the indoor air is moved over the storage medium where it is cooled and then returned to the interior space to provide air conditioning. The temperature difference from the storage medium to the indoor temperature is typically not large, so this system works fairly well as long as the outdoor temperature does not become too warm. Since the cool air at night is

used, the expense is much less than using electrical power to turn compressors in a typical mechanical air conditioning system. The main goal of this type of system is to move large quantities of air throughout the space.

3.7 SOLAR PHOTOVOLTAIC PANELS

Solar PV systems convert sunlight into DC electricity. This section of the chapter will introduce you to all the different types of PV cells. Before you can begin to understand their operation and design, you will need to have a good understanding of basic principles of DC electricity and the basic terms used in electrical systems. In some cases this information will be a review if you have had a previous electricity course. If not you will need this information to fully understand how PV cells are created and connected into a working electrical system. The last chapter in this text will provide all the information you need to fully understand all the electricity used in solar technology. In this section, electrical terms and theories will be introduced where they apply.

3.7.1 Introduction to Voltage, Current, Resistance, and Power

Voltage, current, resistance, and power are all terms used to explain the operating parameters of electricity. *Electricity* is defined as the flow of electrons, also defined as electrical *current*. An electron is the negative part of an atom, and it can be found moving around the nucleus of the atom in an orbit. Each element found on earth has atoms that have a specific number of electrons that orbit around the nucleus of its atoms in specific paths called shells. The electron has a negative charge, and the nucleus of the atom holds protons that are positively charged and neutrons that have a neutral or no charge. Some elements such as copper, gold,

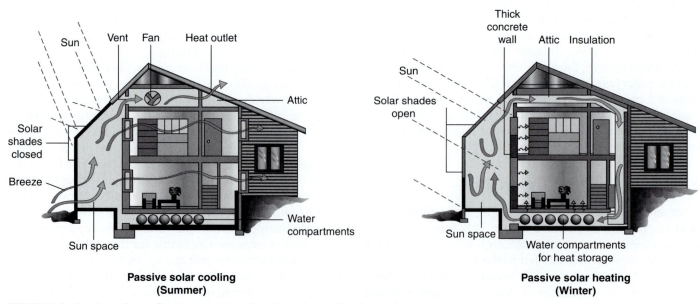

**Passive solar cooling
(Summer)**

**Passive solar heating
(Winter)**

FIGURE 3-9   Passive solar cooling in summer and passive solar heating in winter.

TABLE 3-2	Electrical terms, definitions, units and abbreviations.			
Electrical term	**Definition**	**Units**	**Abbreviation of known values**	**Abbreviation used in Ohm's law**
Voltage	Electromotive forced also called potential that causes electrons to move as current flow	Volts	V	E
Current	The flow of electrons	Ampere	A	I
Resistance	The opposition to electron flow	Ohms	Ω	R
Power	Product of voltage times current	Watts	W	P

silver, and aluminum readily allow the negatively charged electrons in their outermost shell (orbit) to break free of the nucleus that is attracting the electrons with the positive charged protons. The reason the electrons in the outermost shell can break free is that the attraction from the positive proton is weak. When voltage is applied to a copper wire, its charge will cause electrons to break free and begin to move through the wire to a load. A typical electrical load has resistance, such as an electric light bulb or electrical heating element. Electrical *resistance* is defined as the opposition to electrical current and it is measured in Ohms. Electrical current is defined as the fow of electrons and it is measured in amperes, Electrical *voltage* is defined as electromotive force and it is measured in volts. A relationship exists among voltage, current, and resistance in that 1 ampere (A) of current is caused to flow when 1 volt is applied to 1 ohm of resistance.

Electrical power is measured in watts. One watt is defined as 1 volt multiplied by 1 A of current. The units of voltage are volts, the units of current are amperes, and the units of resistance are ohms. Table 3-2 shows all the electrical terms, their definitions, the units for each term, and the abbreviation for the term when the value is known in a problem and the abbreviation of the term when the value is unknown in Ohm's law. Ohm's law states that in a circuit powered with DC voltage, the amount of voltage can be calculated by multiplying the amount of current in the circuit times the amount of resistance in the circuit. When the value of voltage is being calculated in Ohm's law, voltage is represented by the letter E for electromotive force. When the amount of voltage is found and becomes the answer to the problem, voltage is represented by the letter V for volts. When current is being calculated in an Ohm's law formula, it is represented by the letter I for intensity, and it is represented by the letter A for amperes when the answer is known. Resistance is represented by the letter R when

it is used in an Ohm's law formula and the Greek letter "O" (Ω),omega, is used when the answer to the problem is found and the number of ohms is determined. The letter O can be mistaken for the number zero, so the Greek letter omega is used to represent it. Power is represented by the letter P in an Ohm's law formula, and it is represented by the letter W for watts when the answer to the problem is found and the number of watts is known.

The relationship between ohms, volts, and amps and electrical circuit can be mathematically proven with a set of formulas called Ohm's law. The information in Chapter 12 will provide much more information and a more in-depth explanation of this relationship and show various sample problems and solutions that you can follow and try to get a better understanding of these terms. At this point all you need to know is that Ohm's law can be expressed in terms of voltage, current, or resistance, and these formulas can be used to calculate voltage, current, or resistance in any electrical circuit that uses DC voltage any time two or more of the values are known. The formulas for calculating voltage, current, and resistance in an AC circuit vary slightly and will be explained in Chapter 12. Table 3-3 shows the formulas for Ohm's law. In order to understand the formulas of Ohm's law you must first understand that each of the electrical terms, voltage, current, and resistance, is replaced with an abbreviation that is used when each of the values is unknown. For example E is used to represent the voltage in the Ohm's law formulas, and it is derived from the first letter of the term "electromotive force"; I is used to represent current and is derived from the first letter of the term "intensity"; and R is used to represent resistance and is derived from the first letter of the term "resistance." The basic formula for Ohm's law is $E = I \times R$. This means that voltage in any circuit can be found by multiplying the amount of current by the amount of resistance. For example if current in any circuit is 2 amperes

TABLE 3-3	Relationship between ohms and amps in Ohm's law				
Ohm's law units		**To find voltage**	**To find current**	**To find resistance**	**To find power**
E = Voltage I = Current R = Resistance P = Power (in watts)		$E = I \times R$	$I = E/R$	$R = E/I$	$P = E \times I$ or $P = I^2 \times R$

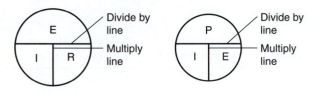

FIGURE 3-10 (left) Ohm's law pie, a memory tool for remembering the Ohm's law formulas for voltage, current, and resistance. (right) Ohm's law formulas for power, which includes voltage, current, and power (wattage).

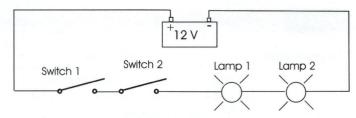

FIGURE 3-11 Two switches connected in series with two lamps that are also connected in series.

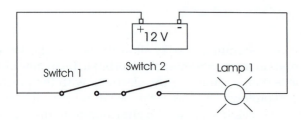

FIGURE 3-12 Two switches connected in series with a single lamp.

and resistance is 6 ohms, the voltage that causes the current flow is 12 V. Since the values in the formula have been determined, a V is now used to represent voltage, so the answer can be represented as 12 V. Remember when the values in Ohm's law are unknown, we will use E to represent voltage. Note that some people may prefer to use V to represent voltage in the Ohm's law formulas when the value of voltage is not known, but the more common standard is to represent the values in the formula when they are unknown with E; V is used once you have solved the problem and know the exact amount of voltage; V is also used on meters when the exact amount of voltage is measured and is known. I is used to represent current and R to represent resistance when their values are unknown in the Ohm's law formula. Once the problem is solved, we will use V to represent voltage, A to represent current (amps), and the Greek letter Ω (omega) to represent ohms. The reason the Greek letter Ω is used instead of the letter "O" is that the letter "O" can be confused for the numeral zero (0). In electricity formulas anytime a letter of the alphabet can be confused with a number or is confusing in any other way, a letter from the Greek alphabet will be used.

If the voltage and current in a circuit are known, the resistance can be calculated with the Ohm's law formula $R = E/I$. If we use the same values from the previous circuit, we would divide 12 V by 2 A and the answer would be 6 ohms. If the voltage and resistance in a circuit are known, the current can be calculated with the Ohm's law formula $I = E/R$. If we use the same values from the previous circuit again, we would divide 12 V by 6 ohms and the answer would be 2 A. The formulas to calculate voltage, current, or resistance using Ohm's law are shown in Table 3-3.

Another way to think about Ohm's law is to use the Ohm's law pie as a memory tool. Figure 3-10 shows an example of this memory tool. It is basically a circle that has a line horizontally across the middle. The term for voltage (E) is placed above that line. A vertical line is placed under the horizontal line, and the term for current (I) is shown in the left panel, and the term for resistance (R) is shown in the right panel. The vertical line is called the "multiply line": in order to find voltage you would multiply the current times the resistance. The horizontal line is called the "divide by line": to find current or resistance you would divide the opposite value into voltage.

3.7.2 Voltage, Current, Lamps, and Switches in a Series Circuit

In order to understand the voltage and current that PV cells produce, you need to understand how voltage and current function in simple circuits. Figure 3-11 shows two lamps connected in series with two switches. The term "series" means that there is only one path for current to flow through as it leaves one battery terminal and returns to the other. Since the current flow is made up of electrons and the electrons are negatively charged, they leave the battery from the negative terminal and return to the positive battery terminal.

Figure 3-11 shows a series circuit with two switches connected in series with two lamps connected in series. In a series circuit, any opening in the circuit wiring or switches will cause current to stop and the lamps will not light. This means that if either switch 1 or switch 2 is opened, the current flow will stop and the lamps will not light. If a hole develops in any of the wiring, the current will also stop and the lamps will not light. The voltage in a series circuit that has two lamps connected in series will split evenly between the two lamps if the lamps are identical. If you measured the voltage across each lamp with a voltmeter, you would measure 6 V across each lamp. Figure 3-12 shows two switches connected in series with a single lamp. Any time either switch is opened, the lamp is turned off.

3.7.3 Voltage, Current, Lamps, and Switches in a Parallel Circuit

The lamps and switches can also be connected in parallel. A parallel circuit has more than one path for current to flow. Figure 3-13 shows switch 1 and lamp 1 in the first parallel circuit, and switch 2 and lamp 2 in the second

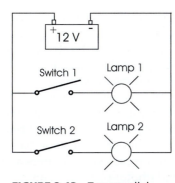

FIGURE 3-13 Two parallel circuits. Lamp 1 is controlled by switch 1, and lamp 2 is controlled independently by switch 2.

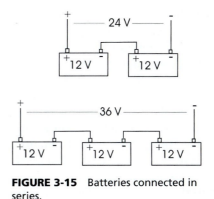

FIGURE 3-15 Batteries connected in series.

parallel circuit. In a parallel circuit the voltage is the same in each branch circuit. Since the power supply is 12 V, the lamp in the first parallel circuit will receive 12 V, and the lamp in the second parallel circuit will also receive 12 V. One of the reason lamps are connected in parallel is that each lamp in each parallel branch will receive the full voltage from the power supply.

Another reason lamps are connected in parallel is that each lamp can be controlled and turned on or off independently from the other. This allows full control over each lamp, as it can be turned on or off regardless of what the other loads in the circuit are being turned on or off.

The circuit in Figure 3-14 shows a lamp that is controlled by two switches that are connected in parallel. In this circuit switch 1 or switch 2 can turn on the lamp. If you need to turn the lamp off, both switch 1 and switch 2 must be turned off at the same time.

3.7.4 Batteries Connected in Series

The next type of circuit that you need to understand connects power sources in series and parallel. The power source in this case is a 12 V battery. Figure 3-15 shows batteries connected in series. In the top part of the circuit two 12 V batteries are connected in series. This is accomplished by connecting the negative terminal of the

battery on the left to the positive terminal of the battery on the right. The total voltage of the two batteries is now added together. If you measured from the positive terminal of the left battery to the negative terminal of the right battery, you would measure 24 V. When batteries or power supplies are placed in series, their voltage is added together.

In the bottom part of the circuit three 12 V batteries are connected in series by joining the negative terminal of one battery to the positive terminal of the battery next to it. When all three batteries are connected in series, again their voltage is added together. It would measure 36 V from the positive terminal of the battery on the left to the negative terminal of the battery on the right. Batteries or power supplies can be put together in series to continue to build the total voltage. The only problem that occurs is that when one or more of the batteries begin to discharge, the amount of total voltage will begin to decrease.

3.7.5 Batteries Connected in Parallel

Batteries can also be connected in parallel rather than in series. Figure 3-16 shows a battery center connected in parallel. In the circuit on the left two 12 V batteries are connected in parallel. The parallel connection is made by connecting the positive terminals of both batteries and then connecting the negative terminals on both batteries. When batteries are connected in parallel, the voltage between the positive and negative terminals

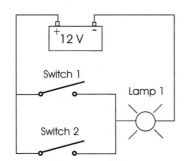

FIGURE 3-14 A lamp controlled by two switches that are connected in parallel.

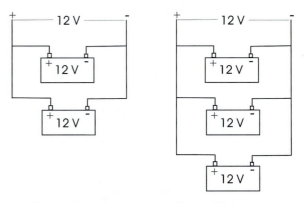

FIGURE 3-16 Batteries connected in parallel.

will remain 12 V. The advantage of connecting batteries in parallel is that the voltage remains the same, but the amount of current the batteries can provide will equal the amount that each battery has available.

The circuit on the right in this figure shows three 12 V batteries connected in parallel. You can see that the total voltage when all the batteries are connected in parallel is 12 V. Again the advantage of connecting three 12 V batteries is that the voltage would remain 12 V, but the available current would equal the total current of each of the three batteries added together.

In the next section we will begin to learn about solar photovoltaic cells and how they are connected. Each solar *photovoltaic cell* is basically a power supply that is similar to a battery because the sunshine that reaches the surface causes it to produce voltage. You will learn about why PV cells are connected in series or parallel to produce a specific amount of voltage or current.

3.8 SOLAR PHOTOVOLTAIC CELLS AND MODULES

Solar photovoltaic systems make electricity when sunlight is directed to their surface. The basic theory of operation of the photovoltaic device is that when certain materials are combined in a specific manner, the energy of photons from the sun strikes the material and causes electrons to move inside the PV cell. Electrons moving in an electrical circuit create an electrical current. In order to make the photovoltaic system provide a useful voltage and current, a large number of photovoltaic cells are connected together in various series and parallel combinations of modules, strings, and arrays.

The photovoltaic cell is also called a "solar cell," and it is the basic building block of all solar PV systems. The cell can be made from any number of materials, including silicon; small quantities of other materials such as boron, phosphorous, gallium, and arsenic can be added to make the cell more or less efficient at converting sunlight into electrical power. Some types of less-efficient solar PV cells

are produced because they are less expensive to manufacture. Each cell develops about half a volt of DC electrical potential (voltage) when light strikes the surface. The maximum amperage of the cell can create is proportional to its surface area and the intensity or amount of light that is striking the surface at any given time. PV cells can produce electricity for approximately 30 to 50 years.

In order to begin to make sense of photovoltaic systems, we will need to agree on some basic terminology. You will find that certain manufacturers and certain companies that install photovoltaic systems may use different terms to describe the same components. In this text we will use the term *photovoltaic (PV) module* or *photovoltaic (PV) panel* to refer to a finished component that can be installed as a unit. A PV module is a grouping of PV cells. Figure 3-17 shows a single PV cell, which has a positive terminal and a negative terminal. Figure 3-18 shows a group of cells that are connected in series (positive terminal of one cell is connected to the negative terminal to the cell next to it), and this grouping is a photovoltaic (PV) module. Typically modules with 36 cells are connected to produce between 18 to 22 V, which will in turn produce a nominal output voltage of 12 V. The reason the voltage output of 18 to 22 V is considered to have a nominal output voltage of 12 V is that the system will not receive the same amount of sunlight at the same intensity, and the output is considered 12 V nominal for design and planning purposes. Some manufacturers will assemble and connect up to 72 PV cells in such a way that the nominal voltage will be 24 V. The higher voltages are used when the voltage is being sent to the grid and is run through an inverter to create a higher AC voltage. Notice that the photovoltaic module has the cells connected in series in such a way that the module has one positive

A Single PV Cell

FIGURE 3-17 A single photovoltaic (PV) cell.

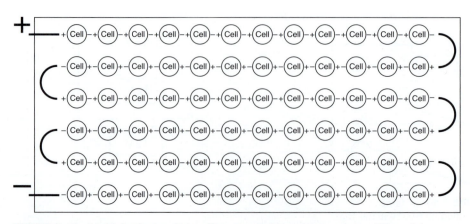

FIGURE 3-18 Multiple photovoltaic cells connected in series to create a PV module.

terminal and one negative terminal that can be used to connect modules together.

At this point it is important to remember that solar energy is used to heat water or air, as well as to produce voltage through photovoltaic cells. The reason that this is important is that some manufacturers will refer to solar panels as systems that use solar energy to create heat, and they will use the term photovoltaic (PV) panel when they are referring to a system that converts light to electrical energy. When this term is used, it is referring to a group of photovoltaic modules that are made of photovoltaic cells.

3.8.1 Solar Photovoltaic Strings and Arrays

A *solar array* is a linked collection of photovoltaic modules made of PV cells. A *string* is a group of modules wired in series. When modules are connecting in series, they will act exactly like batteries that are wired in series and their voltage production will be added together. Figure 3-19 shows three modules connected in series as a string and two strings connected in parallel to make a *photovoltaic array*. Since each module produces 24 V, the three modules together will create a string that produces 72 V nominal voltage. You can see that two of the 72 V strings are connected in parallel at the combiner box to create a PV array, which will increase the amount of current that will be produced. Manufacturers can connect PV modules in a variety of configurations to produce PV arrays that operate at 12 V, 24 V, or 48 V nominal for residential and commercial applications, and up to 600 V for systems that are designed to be connected to the high-voltage grid. The voltage output of the arrays is DC voltage, but it can easily be converted to AC voltage by an inverter so that it can be used for loads that require AC voltage. The inverter provides a second function in that it will act as a voltage regulator to ensure that the voltage level supplied to the appliances and loads is within tolerance of their 40s reading even when the cells began to produce less than nominal voltage. This is accomplished by feeding the inverter a voltage level that is larger than required, so that the voltage regulator will be able to function and provide AC voltage that is slightly lower. Manufactures can produce PV panels that can be connected in a combiner box into strings and produce virtually any voltage by adding the voltages of each panel in series. When additional current is needed, more panels are connected in parallel.

The term "array" describes the whole group of modules in a system. These can be modules connected in series or parallel to create the voltage required by the application. Since the array is created from groups of modules connected together as a string, sometimes the string is called a *sub-array*. The arrays are manufactured in such a way that they are easily mounted on racks that will tilt the surface of the panel the correct amount so that it can absorb the maximum amount of sunlight. The panels can then be connected in parallel through combiner boxes on the rooftop to increase the amount of current the PV system will be able to provide.

3.9 TYPES OF SOLAR PV PANELS

Modern solar PV panels can be made from a wide variety of materials. The most common material used today is silicon, and small amounts of other elements are added to create PV panels for different applications. Two of the important factors that determine which materials will be used are the cost of the raw materials and the cost of the manufacturing process that is used to combine the materials. In some applications the cost is not the most important factor; rather the efficiency and the life of the panel are more important. For example if a photovoltaic cell is being designed to be used in a space exploration application such as the Mars rover, the most important aspects of the PV cell are that it produces sufficient voltage to keep all the systems charged and operational and that it lasts a long time.

In other applications the main goal is to install as many photovoltaic systems as possible at the lowest cost to increase the amount of electrical power production. In these applications lower-cost materials and lowest-cost manufacturing processes are necessary to get as many systems installed as possible. Some combinations of materials and manufacturing processes are being created to provide products that have a specific function. For example

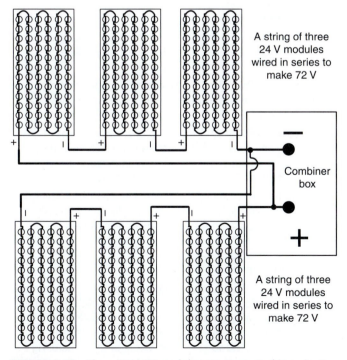

A string of three 24 V modules wired in series to make 72 V

Combiner box

A string of three 24 V modules wired in series to make 72 V

FIGURE 3-19 Three 24 V PV modules are connected in series to create a 72 V PV string. Two of the 72 V strings are connected in parallel to create a 72 V PV array.

thin-film materials are being used in applications where the surface of the panel may need to be flexible enough to fit the outline of a roof panel or other structure.

Advances in *nanotechnology* are also providing a number of new products and manufacturing processes. Nanotechnology is a relatively new entity that is a branch of engineering and science that deals with objects that are smaller than 1,000 nanometers. A nanometer is 0.000,000,001 meter, which can also be defined as 1 times 10^{-9} meters. Nanotechnology is also called nanotech, and it deals with the control of matter or devices at an atomic or molecular level.

This section will introduce all the types of materials that are used to create solar photovoltaic panels. There are many ways to discuss the nanotechnology materials and technology that are used in manufacturing the panels. For example a comparison of cost or the components used in various panels could be used to show the different types. Another way to discuss these products is to compare the manufacturing processes that are used to create them. Table 3-4 shows all the types of solar PV materials that are used. The table also shows the efficiency, the materials that are used, and/or the manufacturing processes and the disadvantages and advantages of each.

TABLE 3-4 Types of PV materials

Types of PV cells and panels	Efficiency	Material/Manufacturing	Disadvantages	Advantages
Monocrystalline silicon panels	15–18%	Created in cylindrical ingots, crystalline silicon produced in large sheets	Must be cut (sawed) from larger cylindrical ingots, which causes irregular form, so corners of rectangular panels are not completely covered	Most prevalent bulk silicon material.
Bifacial mono-crystalline	Up to 20%	Uses mono-crystalline solar cells that have glass on both sides so that they can collect energy from both sides of the solar panel	Requires special mounting and reflectors to get light to be available on bottom of panel	Converts light into electrical power from both sides
Polycrystalline silicon panels	12–14%	Molten silicon in large square blocks	Less efficient than monocrystalline panels	Less expensive to manufacture than monocrystalline panels.
String ribbon silicon	12–14%	Type of monocrystalline silicon	Lower efficiencies than polycrystalline silicon	Lower production costs due to less waste and does not require sawing
Amorphous silicon or thin-film panels	5–6%	No crystalline structure and can be applied as a film directly on different materials	Lowest efficiency of any current photovoltaic technology	The primary advantages of thin-film panels lie in their low manufacturing costs and versatility
Cadmium Telluride (CdTe)	18%	CdTe is easier to deposit and more suitable for large-scale production	Cadmium is toxic, but solar cell has about same amount of cadmium as NiCad battery	Very efficient light-absorbing material for thin-film cells and large-scale cells
CIGS (copper, indium, gallium and selenide) thin-film panels solar cells	19.9%	Copper, indium, gallium and selenide placed on thin soda-lime glass substrate or substrate can be flexible metallic foils, high-temperature polymers, or stainless steel sheets.	Heat retention is a problem because material applied directly to substrate so no space for heat to escape	CIGS cells require approximately 1/50th to 1/100th of the raw materials needed for a typical silicon solar cell; low cost of manufacturing

Types of PV cells and panels	Efficiency	Material/Manufacturing	Disadvantages	Advantages
Group III–V Technologies	25%	Typically combined with gallium arsenide	High cost limits the number of panels that can be manufactured	High efficiencies allow fewer modules to be used.
Building-Integrated Photovoltaics (BIPV)	5–6%	Semitranslucent layer of amorphous silicon into glass used as windows or shingles	Low efficiency	Glass can allow certain amount of light into an area such as a window while making electrical power; also used as shingle roofing material
High Concentration Photovoltaic (HCPV)	More than 25%	Cover standard types of solar panels with optics and lenses to concentrate light	Cannot utilize diffuse light; more complex to install	Create more electrical power from fewer panels
High-Efficiency Multijunction Devices	Up to 40%	Three-layer junction: (1) gallium arsenide; (2) germanium; and (3) gallium, indium, and phosphide	Manufacturing process uses vapor processes	Original research for space applications, such as Mars rover; provides high efficiencies
Dye Sensitive Solar Cells (DSSC)	11%	Ruthenium metalorganic dye on a mesoporous layer of nanoparticulate titanium	Dyes degrade under UV and heat	Manufactured using screen printing process, which makes it cost effective
CIS (copper, indium, and selenide)	13.5%	Material made of $CuInSe_2$ (copper, indium, and $selenide_2$)	High manufacturing costs	Thin-film solar cell application because of its high optical absorption
Gallium Arsenide (GaAs) Multijunction	29%	A triple-junction cell, for example, may consist of the semiconductors GaAs, Ge, and $GaInP_2$	Cost is up to five times that of other solar cells	Very high efficiencies; used for space applications
Non-Silicon Solar Panels Nanoparticle processing; also called quantum dot modified photovoltaics	Up to 42%	Experimental at this point in time; made of carbon nanotubes, also called quantum dots, embedded in conductive polymers called mesoporous metal oxides	Experimental and not ready for production	Very high efficiencies expected
UV solar cells	Still under research	A transparent solar cell that uses ultraviolet (UV) light to generate electricity and allows visible light to pass through it	Experimental and not ready for production	Could be applied to a window as a coating that would produce electricity as well as allow light through it

3.9.1 Monocrystalline Silicon Solar Panels

Created in cylindrical ingots with crystalline silicon produced in large sheets, the monocrystalline silicon module is one of the most efficient types of solar panels. Each of the solar cells use silicon as its main material and it is cut from high-purity single crystal silicon rods. Advanced machining processes are used to cut the rod into discs or wafers that are 0.2 to 0.4 millimeters (mm) thick. The silicon is then processed into photovoltaic cells that are connected into solar strings that are then connected to make a completed solar panel. Monocrystalline solar panels have the highest output per cell area, making the most use of available roof space. They also perform better in high temperature or low light conditions than the multicrystalline or thin-film modules. Some manufacturers use a cutting process that cuts each wafer from a larger cylindrical ingot. This process causes the individual wafer to be round in shape, which means that there may be slight gaps when they are put together in a rectangular panel since the corners of rectangular panels are not completely covered. The monocrystalline silicon solar panel is the

FIGURE 3-20 Monocrystalline silicon solar panels. (iStockphoto.com)

most prevalent bulk silicon material. When the panels are put into production, they tend to be approximately 15% to 18% efficient. Figure 3-20 shows an example of this type of panel installed.

3.9.2 Bifacial Monocrystalline Solar Panels

The bifacial monocrystalline solar panels use monocrystalline solar cells that have glass on both sides of the material so that they can collect energy from both sides of the solar panel at the same time. This type of panel requires special mounting brackets and reflectors that mount the panel in a near vertical position so it can harvest light from both sides of the panel simultaneously.

Typically these panels are installed in a pole-mounted solar array so that ambient light can reach the panel from both the front and the back or the rooftop can be treated with a white matt paint or white paint to allow solar light to be reflected to the back of the panel if the panel is mounted nearly horizontal. Since this type of panel converts light into electrical power from both sides of the material, its efficiency is as high as 20%. Figure 3-21 shows an example of this type of panel installation.

FIGURE 3-21 Bifacial monocrystalline solar panels allow solar energy to be harvested and converted into electricity from both sides of the solar panel. These panels must be mounted in a near vertical position to receive sunlight on both sides. (Courtesy Sanyo)

FIGURE 3-22 Polycrystalline silicon panels. (Courtesy NREL and U.S. Department of Energy.)

3.9.3 Polycrystalline Silicon Panels

Polycrystalline silicon is also called multicrystalline silicon, and it is made from silicon that is produced in a large block of many crystals rather than grown as single crystals. This is what gives it a striking shattered glass appearance. Polycrystalline silicon is made in large square blocks, which makes it less expensive to manufacture than monocrystalline panels. Once the polycrystalline cells are manufactured in large blocks, machining processes are used to slice them into wafers that make up the modules. The modules are wired together as in other solar PV manufacturing processes to make strings and arrays. The polycrystalline silicon panels have an efficiency of approximately 12% to 14%, which makes it less efficient than monocrystalline panels. Figure 3-22 shows a typical polycrystalline solar panel.

3.9.4 String Ribbon Silicon Photovoltaic Cells

String ribbon silicon photovoltaic cells use a manufacturing process that produces multicrystalline silicon strips by pulling high-temperature-resistant wires through molten silicon to form a multicrystalline ribbon of silicon crystals. The ribbon is then cut into lengths that are treated with traditional processes to form solar cells. Problems still exist in the manufacturing process that make it difficult to control the thickness over the entire range of the material. When the thickness varies, it makes sections of the material unusable for photovoltaic cell production. Figure 3-23 shows an example of a string ribbon silicon photovoltaic cell.

3.9.5 Amorphous Silicon or Thin-Film Panels

Amorphous silicon or thin-film solar panels are not made from crystals; rather they use a thin layer of silicon deposited on a metal or glass base material that creates the

FIGURE 3-23 String ribbon type silicon PV cells. (Courtesy Evergreen Solar)

solar panel. The term "amorphous" means that the silicon crystals do not have any defined shape. These amorphous solar panels cost much less than the monocrystalline or polycrystalline solar panels. The energy efficiency of 5% to 6% is also much less than other types of panels, so more square footage is required to produce the same amount of power. In addition to their lower cost, thin-film panels or amorphous solar panels can be made into long sheets of roofing material to cover large areas of a south-facing roof surface.

Variations on this technology use other semiconductor materials such as copper indium diselenide (CIS) and cadmium telluride (CdTe). These materials are connected to the same metal conductor strips used in other technologies, but they do not necessarily use the other components typical in photovoltaic panels, since they do not require the same level of protection needed for more fragile crystalline cells.

Thin-film panels have the potential to grow in use, and they already figure in some of the most exciting enhanced photovoltaic systems, including high-efficiency multifunction devices and building-integrated photovoltaics, which will be explain later. Figure 3-24 shows an example of an amorphous type solar panel.

FIGURE 3-24   Amorphous solar panels are also referred to as thin-film solar panels.   (Courtesy NREL and U.S. Department of Energy.)

FIGURE 3-25   Cadmium telluride solar photovoltaic cells. (Courtesy NREL and U.S. Department of Energy.)

3.9.6 Cadmium Telluride

Cadmium telluride (CdTe) photovoltaic cells use cadmium and telluride instead of silicon as the base material to convert sunlight into electrical power. The process to make CdTe photovoltaic cells is simpler and more suitable for large-scale production, which makes its overall cost of production much less when compared to that for silicon cells. The CdTe cell is an efficient light-absorbing material for thin-film and large-scale cells. Research on CdTe goes back to the early 1950s, but more recently the research has focused on increasing production efficiencies and raising overall conversion efficiency, as well as reducing costs of production.

Among the drawbacks of using CdTe is that cadmium is toxic, but the CdTe solar cell has about same amount of cadmium as a NiCad battery. A disadvantage of using telluride is that it comes from the element tellurium (Te), which is not currently used for many other applications so it is not readily available for large-scale use. Tellurium is a by-product of the copper-making process, so it is limited in its availability. If CdTe cell production is to be increased to the point of providing large-scale production, newer ways to obtain tellurium must be identified.

At the present time the efficiency of CdTe photovoltaic cells is about 18%. Figure 3-25 shows an example of a cadmium telluride solar cell.

3.9.7 Copper Indium Gallium Selenide

The CIGS (copper, indium, gallium, and selenide) solar PV cell is made by placing copper, indium, gallium, and selenide on a thin soda-lime glass substrate or a substrate made of flexible metallic foils. The substrate can also be high-temperature polymers or stainless steel sheets. Heat retention in this PV cell is a problem because the material

FIGURE 3-26   CIGS solar PV cells.   (Courtesy NREL and U.S. Department of Energy.)

is applied directly to substrate and leaves no space for heat to escape when it is absorbing sunlight. The main advantage of the CIGS cell is that it requires approximately 1/50th to 1/100th of the raw materials needed for a typical silicon solar PV cell. The low cost of manufacturing means this type of cell can be mass produced more easily. The efficiency of the CIGS cell is approximately 19.9%. Figure 3-26 shows CIGS solar PV cells.

3.9.8 Group III–V Technologies Solar PV Cells

Group III–V technologies solar PV cells are solar PV cells made from materials that are found in the periodic table of elements in Group III and/or Group V. These PV cells are made from elements that include gallium arsenide (GaAs); gallium, indium, and phosphorous (GaInP); gallium, indium, and arsenide (GaInAs); aluminum, gallium, indium, and antimony (AlGaInSb); and Germanium (Ge). One fairly new Group III composite used is a nitride alloy that is used in the indium, gallium, nitride (InGaN) or in indium, aluminum, nitride (InAlN) PV cells. These alloys can be used in a single cell solar PV, a tandem cell solar PV, or a multijunction.

3.9.9 Building-Integrated Photovoltaics (BIPV)

The building-integrated photovoltaic solar cell is manufactured as a semitranslucent layer of amorphous silicon that is converted into glass used in windows or shingles and can be used on a large scale. It has fairly low efficiency, however, usually in the range of 5% to 6%, so a larger quantity of the material is needed to produce the same amount of electrical energy as more traditional silicon PV cells. Since the BIPV solar cells are designed to work as roofing material or glass, they serve two purposes, which will reduce construction costs for rebuilding slightly when compared to installing large volumes of solar panels on the roof after a traditional roof has been installed. When the BIPV cells are made into glass, the main advantage is the glass allows a certain amount of light into an area such as a window while making electrical power. Figure 3-27 shows an example of BIPV solar cells integrated into roofing shingles on a building.

3.9.10 High Concentrated Photovoltaic

High concentrated photovoltaic solar cells are basically standard types of solar panels with optics and lenses to concentrate light, but they do not utilize *diffuse light* well. These types of photocells are sometimes called solar concentrators, and they must be installed so that they receive direct light rather than diffuse light. Diffuse light is light that does not strike the solar photocell directly. They are also called concentrating photovoltaics (CPV), and they use a large area of mirrors and/or lenses to focus sunlight on a small area of photovoltaic cells. This means a hundred or more times direct sunlight is focused on each individual cell when compared with crystalline silicon panels.

All concentration systems need a single-axis or more often multiple (two axis) tracking systems that move the panels up and down in reference to the sun's rising and setting; it also requires a system that can track the sun as it travels from east to west. The tracking system uses high-precision controls, since the HCPV cells require direct sunlight with an error of less than 3° to produce electrical energy at the most efficient rate. The primary reason HCPV systems are used is that they can produce the same amount of electrical power while consuming smaller amounts of semiconducting materials, which tends to be expensive and is in short supply in some cases. The efficiencies of this type of system are up to 25%.

The main disadvantage of these types of cells is that their use has been limited by the initial and upfront cost of providing expensive tracking, which means the system overall has a longer payback period. Figure 3-28 shows an example of high concentrated photovoltaic solar panel.

3.9.11 High-Efficiency Multijunction Devices Used for Solar PV Cells

The high-efficiency multijunction devices use a three-layer junction to manufacture the photovoltaic cell. Multijunction devices receive their name from their use of multiple layers of cells, each layer acting as a junction where certain amounts of solar energy are absorbed. The first layer of the PV cell is gallium arsenide; the second layer is germanium; and the third layer gallium, indium, and phosphide. Other typical materials used to create the high-efficiency multijunction devices include gallium arsenide and amorphous silicon. In these types of devices, the top photovoltaic layer responds to solar waves that have shorter wavelengths and carry the highest energy. When the cell absorbs this energy, it creates an electrical charge. When other solar waves pass through this layer, they are absorbed by the lower layers and converted into electrical power.

FIGURE 3-27   BIPV solar cells integrated into the shingles on a building.   (Courtesy NREL and U.S. Department of Energy.)

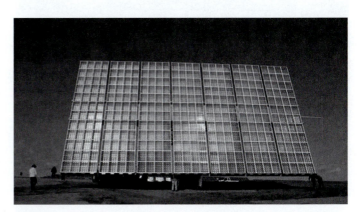

FIGURE 3-28   High concentrated photovoltaic solar cells. (Courtesy NREL and U.S. Department of Energy.)

During manufacturing vapor processes are used to grow the layers into a usable cell. This type of cell is not widely manufactured because of the expense in the manufacturing processes. Its original research and development were for space applications, such as the Mars rover, and it continues to be used in these types of specialized applications. In this application the cells provided high efficiencies up to 40%.

3.9.12 Dye Sensitive Solar Cells

Dye sensitive solar cells (DSSC) are made by applying ruthenium metalorganic dye on a mesoporous layer of nanoparticulate titanium dioxide. These cells are also called Grätzel cells after their inventor, Michael Grätzel. The dyes are a powerful light-harvesting material. These cells can be made into flexible sheets or coatings, and they are usually less expensive to manufacture than silicon solar cells. A low-cost screen-printing process is typically used when they are manufactured.

The major problem with this type of solar cell is that the dyes degrade under UV rays and heat, which are two major components of the sunlight that strike them to make electrical power. This causes their overall efficiency to be approximately 11%, and this efficiency falls off because the dye in the cells degrades over time with exposure to the sunlight. Figure 3-29 shows an example of a dye sensitive solar cell.

3.9.13 CIS (Copper, Indium, and Selenide) Solar PV Cells

CIS (copper, indium, and selenide) solar PV cells are made from copper, indium, and selenide ($CuInSe_2$). This type of solar PV cell has fairly high manufacturing costs, which limits its widespread use in applications. It is basically a thin-film solar cell application with high optical absorption. Its efficiency is approximately 13.5%.

FIGURE 3-29 Dye sensitive solar cells.   (Courtesy NREL and U.S. Department of Energy.)

3.9.14 Gallium Arsenide (GaAs) Multijunction Solar PV Cell

The gallium arsenide (GaAs) multijunction solar PV cell is a triple-junction cell. This cell may be made in combinations of gallium arsenide (GaAs); germanium (Ge); and gallium, indium, and phosphorous ($GaInP_2$). The manufacturing cost is very high, typically up to five times that of other solar cells, so the gallium arsenide multijunction solar PV cell is not widely used in applications. It has very high efficiencies, up to 29% and is typically used for space applications.

3.9.15 Non-Silicon Solar Panels Using Nanoparticle Processing

In the past few years nanotechnologies have provided a number of new manufacturing processes and new products that are now being used in solar panels. These new technologies are still in the experimental stages, but their important feature is that they use products other than silicon. These processes typically use carbon nanotubes, which are also called quantum dots, and they are embedded directly into conductive polymers (plastic materials) or mesoporous metal oxides. The main advantage of this type of solar cell is that the size of the quantum dots can be varied in each cell, which in turn causes that cell to absorb different wavelengths of sunlight. This allows these types of experimental solar panels to achieve efficiencies up to 42%.

3.9.16 Ultraviolet Solar Cells

Typically silicon PV cells use visible and infrared sunlight rays to create electrical power. New technology in Japan has recently found success in the development of a transparent solar cell that uses only the ultraviolet (UV) light to generate electricity. This material allows visible light to pass through it. The advantage this technology provides is that it can be used as a window coating, which would allow visible light to shine through, as opposed to silicon PV cells that block most of the visible light when used as a coating on windows. This new material would allow visible light into the room for passive solar lighting and any heating capability the sunlight would provide in the cooler winter months, at the same time providing electrical power. This technology is still in the research stages, so reliable efficiency data has not been determined.

3.10 MOUNTING SOLAR PANELS

Solar panels can be mounted in a variety of ways to ensure that they harvest the maximum sunlight. This includes mounting them on rooftops as flat panels or sawtooth panels, mounting them on the ground in the same manner, or mounting them on poles. Sawtooth panels are a series of solar panels mounted on an angle and off set from each other so each panel can receive the full amount of sunlight.

Pole mounts are further divided into three subcategories that include top-of-pole mounts, side-of-pole mounts, and pole-tracking mounts. These pole mounts are differentiated by how the panels are positioned on the pole.

Another part of the mounting system is a tracking system. Some solar panels require a tracking system in order to optimize their efficiency. The tracking systems may include hardware mechanisms that allow the panel to track the sun as it moves through the sky from sunrise to sunset so that the maximum amount of solar panel surface is receiving the maximum amount of solar energy. Additional tracking mechanisms may track the sun from north to south. This section of the chapter will discuss all of these mounting applications.

3.10.1 Ground-Mounted Systems for Solar Panels

Solar panels can be mounted on the ground typically on poles. The panels can be mounted directly on top of a pole or on the side of the pole. Figure 3-30 shows a solar panel mounted on top of a pole. The panel is tilted slightly, and it must be positioned so that it picks up the maximum amount of sunlight. The position and direction the panel is mounted may vary slightly for different regions of the world. The panels are mounted facing a particular direction so that they will pick up the sun as the earth rotates and causes the sun to move across the sky from east to west. A variety of pole and mounting hardware are available for various sizes of solar panels.

Another way to mount a solar panel is on the side of the pole, as shown in Figure 3-31. This type of mounting application is especially workable with street lighting systems.

A pole-mounted solar PV panel can be made more efficient by including a tracking system, such as the one

FIGURE 3-31   Sign pole mounted solar panel.   (Courtesy NREL and U.S. Department of Energy.)

FIGURE 3-32   Solar panel mounted on a pole with tracking system.   (Courtesy NREL and U.S. Department of Energy.)

shown in Figure 3-32. The tracking system has sophisticated sensors that determine the direction from which the strongest sunlight is coming and then energizes the tracking motors to move the panel to harvest the strongest light. The tracking systems tend to be complicated and expensive, but in the long run the payback makes the system effective since it raises the overall efficiency of the solar panels.

3.10.2 Roof-Mounted Systems for Solar Panels

Another way to mount solar panels is to position them on hardware that has been mounted to a rooftop of a residence or a commercial or industrial location. One of

FIGURE 3-30   Solar panel mounted on top of pole.   (Courtesy NREL and U.S. Department of Energy.)

the problems with mounting solar panels on rooftops is that additional study must be completed to ensure that the roof is designed to handle the extra load from the weight of the solar panels. Also it is important to be able to determine if the building construction and design are capable of holding the roof mounting hardware. Yet another problem exists with roof-mounted solar panels and that is that care must be taken to limit the amount of damage to the roof from technicians walking on it to install the panels. This includes additional problems caused by penetrating the roof membrane with mounting hardware such as bolts and screws. Each time a bolt or screw is used to mount hardware to the roof and it penetrates the roof material or membrane, the possibility of leaks increases. This is especially problematic in areas of the country where rainfall occurs several times a month. One way to work with this problem is to install the roof mount hardware and then install the roofing material over the hardware so that no openings result in the roofing material or membrane during the installation of the mounting hardware.

3.10.3 Flat Panels Mounting Hardware for Solar Panels

Solar panels can be mounted on the roofs of residential, commercial, and industrial buildings. Flat panels can be mounted on a flat roof with a slight pitch. The panels are mounted on hardware that is mounted into the roof. The hardware is designed to hold the panels in place so that they are directed to the sunlight and to ensure that the panels do not move or come loose when the wind blows.

Figure 3-33 shows flat panels mounted to a roof that has a large pitch. In some regions of the country where snowfall is excessive, the roof of homes and commercial establishments must have a larger of pitch to ensure that

the snow does not accumulate on the roof. If snow is allowed to accumulate on the roof, it becomes a hazard due to the amount of weight that it puts on the roof structure. When the pitch of the roof is increased, such as is shown in the figure, snow is not able to build up on the roof; rather it will fall off due to gravity and/or wind. When mounted on a roof that has a higher pitch, the solar panels must be placed on the side of the house that receives the most sunlight—usually the south-facing side. Since the roof is already pitched, the solar panels will have a similar pitch with their mounting flush to the roof.

3.10.4 Sawtooth or Angular Mounting Hardware for Solar Panels

Solar panels may also be mounted on the roof or on the ground at an angle, which gives the panels a sawtooth appearance when multiple panels are mounted. Figures 3-34 and 3-35 show a number of panels mounted on an angle on a rooftop. The angular mounted solar panels give the appearance of a sawtooth pattern. The sawtooth or angular pattern is designed to ensure that the top of one panel does not create a shadow over the panel behind it. This means the panels must be spaced adequately to ensure that full sunlight is absorbed by each panel. Solar panels can also be mounted on angular brackets that are mounted on the ground. Again if multiple sets of these solar panels are installed together, they must be spaced so that panels do not create shadows in front of the panels that are directly behind.

3.10.5 Tracking Systems for Solar Panels

One of the ways to increase the efficiency of a solar panel is to mount it on a tracking system. The tracking system can have one or multiple axes. The single-axis tracking system typically changes the pitch of the solar panel with respect

FIGURE 3-33 Flat-panel rooftop-mounted solar panels. (Courtesy NREL and U.S. Department of Energy.)

FIGURE 3-34 Sawtooth-mounted roof panels. (Courtesy Ron Swenson, www.SolarSchools.com)

FIGURE 3-35 Sawtooth-mounted solar panels. (Courtesy Ron Swenson, www.SolarSchools.com)

to the sun. A multiple-axes tracking systems also allow a second axis called the *yaw* axis to be moved. The yaw axis moves the solar panel from side to side, which would allow it to track the sun as it moves from the east to the west. Depending on the location or region of the country, a single-axis tracking system may be sufficient to move the solar panel so that it tracks the maximum amount of sunlight. In regions that are in the extreme northern or southern part of the world, multiple-axis tracking may be needed to continually move the panel so that it receives the maximum amount of light. Tracking systems tend to be expensive, but the expense is usually capitalized in the first few years by the increased efficiency of the solar panel. If tracking hardware is used, it may present a problem in causing more periodic maintenance and testing than would typically be used if the panel were mounted in stationary system.

Figure 3-36 shows a typical tracking mechanism that is mounted on a pole. Notice that the solar panel in this application is positioned so that it is nearly vertical. This

FIGURE 3-36 Tracking systems for solar panels. (Courtesy NREL and U.S. Department of Energy.)

position is used when the sun is first rising in the east or later in the day when it is setting in the west. When the sun tracks nearly overhead during the middle of the day, the panel will lay horizontally so that it is nearly flat and can receive the maximum amount of sunlight. When large fields of solar panel are mounted on poles that require tracking, it is important to provide enough space around each solar panel so that it can move to pick up additional light yet not cause a shadow on the panels that are near it.

3.10.6 Parabolic Trough Solar Collector

The parabolic trough solar collector consists of a large reflective trough. A large pipe is mounted just off the surface of the trough so that the light that is reflected from the surface of the trough is focused directly on the pipe. A type of oil circulates through the pipe and picks up heat that is reflected to the pipe. The temperature of the oil will become hot enough to make the fluid in the second loop of the heat exchanger change state and produce steam. This steam is directed to a large utility-type electrical generation system that has a large turbine wheel that turns a generator shaft when the steam from the oil is directed to it. This type of generating system is similar to those used in nuclear power plants and coal-fired power plants that produce utility-grade power. When the steam passes over the turbine blades, the turbine begins to spin at high revolutions per minute (rpms) and this rotates the generator shaft, so that the generator produces three-phase electrical power. The speed of the turbine blade's rotation is controlled so that the turbine blade turns at a constant rpm, which is the speed the generator needs to rotate in order to produce 60-cycle three-phase voltage. (The speed depends on the design. A specific value need not be assigned.) When AC voltage is produced, it is important to maintain its frequency at exactly 60 Hz in the United States; in some countries in Europe and Asia the frequency is maintained at 50 Hz.

Figure 3-37 shows a parabolic trough solar energy conversion system. The troughs are laid down in long rows and are pointed toward the sun. Each trough has a pipe with the oil in it and each is connected from panel to panel so it becomes hotter and hotter as it moves through the panels. The panels are approximately 12–15 feet high. Figure 3-38 shows an aerial view of a solar farm made up of parabolic trough systems. In this image you can see two large generating stations sitting between the large fields of parabolic troughs. Each generating station is the size of a traditional coal-fired or nuclear-fired electrical generation station. The turbine and generator are housed inside the station, and the heat from the oil is transferred in the heat exchanger. The heat creates steam pressure in the second loop of the system that is directed to the turbine to cause it to turn and produce electric power. Figure 3-39 shows a diagram of all the component parts in the parabolic trough generating system.

The oil from the parabolic troughs moves to a heat exchanger in the generating station. The second loop of

FIGURE 3-38   A large field of parabolic trough solar collectors and electrical generating sites.   (Courtesy NREL and U.S. Department of Energy.)

FIGURE 3-37   A parabolic trough solar energy conversion system. This type of system heats oil to the point of boiling and the steam from the oil is used to turn a traditional turbine, which then turns a large generator.   (Courtesy NREL and U.S. Department of Energy.)

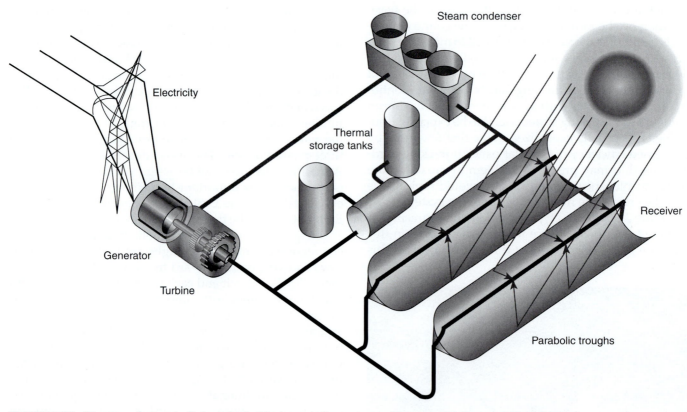

FIGURE 3-39   Diagram of a parabolic trough electrical generating system.   (Courtesy U.S. Department of Energy.)

the heat exchanger has water that absorbs the heat from the oil and turns to steam. The steam is then directed to the turbine wheel, and causes the turbine wheel to turn. The armature of the generator is connected directly to the turbine wheel, so that when steam causes the turbine wheel to rotate, it turns the alternator armature, which produces electricity.. When the turbine wheel converts the steam energy to rotational energy, sufficient heat is removed from the oil in the heat exchanger, so that the oil can be sent back into the parabolic solar array to be reheated. After the steam moves over the turbine wheel, it looses much of its energy, is condensed back to a liquid, and is once again pumped through the heat exchanger to create steam. If a sufficient amount of heat energy is not removed from the steam to get it to condense back to water, a water-cooled condenser must be used to remove the additional heat. As long as the sun is shining, it provides enough solar energy to heat the oil in the tube hot enough to create steam in the heat exchanger. When the sun goes down in the evening, the generating plant can continue running by supplementing heat to the heat exchanger by burning natural gas. This supplemental heat ensures that sufficient steam is always available to turn the turbine wheel. This means that this type of system can generate electricity 24/7 without any interruptions.

Some states do not encourage the utility companies to use hydrocarbon type fuels such as natural gas to provide the auxiliary heating when the sun is not providing heat energy. In these applications the generating station is allowed to shut down when the sun is not providing enough energy. The problem this creates is that the electrical energy supplied by the solar energy is available only during the daylight hours. These states are learning to balance the need for an uninterruptible source of electrical power with the drawbacks of burning hydrocarbon-type fuels. When natural gas is used as the alternative energy for this system, the amount of hydrocarbon pollution that it creates is minimal, when compared to coal.

This type of electrical generating system requires personnel around the clock to ensure that the system runs without maintenance problems. The pumps, motors, and switchgear require maintenance just as that same equipment in a traditional coal-fired or nuclear-powered electrical generation station. The turbine wheel and generator also wear down eventually and need to have their bearings replaced or the turbine itself has to be replaced over time. The major advantage of this type of system is that it can produce large amounts of electrical power that can be added to the grid. Operating systems today range from 50-megawatt (MW) generators to 80-MW generators. Generators in excess of 100 MW are in the design phase and are expected to be up and running and producing power within the next 10 years. These large-scale systems are producing electrical power at costs that are similar to comparably sized coal-fired or nuclear-powered electrical generating stations, which tend to provide the lowest-cost electricity for business, industry, and residential use. The final cost of electricity to the consumer has always been a problem for electricity that is produced by solar or wind energy because of the large upfront cost for capitalization of the equipment and the nature of the low efficiencies of the systems. The parabolic trough type system has a large installation cost, but once installed the solar energy portion of the system is relatively free.

One of the disadvantages of this type of generating system is that it can be installed only in the Southwest United States, where the amount of solar energy is available to provide the higher temperatures more continuously.

These sites are some distance from the urban and industrial areas that need large amounts of electricity; therefore this electricity must be transmitted longer distances across the electrical grid to where it is finally used. Another problem is that in order to get the steam to condense back to liquid completely, large water-cooled condensers may be required. The water-cooled condensers cause large amounts of water to be evaporated into the atmosphere and this water could be used for agricultural purposes instead. The issue of water usage is an important one that must be considered as additional systems are put into production. Another problem with the parabolic trough system is that it must be placed over large areas of desert or land areas that are protected by environmental laws. This means that studies must be undertaken to ensure that the systems do not destroy the protected lands or disturb the environmental balance of all the systems that use the land.

3.11 CONCENTRATED SOLAR ENERGY SYSTEMS AND THE STIRLING ENGINE

Another way to produce electricity from solar energy is to use large concentrator panels that are shaped like a dish and concentrate the reflected sunlight to a thermal receiver. Figure 3-40 shows an example of the dish for this system. This technology is called concentrating solar power (CSP) technology, and it produces a small amount of electricity in the range of 3 to 25 kW. The thermal receiver is filled with hydrogen or helium in a sealed system. When the solar energy from the sun is concentrated on the thermal receiver, it turns the hydrogen or helium into a gas or vapor. When the hydrogen or helium is in the gaseous state, it is directed to a Stirling engine that is similar to a piston engine in an automobile. The difference in the Stirling engine is that the gas that enters the pistons and causes them to move is not allowed to escape; rather it is recaptured and condensed and returned to the thermal

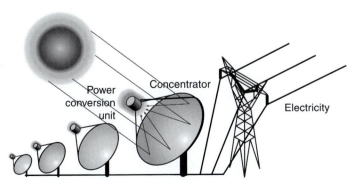

FIGURE 3-40 A solar concentrator dish used to focus solar energy to a receiver that is filled with helium or hydrogen. The heated gas is sent to a Stirling engine that converts the energy in the gas to rotary energy that turns the shaft of a generator. (U.S. Department of Energy)

receiver. The energy that the gas exerts on the pistons in the engine causes them to move in a downward direction, thereby turning the crankshaft. The generator for the system is connected to the crankshaft and will turn as long as the heat energy in the gas causes the pistons to move. Since the hydrogen or helium is maintained inside the system, the engine can continue running as long as solar energy provides the heat to the thermal receiver. At night when the sun's energy is no longer sufficient to heat the thermal receiver, the hydrogen or helium slowly cools and returns to a liquid state and is ready for the next sunrise to heat it up and start the engine again. This type of system is designed to automatically shut down the generator when the engine stops turning and bring it back up when the engine begins to turn again.

This type of solar energy conversion to electrical energy is quite useful for applications in remote areas. This type of system would provide sufficient power to a small village or small ranch that is too far from traditional power lines.

Questions

1. Identify ways that solar energy is used to provide heating and lighting and to produce electrical energy.
2. Explain the differences among a photovoltaic cell, a photovoltaic panel, a photovoltaic module, and a photovoltaic array.
3. Explain the operation of the parabolic trough electrical generating system.
4. Identify five types of solar PV materials and/or processes used to manufacture PV material.
5. Identify advantages and disadvantages of five types of solar PV materials and/or processes used to manufacture PV material.
6. Explain the difference between active and passive solar energy systems.

Multiple Choice

1. The main difference between an active and passive solar system is _____
   a. the passive solar system uses more energy than the active system.
   b. the active solar system uses more energy than the passive system.
   c. there is no difference between active and passive solar systems.
2. When you have a photovoltaic cell, a photovoltaic panel, and a photovoltaic array, which is the smallest unit?
   a. The photovoltaic cell.
   b. The photovoltaic panel.
   c. The photovoltaic array.
3. A parabolic trough solar energy system _____
   a. uses concentrated solar energy to create electrical energy through PV cells.
   b. uses concentrated solar energy to heat oil to make steam that turns an electrical generator.
   c. uses concentrated solar energy in small quantities to heat a home.
4. The most efficient solar PV cell is_____
   a. Bifacial monocrystalline PV cell.
   b. Amorphous silicon thin-film panel.
   c. String ribbon silicon cell.
5. Solar energy that is used to heat water as part of the hot water system or as part of the main heating system for residential applications _____
   a. needs to circulate a glycol solution to prevent the solution from freezing.
   b. needs to circulate a glycol solution because glycol should be added to the drinking water and potable water.
   c. does not need a heat exchanger as it is okay for the glycol and potable water to mix.

4

Solar Installations and Solar Farms

OBJECTIVES FOR THIS CHAPTER

When you have completed this chapter, you will be able to

- Explain the term *turn key*.
- Identify four types of site issues that you may encounter when you are selecting a location for a solar energy system.
- Explain why a small residential solar energy system may need a power electronic frequency converter (inverter).
- List three types of commercial applications of solar energy.
- Identify the five largest solar energy farms in the United States as of 2010.
- Identify the total amount of electrical energy produced by solar energy worldwide as of 2008.
- Identify two solar energy farms that are located outside the United States.

TERMS FOR THIS CHAPTER

Capacity Factor
Gigawatt (GW)
Kilowatt Peak (kWp) Power
Megawatt (MW)
Net Meter
Peak Sun Hours

Power Electronic Frequency Converter
Project Development Plan
Residential Solar Energy System
Solar Energy Farm
Solar Site Assessment
Turn Key Project

4.0 OVERVIEW OF THIS CHAPTER

This chapter will help you get a better understanding of solar energy installations, such as large solar energy farms and smaller installations of two or three solar energy systems, including photovoltaic (PV) panels or solar heating for a residential or commercial establishment. This chapter contains information to help determine where individual solar energy systems and larger solar energy farms should be located. You will also learn about the issues regarding the placement of solar equipment that include project development, solar site assessment, the placement of individual solar panels, as well as the problems you may encounter when trying to determine the best location for larger solar energy systems.

Information in this chapter will also include ways to determine the best location to place a small *residential solar energy system,* homemade solar energy systems, commercial solar energy systems, and finally solar energy system farms to make them most efficient. If the location where solar energy equipment is to be installed is not studied correctly, the equipment may not work to its optimum efficiency. This information will cover solar heating and solar photovoltaic (PV) systems that produce DC electricity and help you select the best location to place solar energy equipment. You will also learn about where the largest solar energy farms are located in the United States and worldwide, and where the greatest amount of growth in new installations of solar energy farms will be. You will also learn about how some countries that are located outside the United States have learned to develop and place large solar energy farms in locations where they will be most efficient.

If the solar energy system is installed on the rooftop of a residential, commercial, or industrial building, the construction codes, building codes, and zoning codes must be understood completely and followed so that the system can pass inspection. Additional information may be required to ensure that the building has a proper design to support the excess weight of the solar system. For example if a solar heating system that includes a water collection system will be located on a rooftop, the weight of the water must be accounted for in addition to the weight of the equipment when determining if the roof can support this type of system. You will learn about all these factors when you are determining where solar PV systems or solar heating systems should be located. You will also learn about project development and site assessments.

4.1 PROJECT DEVELOPMENT

When you are analyzing information about the best locations to place a solar energy system, you will need to create a *project development plan.* This plan will include selecting the proper site for the solar energy system or solar energy system farm; identifying available solar energy resources; predicting energy output for different-sized solar energy systems; identifying landowners and developing landowner agreements; identifying grid interconnection if necessary and utility companies that control the grid; identifying government agencies that control the country, state, county or city regulations that will affect your selection; and finally in developing an environmental plan to ensure that the solar energy system conforms to all requirements. If the solar equipment is being installed on a rooftop or in another integral part of the building, the building and zoning codes will have to be identified and followed for this type of installation.

A number of companies will complete the solar project development process for you on a fee basis. The fee may be included in the cost of the solar energy system installation if you select the company's equipment. Other companies do not sell solar energy system equipment directly; rather they will recommend one or more brand names and construction companies will help with the installation and startup process. Some of these projects are under the category of *turn key projects.* All aspects of a turn key project are taken care of from initial design and equipment installation all the way through startup, and then the project is turned over to the owner. A turn key project may be more expensive, but all of the details identified in a project development plan will be taken care of for the owner of the project.

4.2 SOLAR ENERGY SITE ASSESSMENT

A perspective customer interested in installing a solar energy system will want to know the return on investment and how much energy can be expected from the system. A solar energy site assessment is needed to predict how much energy can be harvested at the locations or sites.

The *solar site assessment* includes gathering data about the amount of solar energy that is available at a particular site, which direction solar energy panels need to be facing, and if a tracking mechanism is needed. You will also need to gather information regarding the number of hours per day that solar energy is available and these data should be documented by the month, day, and hour to account for changes in the position of the sun in relationship to how strong the solar energy is at that location. A comparison of the differences of the amount of solar energy that is available at several sites will help you determine the best location for the solar equipment. It is important to understand that the amount of sunlight will vary from site to site and that the amount of sunlight that can be converted into electrical energy will vary from panel to panel; therefore similar panels need to be used when making comparisons of different locations.

In order to make a valid comparison of different sites for a solar energy installation, a standard needs to be used that includes the amount of solar energy that is available and can be produced. One method of measurement for this comparison is *kilowatt peak (kWp) power.* Kilowatt peak refers to peak power. This value specifies the output power achieved by a solar module under full solar radiation (under a set of standard test conditions). Solar radiation of 1,000 watts per square meter (w/m^2) is the measure used to define standard test conditions. The higher or larger the amount of peak power that a solar module has, the more efficient it is, so this value can be used to compare solar energy modules that are manufactured by different companies and solar panels that use different manufacturing technologies and different materials. Another important part of the comparison is the *peak sun hours,* which will differ from area to area. The peak sun hours are the equivalent number of hours per day when solar irradiance averages 1,000 w/m^2. This means that if the sun shines off and on all day long and the intensity changes from time to time during the day, and 6 peak sun hours is the amount of energy that is received at the panel during total daylight hours, that equals the energy that would have been received if the irradiance had been steady at 1,000 w/m^2 for 6 straight hours. Since the amount of sunshine is not constant, the number of peak sun hours is a better way to make comparisons among sites where solar panels might be installed.

An efficient panel may be able to produce the same amount of power in an area that does not have as much peak sun hours, as does a less inefficient panel that has more peak sun hours. Another issue that comes into play in the site selection is the state and local enticements that are offered, such as rebates, tax credits, or tax incentives, so it is possible to get more dollars of return from a panel that has a state and local enticement, even though it is in a site location that does not have a large number of peak sun hours. Typically the site with the highest number of peak sun hours will be the best site to select, but you do have to be aware of other issues that may come into play.

Peak power for a solar PV panel is also referred to as "nominal power" by most manufacturers. It is one of several values that can be used to compare solar energy units. Since it is based on measurements under optimum conditions, the peak power is not the same as the power produced under actual radiation conditions. It is also important to understand that in practice, the peak power will be approximately 15% to 20% lower due to the considerable heating of the solar cells and other issues that affect the conversion of sunlight to electricity.

Companies will use the data that have been recorded by data loggers in conjunction with data from instrumentation that measures the amount of solar energy. This instrumentation also measures the variations in the solar energy for a given site based on the earth's position and the season during which the energy is being measured, as well as other important information such as the length of days and times that daylight is available. The data loggers will compile the information from a number of these instruments spread across a site and pointed in the directions to obtain the highest solar energy readings. Solar energy site assessments will usually include GPS (global positioning system) information that will provide precise locations where the instruments have been located and where the solar energy data information has been gathered. Other useful information can be obtained from aerial photographs or satellite images that will show all the necessary information about the location where solar panels will be installed and other information about the surrounding environment such as the location of woods and rivers and the possibility of flood plains or areas that may flood if rainfall is excessive.

Other important issues that must be taken into consideration involve the geology of the land where larger solar energy systems will be located. Problems may exist with soft soil or wetlands that do not drain well, which will restrict the size and weight of the solar energy system and the equipment that is used to install it. Other issues may limit the placing of large solar energy systems in close proximity to cities or residential areas. Some cities in other areas, such as resort areas, may enact legislation to prohibit the location of solar energy systems in their areas that may be considered scenic areas.

If the solar energy system is being connected directly to a residential or commercial building, local and state building codes and zoning codes will have to be followed. This means that you will need to know the weight and structure of the solar energy equipment that you consider mounting on the roof or on other structural parts of the building. An architect or building engineer may be needed to identify problems inherent in installing equipment on a rooftop that was not specifically designed for this purpose. If a building is being designed with the solar energy system built into it, then the designers and engineers will need to be familiar with the problems of mounting solar energy equipment on a rooftop and providing access for air ducts or water lines to and from the roof. If the system is a PV electrical system, the design will have to include locations for installing electrical components such as the combiner box and inverter on the rooftop or somewhere within the building. If the system is designed to have a connection to the grid, all state and local codes, electrical codes, and requirements of local power utility companies will have to be integrated into the system design.

In some states, government agencies will help with the site assessment process, in order to entice more solar energy systems to be placed in their area. Some states have also provided financial incentives for solar energy systems placed in their states. In some cases you will find that other solar energy systems have already been placed in nearby locations, which may make it simpler to complete the solar energy site assessment because information such as grid connection, utility company contacts, state forms, and other common information may already be available and completed from previous projects. In these instances the information from the original installations will be specific to the site and will include the amount of solar energy availability at the original site and the amount of solar energy availability at the new location.

4.3 SITE ISSUES

You will become aware of a number of site issues as you begin to select a location for solar energy systems. The site issues can be grouped into the following categories: environmental issues, visual issues, zoning and code issues, and electrical power and connection issues.

Environmental issues may include how well the solar energy system fits in with the local environment. For example if the solar energy system is located near a wildlife area such as a desert, there will be concern about any species of plants and animals that have been displaced or had their habitat disturbed, which causes a change in their numbers. Other environmental issues may include impact on the environment where roads must be created to move the solar energy equipment to the solar energy system site. There are also concerns for any environmental damage due to installing power lines and power transmission towers near critical environmental locations.

Another issue that may affect the placement of solar energy systems is their visual impact. Since solar energy panels and equipment for solar energy farms may be the only structures when they are located in remote areas, people may have concerns that the equipment and structures will have a visual impact on surrounding areas, such as large prairies, deserts, or residential developments. An agricultural area may already have a variety of equipment and existing structures, so these issues may not be as much of a concern as they would be in residential or other locations where they do not match the local environment.

Another problem for solar systems is zoning and code restrictions. All kinds of solar systems are included in many zoning laws that may prohibit the location of solar energy systems near residential locations. These restrictions may include how close a solar energy system can be to the residential area, or they may restrict the size and location of the system that would be permitted near residential areas.

Other concerns when solar energy system sites are being planned include maintaining the solar energy system site and removing older equipment as it wears out. Many communities are concerned that companies will abandon older equipment and leave it in place where it may become an eyesore. Some regulations require companies to establish escrow accounts for the purpose of removing older equipment in the event these companies go bankrupt or abandon the site area.

The final site issue involves the location where the solar panels are connected to the power grid. Some electrical companies are concerned about solar systems that are connected to the grid. It is important that these solar systems have the correct disconnect switch and controls that ensure that the voltage that is put back into the grid from the solar system matches the voltage and frequency of the grid voltage.

4.4 VISUAL AND LANDSCAPE ASSESSMENT

Since solar energy systems can be placed on building rooftops or on the ground, they will generally be visible for quite a distance. If the location of the building or the site where the solar equipment is being placed is within sight of other residences or commercial establishments, landscape codes and zoning codes may need to be evaluated so that the project can be inspected and approved. In some cases residential and commercial buildings may be part of a larger historical conservation area where they

have been deemed a historical asset, and this may limit the types of equipment that can be exposed on the rooftop or in the yard near this area. If the solar energy system is to be installed in rural agricultural or farmland areas, these issues may not occur. Many state and local codes have begun to address these issues and weigh the value of installing renewable energy sources and comparing that to the need to maintain certain code restrictions. Newer solar energy systems tend to be larger than previous ones because more solar panels are used, which tends to compound visual issues. Some cities and areas of the country consider the appearance of solar energy equipment on residential and commercial buildings a "badge of honor" and have amended their codes to ensure that solar energy systems can be easily incorporated into new and existing buildings.

Other issues that impact visual and landscape concerns include the construction of large electrical power transmission towers to get the electricity that the solar energy system farms produce and send to the grid. There are basically two ways to work around this problem. One way is to select a site for the solar energy system farm that is close to existing power lines, so that a minimum number of new power lines will need to be installed to get power into the grid. Another way is to install as much of the power transmission cables under the ground as possible. Under the ground transmission lines are electrical power lines that are buried 3 to 4 feet under the ground and provide electrical power connections at transformers or switch gear that are above the ground. On new solar energy farms these underground transmission lines may be capable of transmitting power several miles to a location where the construction of larger power transmission towers does not create an issue.

Other ways to head off problems with visual concerns is to hold meetings with residents and businesses in the area where the solar energy system installation is planned. These meetings are designed to get input from area residents and businesses and have a thorough discussion of pros and cons for the site. At these meetings residents can express concerns, and many times these concerns can be integrated into the plan and will head off problems in the future. Also government personnel can help explain the code requirements that must be met.

4.5 SMALL RESIDENTIAL SOLAR ENERGY SYSTEMS

Small residential solar energy system systems that use photovoltaic panels to produce DC electricity are available for purchase and installation in a variety of sizes. In order to understand what size is needed, you need to start with how much energy the residents need on a day-by-day basis over a month. The average monthly consumption in kilowatt-hours (kWh) ranges from a low of 521 kWh in households in Maine to the highest consumption of 1,271 kWh per month in households in Alabama. If you divide these values by 30 to get the average daily consumption, they will need from 17.6 kWh to 42.4 kWh per day. Once you have established how much electricity is used, the next step is to determine what size PV panel will be needed to create that amount of electricity. Basically 60 square feet of high-efficiency monocrystalline cells will produce 1 kW per hour. PV cells that are less efficient will need up to 130 square feet, and some of the least expensive types of PV cells will need more than 300 square feet of cell surface to produce 1 kW. Another factor in deciding how large the system is depends on whether you are going to store electrical energy during the daylight hours for use when the cells will not be producing electricity. In this case you can figure that you will need approximately 30% to 40% more electricity to be stored during the day to have the excess to use at night. The reason you do not need to double the amount for nighttime use is that most of the night the electrical energy usage will be less because people are sleeping and most loads are turned off. If larger loads will be running during the night like air conditioning systems, you will need to add more storage.

Since it may not be possible to have enough solar PV panels and battery storage to produce the amount of electrical energy the residence uses, an alternative plan would be to have the solar panels produce as much as possible and use the electrical grid to supply the remaining amount of electrical power needed. This type of system is a grid-tied system, and it allows the residents to use electricity from the grid when the PV system is not providing voltage. Another advantage of this type of system is that it does not need battery storage capacity, which tends to be a little more expensive. If the system does not use battery storage, then it must have an alternative source of electricity when the solar panels are not producing.

Another way a solar PV system can be designed is to use more P panels than are needed to provide electricity for the residence and connect the system to the grid. This type of grid-tied system allows the residential customer to sell back to the grid any excess electricity that is generated by the solar panels during the day. In order to do this the size of the solar array must be large enough to provide the voltage used by the household plus produce enough extra electrical power to sell to the grid. The idea is to sell enough excess electricity back to the grid to cover the cost of the electrical power that is used during the night or when the PV cells are not producing. In most cases the price of the electricity that is sold back to the grid may be substantially less than the rate the customer pays to the utility for the electricity it uses from the utility. It is also important to take into consideration that the amount of electrical power that is used by the residents is not constant throughout the day. For example washing machines, clothes dryers, and other equipment that use large amounts of electrical power may run during one hour and not the next, whereas other large energy users such as air conditioners may run constantly during certain months or on certain days of the month depending on the outside temperature. This type of electrical power usage that does not constitute a constant load makes it more difficult to design a system that has the amount of power required for peak usage times, yet has enough excess capacity to sell electrical power back to the grid to cover the cost of the power that will be used when the PV cells are not producing. In some of these cases the solar energy system will be considered an auxiliary power source rather than the main power source, and the customer will be satisfied with offsetting some part of the amount of the traditional electrical bills.

Another problem that may arise when you are considering installing a residential solar energy system is determining if the residential building has sufficient rooftop space that is pointed in the right direction to support the number of panels needed to produce the required energy. Some two-story homes have less than half the roof space of a single-story home, and the slope of the roof may be pointing in a direction that will not produce the maximum amount of solar energy from the PV cells. These conditions may limit the number of solar panels that can be mounted on the roof to provide the necessary electrical. In some applications solar panels may be mounted on the ground instead of the rooftop to provide sufficient numbers of panels to generate the power needed by the residents. All these factors must be taken into consideration when determining whether a solar PV system or solar heating system can be used on the rooftop of a residential structure. If the home is new construction, some of these issues can be accounted for during the design and building processes. If the solar energy system is to be added to an existing residence, some of the factors that limit the amount of practical power that can be generated may not be able to be overcome.

The total amount of electrical power from the solar energy system that will be available will depend on the amount of solar energy available at that site and the number of hours the solar energy can be converted by the PV cells each day. The DC current produced by the PV can be stored in batteries or used directly in the home by sending the DC power through an inverter. Since the loads in most homes are powered by AC voltage, the inverter, which is also known as the power electronic frequency converter, will be needed to take the DC electricity produced by a solar energy system and convert it to AC electricity with a frequency of exactly 60 Hz. The inverter can convert the DC power directly from the solar panels or it can take the DC power from the batteries. This also allows the electrical power to be integrated directly with the power coming from grid. If excess electrical power is created, it can be returned to the grid. If the system is being designed to send power to the grid, permission is needed from the power company, which will work with you to ensure that you have the correct switch gear and meter for this application and that the installation follows its guidelines and codes. The switch gear includes an electrical control circuit that ensures that the voltage matches the grid voltage exactly before the residential solar energy system is connected to the grid.

Another component that is needed with this type of system is called a *net meter*, which is an electrical power meter that can measure the amount of electrical power that flows through it in either direction. This means that the meter can measure electrical power coming from the grid to the user, and it can also measure electrical power that is generated by the PV cells and then sent back into the grid. Some power companies will provide the net meter, and others will require the residential or commercial site to purchase it. Figure 4-1 shows an electric meter, an inverter, and a net meter for a residential system.

Figure 4-2 shows an example of a residential solar photovoltaic energy system that is located on the roof of a home in a suburban area. Some suburban areas may have code restrictions and other zoning restrictions that may restrict the size and location of the solar energy systems used on homes in their area. If the solar PV system is a stand-alone system and is not connected to the grid it is often referred to as an "off grid system." These types of systems may also be used in cabins or other remote areas where traditional grid power electricity is not available. In these systems the solar heating or solar hot water heating system may also be used in addition to solar PV panels, and the total energy from the PV panels is stored in large banks of batteries. When the residents need to use power, the circuits in the home are connected through the inverter, and the power put into the batteries from the solar energy system is used to provide the limited amount of power for which the inverter is rated. If the solar PV system is large enough to provide all the power needs for the home at peak power usage times, the inverter must be large enough to convert all this power. If the inverter is too small, it will limit the amount of

FIGURE 4-2 Solar energy system on a home in a residential setting. Zoning laws or codes may limit the types of solar energy systems that can be installed. (Courtesy NREL and U.S. Department of Energy.)

electrical power that can be converted from the PV cells or the battery. In these applications the battery acts as a buffer and allows voltage created from the PV cells to flow directly through it, and excess power will be stored if electrical loads in the home are not turned on. In other words the batteries and the inverter must be large enough to provide all the power needed at peak times, or they will be a limiting factor in which loads in the home can be turned on and energized at the same time. This is one of the inherent problems in using solar energy, as most individuals are used to turning the switch on for whatever electrical load they expect to use, and they may not have ever experienced having to shed certain loads in order to have enough electrical power to run other loads.

Figure 4-3 shows a small solar PV energy system that has the PV panels mounted on the ground in the backyard. The reason the panels are mounted on the

FIGURE 4-1 Solar meter on top left, electric meter below, and inverter for the solar electrical system on right. (Courtesy Jim Richmond, Granite Viewpoint Blog.)

FIGURE 4-3 A small solar PV energy system on the ground in the backyard of a home. This home is heavily shaded by trees and does not have sufficient roof space for enough PV panels. (Courtesy NREL and U.S. Department of Energy.)

ground is that the house is in a heavily shaded lot or there is not sufficient roof space to mount the PV panels so that they are pointed in the correct direction. Since this home is in a mostly rural area, the zoning and other regulations are more lax, so the panels can be placed anywhere on the property to ensure they get enough sunlight.

4.6 HOMEMADE SOLAR ENERGY SYSTEM SYSTEMS

Another type of small solar PV system is a homemade or home owner installed system. Figure 4-4 shows an example of this type of small system. These systems usually have a small kW output that has sufficient power to charge batteries and store the energy needed by the home. These types of smaller systems are mainly designed to offset part of the total electrical usage. In some cases such as a rural location or at a vacation cabin, the unit is sized so that it provides sufficient electricity for essentials such as lights, small appliances, and a small refrigeration unit. Today all the parts for this type of system are available for purchase through the Internet or at energy stores that sell the small homemade systems. It is also possible to purchase the inverter, battery charger, and batteries locally and order the solar PV panels to complete the system. This allows the system to be built completely from scratch. For some people it is important to size, locate, and assemble the entire system so they are thoroughly familiar with its operation in case it breaks down or needs periodic maintenance. In order to make the electrical connections from the PV panels to the charge controller and then to the batteries and on to the inverter, a person must know a little about electrical wiring and be able to follow the directions that are provided. If the system is wired into the existing electrical system that is tied to the grid, an electrician will

be needed to ensure that the transfer switch is connected correctly, tested, and meets the standards and inspection of the electric utility. In many cases the homemade or home-built system provides separate circuits that are not tied to the existing electrical system, so a novice can make the connections among the PV panels, the batteries, and the inverter and provide several outlets for use in the home. Another way to do this is to use the PV panels, and a battery with a battery charger and provide the electrical power in the home as DC power; this will necessitate buying electrical appliances that use DC power. Since a large variety of DC-powered electrical appliances are available for RVs and campers and truck drivers who spend a significant number of hours in a DC-voltage environment, it is easy to locate these appliances and use them on the DC circuits in a residence.

When batteries are used to store the generated DC electrical power, two additional issues must be taken into consideration. First an electronic control must be added to the batteries' circuit to ensure that they do not become overcharged. If the battery charge level is not controlled, the solar energy PV system can possibly continue to add electrical power to the batteries after they are fully charged, and this overcharged condition causes the cells to dry out and become damaged. The second issue that must be controlled occurs when the solar energy system has not run for some period of time due to darkness, storms, or cloudy days, and the batteries are drained down due to electrical power usage in the residence. If the electrical system needs any amount of voltage and the battery is nearly depleted, the power will not be available since the voltage level of the batteries is too low to provide power to energize any of the items that need the power. For this reason it is important that an electronic circuit control is provided to ensure that the batteries never drain down to a voltage level at which the power system becomes unusable. This electronic cutout switch disconnects the batteries before they are drained below usable levels and then reconnects them when the solar energy system is producing electrical power again. In some cases the overcharge control and the low-voltage control are built into the electrical frequency converter (inverter).

If the output from the batteries is converted to 110 volts, a *power electronic frequency converter*, also called an inverter or electronic inverter, will be needed to produce the single-phase AC voltage with a controlled 60 Hz frequency and a regulated voltage level. The batteries for these types of applications will require significant inspection and maintenance so that they do not run dry or become overcharged or undercharged as electrical power is constantly charged and discharged on a daily basis. Another issue with batteries is that they will eventually fail completely and need to be replaced. Battery life runs from 4 to 10 years, so this is a cost that will be repeated from time to time. Newer batteries made with materials

FIGURE 4-4 Smaller home-installed solar energy system. (Courtesy NREL and U.S. Department of Energy.)

that will extend their life are constantly being brought to market, and the research on these products is continually bringing about new discoveries in battery technology.

Smaller, homemade photovoltaic systems are constantly evolving so that people who want to install a solar energy system to provide electrical energy can do so on their own or with the help of a few people. These small systems will help offset the electric bill but typically will not replace the electric power company completely. The smaller systems will usually require more inspection and minor maintenance such as checking electrical connections, cleaning the outer glass on the PV panel, and checking batteries than some of the larger systems designed for commercial use. This means that the person who installs this type of system needs to be able to perform the inspections and minor maintenance or have someone who will be willing to do the maintenance to keep the system operating. If you have to pay for the maintenance, the cost will typically consume the benefits gained by offsetting the electric bills. When you are deciding on the feasibility and cost of installing and operating a small home solar system, you must include all of these factors in your costs analysis.

4.6.1 Residential Solar Heating and Solar Hot Water Heating Systems

Another way to use solar energy for a residence is to use the solar energy to heat the building and to heat water in place of the hot water heating system. When solar energy is used to heat water, the hot water can be used to heat the home or it can be used to heat the potable water that is used in the residence for bathing, washing clothes or dishes. The solar water heating system includes a collector system on the roof and a storage system for the water and pumps and piping to move the water that is warmed in the collector to the storage tank. The warmed water can be circulated throughout the house where radiators radiate heat into the rooms. The heated water in the storage tank can be used for showers, washing clothes, or other typical uses in addition to being circulated through out the rooms for heating.

In some applications the heated water is circulated through a coil that is located in a furnace, and the air from the furnace is circulated over the water coils; heat is transferred to the air, which circulates throughout the house much like any other gas or electric warm-air furnace.

A second type of solar heating system uses a sealed liquid medium to circulate through the solar collector on the roof, and this medium is used to transfer heat to potable water that is used in the house. The liquid medium may have antifreeze in it so that it does not freeze when the outdoor temperature drops below the freezing level. The antifreeze inside the solar collector can be circulated even when the outside temperature is cold, because a small amount of solar energy can still be

FIGURE 4-5 A solar collector mounted on a residential roof. The water that circulates through the solar collector heats water that circulates to a storage tank on a lower level. (Courtesy NREL and the U.S. Department of Energy.)

harvested from the sun. The antifreeze is needed when the circulation pumps are turned off and the liquid solution is not circulated. If the solution sets in the collectors, there is a chance it would freeze if it did not have antifreeze in it. Figure 4-5 shows a residential solar collector that is mounted on a rooftop and has water circulated through it to collect the solar energy. Figure 4-6 shows a diagram of how the water is used for hot water needs and for heating the home.

4.6.2 Residential Wind Energy Systems Combined with Solar Energy Systems

Solar energy systems have a drawback in that they do not produce electrical power during the night or on cloudy days. One way a home owner can offset this problem is to add a small residential wind turbine that can produce electrical energy during the night or on cloudy days if the wind is blowing strongly enough. In this type of system a small residential wind turbine is mounted on a roof with solar heating panels. The blades of the wind turbines are allowed to turn at any speed the wind will move them, so its output power will not be controlled for voltage level or frequency. In many residential wind turbines the electrical power output will be DC voltage instead of AC voltage, so it can be stored in batteries and then used in the residence with DC-powered appliances, or an inverter can be used to convert the DC power stored in the batteries into AC electricity at the correct voltage and at exactly 60 Hz. The same battery controls and interconnection switches must be used if the AC voltage is put into the same electrical system that is supplied by the electrical grid. If the wind is blowing at sufficient strength during the day, the electrical power that is produced by the wind turbine can be added to the battery storage, and it can be used from the battery at night or

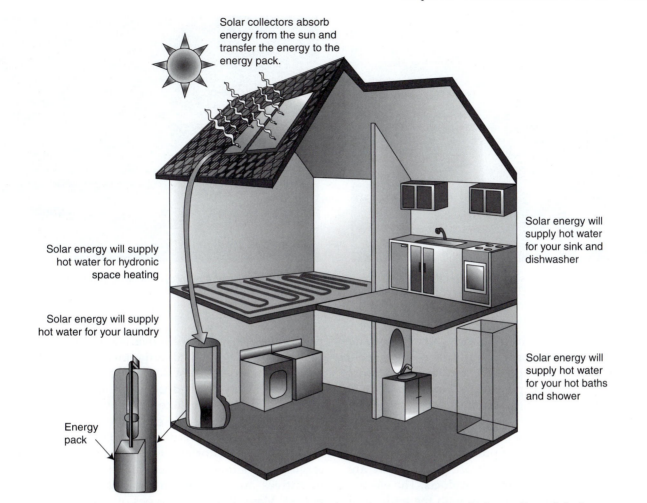

Solar collectors absorb energy from the sun and transfer the energy to the energy pack.

Solar energy will supply hot water for your sink and dishwasher

Solar energy will supply hot water for hydronic space heating

Solar energy will supply hot water for your laundry

Solar energy will supply hot water for your hot baths and shower

Energy pack

FIGURE 4-6   A residential solar water heating system. The heated water is stored on the lowest floor of the home. From there the water is used for traditional hot water uses such as bathing and washing clothes. Additional hot water is circulated through the floors or through radiators in the cold months to heat the house. The hot water can also be circulated through heating coils located in a furnace.

when wind production and solar power production become too low to meet the power requirements of the house.

Some wind turbines use a generator that needs voltage to flash its fields to create the initial magnetic field so it can begin to produce electricity when the turbine blades rotate the generator. Flashing the field is accomplished by applying an external electrical voltage to the generator field coils. If the residential wind turbine uses this type of generator, the generator must be connected to the battery system to receive the voltage to create this magnetic field. The voltage level for the battery system must be protected so it does not discharge to a voltage level that is too low to provide the voltage to flash the electrical fields in the generator to start the wind turbine producing electrical power. The control circuit will isolate the batteries when the voltage comes near the point that is too low to be used to provide the magnetic field in the generator. If the household needs additional voltage after this point is reached, it will not be able to use the batteries and drain them any further down.

4.6.3 Commercial Solar Energy Systems

Small- and medium-size solar energy systems from 4 to 40 kW are available to be installed right on the property of a commercial establishment. The small business or has a solar energy system and either uses the electrical energy it produces directly to offset the energy it purchases from the electric utility or it can sell the electricity back to the grid. Small- to midsize commercial establishments are finding that they generally have room on their rooftops or on their property to install a solar energy system, and they use the electrical energy it produces to offset some of power they previously purchased from the electric utility. In most cases these midsize solar energy systems are grid tied and have a net meter so they can sell any excess electrical power back to the utility. Typically these commercial applications will not use the same amount of electrical power at every hour during the day, so there will be times when the solar energy system is providing maximum energy and excess energy can be sold back to the electric utility. Figure 4-7 shows a solar

FIGURE 4-7 A solar energy system located on the front awing of a restaurant. (Courtesy NREL and U.S. Department of Energy.)

FIGURE 4-8 A solar energy system installed on the rooftop of a public entertainment building.

energy system located on the front awing of a restaurant. The solar energy system for this application works well, because the facility requires power approximately 12 hours during the day. The solar panel array on the awning is large enough that on the very sunniest days, the system may be capable of producing excess power that can be sold to the grid. Also during the winter months, the electricity the solar energy system produces can be used for providing power to electrical heating coils that are mounted directly into the furnace air ducts. The electric heating coils are usually used along with the traditional heating system, such as natural gas or a heat pump. This type of system works better in some locations where it is not as cold since the amount of heat needed in the home would be less than if the building were located where the outside temperature was below 20° and the heat load of the building would be very large. When the solar electrical system is generating electrical power, it provides electricity to the electric heating coils in the furnace and the traditional natural gas or other heating fuel is used only if the electric heating coils cannot produce enough heat energy to keep the building at the set point temperature. It is important to remember that if the solar energy system is connected to the grid, it will require safety disconnect switches and transformers to ensure that the system connection meets the codes for the grid.

Figure 4-8 shows a solar energy system installed on the rooftop of a building that is used primarily as entertainment space. The front of the building has a large number of mirrors that do not serve any purpose other than to make the façade of the building match the solar panels on the roof more closely. The solar energy system in this application can produce the electrical power the building needs for lighting and ventilating. Any excess electrical power the solar energy system produces is used in other buildings on the property or is sent back to the grid and sold to the electric utility. Another advantage of

using solar energy or other green energy applications on a commercial building is that the installation of the green energy or solar energy system will provide green energy credits that may be used to help obtain government or other contracts. Many states have systems of credits set up that encourage the use of green energy and solar energy, and these credits can be used to help the company move up the list for contracts to sell their goods and services. Sometimes the use of these systems will lead to incentives and grants that will shorten the payback period. Many times these types of systems are installed by companies that have an environmental or green energy philosophy and policies that require them to utilize all the possible technologies to limit pollution and help the environment.

Figure 4-9 shows a solar energy PV system on a school building and Figure 4-10 shows a PV system on the rooftop of a government building. The solar energy system in these applications is large enough to provide electrical

FIGURE 4-9 Solar energy PV panels mounted on the rooftop of a school. (Courtesy of NREL and U.S. Department of Energy.)

FIGURE 4-10 Solar energy PV system mounted on the rooftop of the government building.   (Courtesy NREL and U.S. Department of Energy.)

FIGURE 4-11 Solar PV array is used to produce electricity to pump water for livestock.   (Courtesy NREL and U.S. Department of Energy.)

power required to operate all the lighting that is required for the buildings, since they primarily use the lighting during the daylight hours when the sun is shining. Since these types of applications consume major amounts of electrical energy only five days a week, they can sell excess power back to the grid on the weekends when their usage is minimal. Since the rooftops of some of these buildings are extremely large, the solar system can be designed to be pointed in the correct direction and to be large enough to provide power for the building and have excess remaining to sell back to the grid on daily basis. Schools that are in session only from September to May are able to sell the majority of the power produced during the summer months when energy usage is smaller, back to the grid and bank the money against future bills during the year.

4.6.4 Other Applications Using Solar PV Panels

Solar PV panels can be used in rural or agricultural settings to provide sufficient amounts of electricity to pump water or to provide other services to feed cattle or to monitor herds. The solar PV array is located in an agricultural setting, and the electricity from the solar panel is used to pump water for cattle and livestock. Figure 4-11 shows solar panels used in an application to provide electrical power to pump water in a remote range area for cattle. Notice there are windmills in the background, but sometimes the wind does not blow continuously enough to provide the water. Also the windmill mechanically turns a shallow pump that pulls water from no deeper than 30 feet, which means the well might dry up in years with minimal rain fall. If an electrical deep well pump is used at the bottom of the well, the well can be more than 200 feet deep and will seldom run dry. The solar PV cells can charge batteries and the

power from the batteries can be used for a water pump that has a DC motor, or the electrical power from the batteries can be run through an inverter and changed to 60 Hz AC voltage so the electric motor for the well pump can be an AC motor. The electrical cables for the pump motor run from the batteries or inverter down the well casing to the electric motor for the pump at the bottom of the well.

Another type of application for solar PV cells is to mount the PV array directly on a light pole that provides illumination for a parking lot at night. Figure 4-12 shows an example of a solar array mounted on a light pole. A battery and inverter are provided in the system, and sunlight causes the PV cells to charge the batteries; the voltage from the batteries runs through the inverter to provide 60 Hz AC electricity to power the lighting. This type of system is also tied to the grid, in case of long

FIGURE 4-12 Solar PV array mounted on a light pole. A battery and inverter are used to provide 60 Hz voltage for lighting. The system is also grid tied so light is always provided even if the batteries are not fully charged.

periods of cloudy weather or other conditions that may limit the amount of power the PV cells provide. Since the lighting is critical for safety, it needs to be backed up by voltage from the grid. This type of solar PV installation will provide sufficient power to allow cities to lower their lighting bills. There is an initial cost for the extra equipment during the installation portion of the project, but the offset of purchasing grid electricity from the power company will help pay for the project many times over. These types of PV cells are mounted in one direction to harvest the maximum amount of sunlight throughout the day.

4.7 SOLAR ENERGY FARMS

A *solar energy farm* is a group of solar energy systems that can range from two to more than a thousand solar energy modules placed close together so they can share the electrical grid connection, transformers, and maintenance equipment. In some cases several hundred solar energy systems will be placed in the same location. There are many advantages to grouping solar energy panels in close proximity. One is a single interconnection to the grid, through either electrical switching or a substation. This also allows the multiple solar energy systems to share larger transformers that are used to increase the voltage to the amount needed at the grid. Another advantage of placing multiple larger solar energy systems on the same location is that they can all be serviced and inspected at nearly the same time.

Once a solar energy farms is established, it becomes less expensive to install additional solar energy systems, because much of the work for this site application has been completed on previous installations. Also if any zoning restrictions need to be negotiated, this can be done for all the solar energy systems on the site, which makes it easier than negotiating a dozen sites that may run across different state, county, and other jurisdictions. Typically complaints about visual appearance are minimized because the solar energy farm will produce a tax base for the area and also provide additional electrical power, which most industrial regions and states are looking for to entice additional industries into their areas. When there are large areas of unused land that are perfect sites for solar energy systems, solar energy farms serve the purpose of producing energy and using land that would not be used for any other purpose. It is important that the location that is used for equipment on a larger

farm has been analyzed to ensure that it will produce sufficient solar energy throughout the year.

The newest trend in solar energy farm system installations includes larger installations that can provide half the amount of electrical power that a small coal-fired plant would provide. Table 4-1 shows the amount of electricity produced from solar energy in the United States from 2004 through 2007. These solar energy farms are producing sufficient amounts of power to offset construction of other types of electrical power plants. Some of the larger energy producers such as BP, General Electric, and Siemens are involved in large solar energy projects in the United States. Other types of energy companies and utilities have also proposed large solar energy farm projects throughout the United States. Table 4-2 lists some of the largest solar energy farms in the United States as of 2008. The names on the list are not as important as the fact that large numbers of solar energy system farm projects have recently been completed or are currently under construction. It is also important to note that at one time most of these larger solar energy farms were located in California and the Southwest, whereas today they are being placed in nearly every state including Florida, where solar energy power production is sufficient to operate large solar energy systems throughout the year.

One of the terms used when comparing solar energy farms is *capacity factor*, which is equal to the energy generated by any power plant during the year divided by the energy it could have generated if it had run at its full capacity throughout the entire year for a total of 8,760 hours.

Table 4-2 lists the name of the project and its location, the amount of the DC power rating for the solar farm, and the power that can be generated per year in *gigawatts* (GW). A gigawatt is 1 billion watts. A capacity factor is also provided to show the amount of power output the system produces as a percentage of what it could produced every hour of the year. The projects that are completed and are currently producing solar energy are located at the top of the table. Projects that are under construction or in the planning phases are listed near the bottom of the table. One of the largest solar power projects is in Florida and it is rated for 25 megawatts (MW). The next largest solar project in the United States is located at Nellis Air Force Base in Nevada and it is rated at 14 MW. New projects are coming online each year and they are becoming larger and larger, so the actual list of the largest farms may change each year. One of

TABLE 4-1 Electrical energy production from solar energy system in the United States from 2004 through 2008 (in thousands of kilowatt-hours)

Year	2004	2005	2006	2007	2008
Amount of solar PV in thousands of kWh	575,155	550,294	507,706	611,793	843,054

Source: U.S. Department of Energy

TABLE 4-2 Largest PV solar energy farms in the United States

Name of PV power plant	Location	DC Peak power (MW)	GWh /year	Capacity factor	Notes
DeSoto Next Generation Solar Energy Center	Arcadia, DeSoto County, FL; Florida Power & Light (FPL)	25	42	0.19	Completed October 2009
Nellis Solar Power Plant	Nellis Air Force Base, NV Nevada	14.02	30	0.24	70,000 solar panels currently in production
Alamosa Photovoltaic Power Plant	San Luis Valley, CO	8.2	17	0.24	Completed December 2007
Rancho Cielo Solar Farm	Belen, NM	600			Project not completed; Thin film silicon from Signet Solar
Topaz Solar Farm	San Luis Obispo County, CA	550	1,100	0.23	Project not completed; Thin film CdTe from First Solar
AV Solar Ranch One	Antelope Valley, CA	230	600	0.30	Project not completed
KCRD Solar Farm	California	80			Scheduled to be completed in 2012
Davidson County Solar Farm	Linwood, NC	21.5			Project not completed; 36 individual structures
Kennedy Space Center	Florida	10			To be constructed by SunPower for FPL Energy, completion date 2010

the latest solar farms to come online in Ohio consists of 159,200 solar panels built by First Solar Inc., Perrysburg, Ohio, and produces approximately 12 MW of electricity and supplies power to more than 1,400 residences. A megawatt is 1 million watts. The farm is located on about 80 acres about 40 miles north of Columbus, Ohio.

Other projects that are under construction include a 600 MW solar farm in New Mexico and a 550 MW farm in California. A 230 MW farm and an 80 MW farm are also planned for California. All these farms will be larger than the current largest farm in the United States. This table shows how quickly solar energy installations are occurring and future installations are being planned in the United States. This represents a large investment and an opportunity for a large number of jobs in the solar energy industry. The next sections will provide more in-depth details about each of these large projects.

4.7.1 DeSoto Next Generation Solar Energy Center

The DeSoto Next Generation Solar Energy Center is located in Arcadia, DeSoto County, Florida. DeSoto County is southeast of Tampa and northeast of Port Charlotte. This project is operated by Florida Power & Light (FPL). The construction for this project was started in late 2008 and was completed in October 2009. This solar farm is rated for 25 MW and is currently the largest photovoltaic plant in the country. This project is estimated to generate approximately 42,000 kWh per, year which is enough power for approximately 3,000 homes. The plant consists of more than 90,000 solar panels.

The project provided approximately 400 jobs during the peak construction. It will provide approximately $2 million in additional property tax revenue for DeSoto County through the end of 2010. This project represents one of the largest private capital investments in the county. The electricity provided by the solar panels will offset the need to produce electricity through traditional means. If this electricity were produced by a coal-fired plant it, would produce 75,000 tons of greenhouse gases. If this amount of electricity were produced by a power plant that uses natural gas, it would use approximately 7 billion cubic feet of natural gas. If it were produced by a power plant burning oil, it would use 277,000 barrels of oil. The most important feature of this project is that it allows Florida to have a stable source of electrical power without fear of cost increases from oil, gas, or coal. It also allows the project to continue for a number of years without concerns of environmental issues in additional charges or taxes for causing pollution.

4.7.2 Collaboration Among the United States Postal Service, the Department of Energy, and the Federal Energy Management Program

Another way that large installations of solar panels have taken place is the collaboration between government agencies. One such project is a collaboration among the United States Postal Service (USPS), the U.S. Department of Energy (DOE), and the Federal Energy Management Program (FEMP). This collaboration installed a photovoltaic distributive energy resource (DER) system at the USPS Marina Processing and Distribution Center in

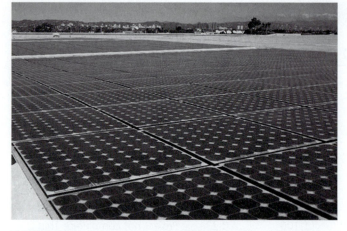

FIGURE 4-13 Solar panels at the USPS Marina Processing and Distribution Center in Marina del Rey, California. (Courtesy NREL and the U.S. Department of Energy.)

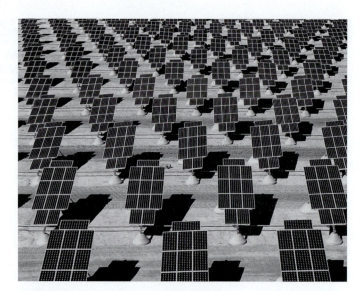

FIGURE 4-14 Solar array at Nellis Air Force Base. These panels track the sun in one axis. (US Air Force-digital Version c/Science Faction/CORBIS All Rights Reserved)

Marina del Rey, California. The agencies were assisted by a national laboratory, the local utility, and private-sector companies. The end result is a 127-kW rooftop PV system—the largest roof-integrated federal system in the nation. The project currently generates 127 kW of electricity and is expected to save approximately 10% off the facility's 1.2-MW peak power demand and $25,000 per year in utility costs. Figure 4-13 shows a picture of this installation.

4.7.3 Nellis Solar Power Plant

The Nellis Solar Power Plant is the second largest solar photovoltaic system in North America and is located on Nellis Air Force Base in Clark County, Nevada. Clark County is on the northeast side of Las Vegas. The Nellis system will generate more than 25% of the power used at the base, which is in excess of 25 million kWh of electricity annually. Construction started on this project in April 2007, and in October 2007 the first 5 MW became operational. The system was brought under full power in December 2007. It currently is in full operation and produces approximately 13 MW from the array. Figure 4-14 shows the entire array.

The entire project occupies slightly more than 140 acres, and this land is leased from the Air Force. The system consists of individual ground-mounted solar systems that use an advanced sun-tracking system. Figure 4-14 shows these individual systems and how each module can track the sun independently. The system is designed and deployed by SunPower. It consists of approximately 70,000 solar panels, and the system has a design rating of 14.2 MW. At its peak generating capacity the plant is producing approximately 13 MW AC. This means the capacity factor (the ratio of average output to rated peak output) is approximately 20%. The total output of electrical energy will support more than 12,000 civilians and military personnel at Nellis.

4.7.4 Alamosa, Colorado, Photovoltaic Power Plant

Alamosa Photovoltaic Power Plant is rated at 8.22 MW and is located in the San Luis Valley, an alpine valley in Colorado. The power plant construction was completed in December 2007 by SunEdison for Xcel Energy. The solar arrays use approximately 83 acres, and they provide enough power to supply 1,400 homes. This made it the largest producer of photovoltaic power in the United States at the time it was brought under power. Colorado Xcel customers pay about 2% more for electricity to subsidize renewable energy projects. Colorado has combined solar energy projects with wind energy projects to increase the amount of electricity that is provided by renewable resources.

The following projects are either at the end of their construction phase or are being designed and will be brought online in the near future. These projects are much larger than existing projects and represent a commitment to use solar energy in the future.

4.7.5 Rancho Cielo Solar Farm in Belen, New Mexico

The Rancho Cielo Solar Farm in Belen, New Mexico, is the largest proposed solar farm in the United States. It is being constructed on 6,000 acres in Belen, which is south of Albuquerque. The site is expected to provide the majority of the power for the community when it is under full power. The projected cost is approximately $840 million and it is projected to provide 600 MW. The solar array will cover more than 700 acres. The solar panels for this project will be thin-film silicon panels, and they will be built locally in New Mexico by Signet Solar, which is opening a new PV manufacturing facility in New Mexico.

4.7.6 High Plains Ranch II in California

High Plains Ranch II, LLC, owned by SunPower of San Jose, California, has proposed the California Valley Solar Ranch. This solar farm will provide approximately 250 MW of electrical power from solar voltaic cells, and it will be built in the Carrizo Plain, northeast of California Valley. The Carrizo Plain is a large enclosed plain in southeastern San Luis Obispo County, about 100 miles northwest of Los Angeles.

To get the project underway in 2008, Pacific Gas and Electric announced an agreement to buy all the power from the power plant. This project will be combined with a nearby 550-MW project to provide a large increase in electrical power. The project will utilize approximately 1,966 acres for solar power generation. The PV panels used for the project will use high-efficiency, crystalline PV panels designed and manufactured by SunPower. A total of 88,000 solar tracking devices will hold PV panels, and these devices will allow the panels to track the sun as it moves across the sky.

The solar farm is projected to deliver approximately 550 gigawatt-hours (GWh) annually. This will be the equivalent of 250 MW of electrical power. The plant will have a capacity factor of 25%, which means it will be able to produce 25% of its rated power out. Since the most solar power is generated during the middle of the day, it will be available when the demand for electricity is much higher than it is at night. The project is expected to begin delivering power in late 2010and should be fully operational by 2012.

4.7.7 AV Solar Ranch One in Kern County, California

AV Solar Ranch is a proposed photovoltaic project near Bakersfield, California, which is north of Los Angeles. This solar farm is projected to be rated at 230 MW, and it is owned by NextLight Renewable Power. The photovoltaic array will occupy about 2,100 acres in the Antelope Valley. The project is scheduled to begin construction in 2010 to begin to produce partial power in 2011 and full operational power by 2013. When the system is fully functioning, it will produce approximately 600 GWh of energy each year, which is enough to meet the needs of 70,000 homes. An integral part of the project is an agreement that was signed by Pacific Gas & Electric (PG&E) to purchase from NextLight the power that is produced by the project for the next 25 years at a rate not to exceed 13.3 cents per kWh.

The PV solar panels will be mounted on tracking assemblies that will face south with a tilt of approximately 20° to the horizon. The tracking system will be set up in east to west rows, and they will be powered by drive motors that will have a closed loop control that tracks the maximum sunlight as the sun moves from east to west throughout the day. The tracking mechanism will increase the efficiency of the panels in this installation.

4.7.8 Davidson County Solar Farm in North Carolina

Another solar farm is in the early stages of construction and is being built near Linwood, North Carolina, in the middle part of the state. When it is completed in 2011, the 350-acre Davidson County Solar Farm will be the largest solar farm in the United States. This solar farm will be rated as a 21.5 MW power station. SunEdison from Maryland will build the array of photovoltaic panels, and Duke Energy plans to buy all the output from the solar farm. The first stage of the project began producing power in February 2010 and it will produce approximately 4 MW of electrical power, which will serve about 650 homes. When the complete project is fully operational, it will produce about 16 MW of power, which is enough to provide power for 2,600 homes. North Carolina is one of a number states that are under mandates to produce 20% of their energy from renewable sources in the near future.

4.7.9 The John F. Kennedy Space Center (KSC) Florida

Another project that is under construction is located at the Kennedy Space Center on the east coast of Florida. The Kennedy space Center is one of many NASA facilities across the United States. This photovoltaic farm is located on Merritt Island, Florida, north-northwest of Cape Canaveral. The first phase of the project was brought under power in November 2009 and the electricity that is produces feeds the space center's power grid. Two additional sections of the solar farm will be brought under power in the near future to bring the system to full capacity, when it will provide about 1% of the space center's electrical power use. Figure 4-15 shows individual solar panels used in the system.

FIGURE 4-15 Individual solar panels that are part of the array at Kennedy Space Center in Florida.

The NASA facility has worked closely with Florida Power & Light, which is the local electrical utility and together they are working to build solar power systems in Florida. NASA is doing a feasibility study on utilizing an additional 500 acres of its land for additional solar power systems that will provide additional electrical power to meet more of the needs of the facility.

4.7.10 Other Solar Farms in Florida

Another large solar project in Florida is just west of Jacksonville, in the northeast part of the state on the Atlantic coast. The Jacksonville Solar project will produce about 15.01 MW of DC power when it is in full operation, and it is the result of a partnership between Jacksonville, Florida, municipal authority (JEA) and PSEG Solar Source. JEA is the largest community-owned utility in Florida and the eighth largest in the United States. PSEG Solar Source owns the system, which is located on JEA's property, and JEA has contracted to purchase and use the entire output of the system, including the environmental attributes generated by the facility, through a 30-year power purchase agreement with PSEG.

The project is managed by a company named Juwi Solar Inc., which is a solar energy company located in Boulder, Colorado. This company developed, designed, and obtained the permits for the 15.01-MW (DC) solar farm at the Jacksonville Solar Energy Generation Facility, and it is providing the engineering, procurement, and construction services and will also be providing the initial operation and maintenance services. Juwi Solar will also operate, maintain, and monitor the project on behalf of PSEG Solar Source during the initial years of the project.

PSEG Solar Source is a subsidiary of Newark, New Jersey–based Public Service Electricity & Gas (PSE&G). The Jacksonville Solar project is one of three solar projects the company currently owns, totaling 29.2 MW. These are the first in a planned portfolio of solar facilities throughout the United States to be developed, constructed, owned, and operated by PSEG Solar Source.

The Jacksonville Solar Project will have a capacity 15-MW DC power, and the annual generation output will be 22,430 MWh AC power. The solar panels used in this project consist of 200,000 fixed ground-mounted thin-film (cadmium teluride) solar panels that are provided by First Solar. The panels are situated on 100 acres of land owned by Jacksonville Electric. The project broke ground in October 2009 and began to deliver power in May 2010.

4.7.11 Solar Farms Projects Between Industry and Electrical Utilities

Some solar farms have been specifically created as a partnership among an industry or business, a solar energy company, and the local electrical utility company that serves them. One such project is located in Hackettstown, New Jersey, and it is a solar garden (farm) that is composed

FIGURE 4-16   Aerial view of the PSEG solar farm at Hackettstown, N.J.   (Courtesy PSEG)

of more than 28,000 ground-mounted solar panels on 18 acres. Figure 4-16 shows an aerial view of this installation. The solar panels are mounted in an area adjacent to the headquarters of Mars Chocolate North America in Hackettstown, where more than 1,200 associates work and M&M's® Brand Chocolate candies are manufactured. The solar installation provides 2 MW of power during peak hours, which is equivalent to approximately 20% of the plant's peak energy consumption. It will reduce carbon dioxide emissions by more than 1,000 metric tons, equivalent to removing 190 vehicles from the road each year. The remainder of the electrical power comes directly from the company's grid connection to the local utility.

Public Service Enterprise Group (PSEG Solar Source) owns the system, and Mars has contracted to use the entire electrical output of the system. Juwi Solar Inc. performed the engineering, procurement, and construction services for the system and will also be providing the initial operation and maintenance services. A long-term partnership between Mars Chocolate North America and PSEG Solar Source will ensure the solar garden's success. Thin-film panels are used on this project and they are provided by First Solar.

4.8 LARGEST SOLAR FARMS IN THE WORLD

Solar energy has been installed in a large number of sites around the world. Germany has led the world in the production of solar PV panels, and Spain and Germany are leading in the installation of solar panels and the production of electrical power from PV cells. Table 4-3 shows the top 10 solar photovoltaic farms in the world as of January 2010. The size of the farms ranges from 30 MW to 60 MW output.

Table 4-4 lists the five top solar energy producing countries in 2008. Germany leads all countries by producing 3.5 GWh annually. The United States produced 0.843 GWh in 2008, China produced 0.14 GWh, Canada produced 0.017 GWh, and Japan

TABLE 4-3	Largest solar energy farms currently producing, proposed, or under construction outside the United States as of January 2010.				
Name of PV power plant	**Country**	**DC Peak power (MW)**	**GWh /year**	**Capacity factor**	**Notes**
Olmedill Photovoltaic Park	Spain	60	85	0.16	Completed September 2008
Strasskirchen Solar Park	Germany	54			
Lieberose Photovoltaic Park	Germany	53	53	0.11	2009
Puertollano Photovoltaic Park	Spain	50			2008
Moura Photovoltaic Power Station	Portugal	46	93	0.23	Completed December 2008
Waldpolenz Solar Park	Germany	40	40	0.11	550,000 First Solar thin-film CdTe modules; completed December 2008
Arnedo Solar Plant	Spain	34			Completed October 2008
Merida/Don Alvaro Solar Park	Spain	30			Completed September 2008
Planta Solar La Magascona & La Magasquila	Spain	30			
Planta Solar Ose de la Vega	Spain	30			

TABLE 4-4	Top five countries producing solar energy in 2008
Country	**Production in 2008 in GWh**
Germany	3.5
The United States	0.843
China	0.140
Canada	0.017
Japan	0.002

produced 0.002 GWh. Even though Spain has several of the largest solar energy farms, it was not in the top five countries for producing solar energy in 2008.

4.9 ELECTRICAL ENERGY PRODUCED BY PARABOLIC TROUGH SOLAR COLLECTORS

Another way to produce electrical power from solar energy is through the use of parabolic trough solar collectors. The collectors for the parabolic trough are filled with oil or a similar substance that will easily absorb heat from the solar collectors. The surface of the solar collectors is set at a specific angle to take the sun's rays and focus them on the tube that has the oil in it. As the sun's rays intensify on the surface of the tube, the oil that is flowing inside the tube that runs through the parabolic trough is heated to the temperatures up to 800°F. This oil flows through the parabolic trough and on to a water tube heat exchanger where the oil heats the water that flows through the second loop of the heat exchanger and the water turns to steam. The steam is then directed onto a turbine wheel that is similar to the turbines that are used in a coal-fired or nuclear-fired power plant. The shaft of the steam turbine is

connected directly to the armature of a generator. When the steam is directed onto the veins of this turbine wheel, it makes the turbine wheel spin at a high rate of speed. As a turbine spins the shaft on the generators also spins and causes the generator to produce AC electricity. The amount of steam that is allowed to reach the turbine veins is closely regulated and controlled to ensure that the turbine wheel spins at an exact speed to cause the generator to produce electricity at 60 Hz. Typically the generator will be a three-phase generator just as in a coal-fired or nuclear-powered generating station.

The main energy source that heats the water to create the steam for the traditional electrical power station is usually coal, nuclear, or natural gas fired. Since the parabolic trough system can produce electricity only when the sunlight is focused on the tubes, it is considered to be the backup or auxiliary power source for the power plant. The parabolic tube system does not produce electricity at night, so it needs steam from another energy source such as coal, nuclear power, or a natural gas system like a traditional power plant when the sunlight is not available.

The parabolic trough system has some issues and limitations. One of the ways the cost of the parabolic trough system is reduced is to receive tax credits. If the system is to receive credits as a renewable energy system, the amount of energy from coal, nuclear, or other sources is limited to 27% of the system input; the remaining 73% must come from solar energy. Another issue that occurs with the parabolic trough system is that its reflectors and collectors may be mounted stationary and do not move, which limits the efficiency of the system. The reflective panels in the parabolic trough system keep plumbing in the piping system mounted rigidly and they provide a reflective surface to ensure the heat energy from the sun is focused directly on the piping. The tracking systems

TABLE 4-5   Parabolic trough solar energy systems data for the United States (1 bar = 14.5PSI)

Plant name	Location	First year of operation	Net output (Mwe)	Solar field Outlet (C)	Solar field Area (m²)	Solar turbine Effic. (%)	Power Cycle	Dispatch ability provided by
Nevada Solar One	Boulder City, NV	2007	64	390	357,200	37.6	100 bar, reheat	None
APS Saguaro	Tucson, AZ	2006	1	300	10,340	20.7	ORC	None
SEGS IX	Harper Lake, CA	1991	80	390	483,960	37.6	100 bar, reheat	HTF heater
SEGS VIII	Harper Lake, CA	1990	80	390	464,340	37.6	100 bar, reheat	HTF heater
SEGS VI	Kramer Junction, CA	1989	30	349	188,000	37.5	100 bar, reheat	Gas boiler
SEGS V	Kramer Junction, CA	1988	30	349	250,500	30.6	40 bar, steam	Gas boiler
SEGS III	Kramer Junction, CA	1987	30	349	230,300	30.6	40 bar, steam	Gas boiler
SEGS IV	Kramer Junction, CA	1987	30	349	230,300	30.6	40 bar, steam	Gas boiler
SEGS II	Daggett, CA	1986	30	316	190,338	29.4	40 bar, steam	Gas boiler
SEGS I	Daggett, CA	1985	13.8	307	82,960	31.5	40 bar, steam	3-hrs TES

move the reflectors in sync with the movement of the sun to ensure that the panels receive the most heat energy available. The tracking system makes the installation more expensive but more efficient.

One of the largest parabolic trough solar collectors is at Kramer Junction in the Mojave Desert in California. The Solar Energy Generating Systems (SEGS) at Kramer Junction is the world's largest solar energy generating facility, which consists of nine solar power plants. The plants are all located in Mojave Desert, and they generate a combined total of about 355 MW. The plant at Kramer Junction produces about 150 MW, the plant at Harper Lake produces about 160 MW, and the plant at Daggett about 44 MW. Figure 4-17 shows the parabolic trough solar collectors at Kramer Junction. Table 4-5 shows the data for the largest parabolic trough solar systems in the United States, including the name of the plant, the size of the generating capacity, and the size of the solar field capacity.

4.10 PRODUCTION OF SOLAR PANELS AND SOLAR EQUIPMENT

Germany produces more than 40% of all the solar energy that is produced in the world. One reason is that a large number of companies that produce solar energy products are located in Germany. The German government took a unique approach to developing solar energy, heavily incentivizing it while punishing companies that did not use clean energy. As a result Germany's solar energy

FIGURE 4-17   Parabolic trough solar collectors at Kramer Junction in the Mojave Desert in California.   (Courtesy NREL and the U.S. Department of Energy)

market, which includes solar photovoltaic panel production and the production of electrical power from solar energy, has grown at a much faster pace than in any other country in the world. In fact the next closest country produces less than 20% of the world's solar electrical power output. The United States and other countries are beginning to catch up with the amount of electrical power that is produced from solar panels. The United States will begin to draw nearer to Germany's output when the solar projects that are in development begin to be installed and start to produce electricity. One of the reasons Germany has become a dominant force in producing solar energy is that it adds a tariff to all electricity consumed by all customers, and the extra income is used to subsidize solar energy at approximately 50 cents per kWh, which makes the electricity generated from solar energy equipment nearly the same cost as electrical energy produce from coal-fired and nuclear plants. Some of the money from selling electrical power is returned to a number of institutes and universities dedicated to studies for increasing the rate and efficiency at which the world installs solar energy products and produces electrical power from them.

4.11 LARGEST SOLAR ENERGY COMPANIES AROUND THE WORLD

Over the past 20 years the number and size of solar energy companies have continually increased. Solar energy companies include companies that manufacture solar energy equipment, install or maintain solar energy equipment, and companies that own or lease large solar energy fields. This information will give you some idea of all the companies that are involved in solar energy one way or another and they are also the larger companies that will be hiring solar energy technicians in the future.

TABLE 4-6	Some of the larger solar energy companies
Company name	**Location**
British Petroleum (BP)	United Kingdom
Chevron	California, the United States
First solar	Perrysburg, OH
General Electric	United States
GT Solar International, Inc.	Merrimack, New Hampshire
JA Solar	China
Nano Solar	Switzerland
Siemens Solar Energy	Germany
Solar Enertech Corp	Menlo Park, CA
Sun Edison	Maryland, the United States
Sun Tech power	China
Sunergy	China
SunPower Corporation	San Jose, CA
Suntech Power Holding Ltd	China

Note: Since solar energy companies are continually changing in size, purchasing companies, and expanding to increase their size, this list does not indicate the size of the company.

During this time various governments around the world have created incentives to increase the number and size of solar energy products. The governments have provided financial incentives and a reduction in taxes, as well as making financial aid available through loans so that projects can be fast tracked. Table 4-6 shows a list of companies around the world that produce solar energy products and also manage solar energy projects. The companies are listed in alphabetical order and do not represent any value or size.

Questions

1. What does the term *turn key* mean?
2. Identify three types of sight issues that you may encounter when you are selecting a location for a solar energy system.
3. Explain why a small residential solar energy system may need a power electronic frequency converter (inverter).
4. List three types of applications of solar energy systems.
5. Identify the five largest solar energy farms in the world as of January 2010.
6. Identify the total amount of electrical energy produced by PV solar energy systems in the United States in 2008.
7. Explain how the parabolic trough solar energy system produces electricity.
8. Identify two PV solar energy farms that are located in Europe.
9. Identify two factors that limit the amount of electrical energy produced by PV solar systems.
10. Identify two sites that utilize parabolic trough solar energy systems.

Multiple-Choice Questions

1. The three categories for site issues are _____
   a. environmental issues, visual issues, and connection issues.
   b. environmental issues, time-of-use issues, and noise issues.
   c. environmental issues, height issues, and noise issues.
   d. All the above.

2. A small residential solar energy system may need an electronic inverter if _____
   a. the solar energy system is used to charge batteries in the home and all the loads in the home require AC voltage.
   b. the solar energy system produces AC voltage that will need to be converted to DC voltage.
   c. the solar energy system produces hot water that needs to be converted to heat energy to augment the heating system for the home.
   d. All the above.

3. The three largest PV solar energy farms rated by DC peak power that are currently in production in the United States as of 2009 are _____
   a. Kramer Junction California, Desoto County Florida, and Nellis AFB Neveda
   b. Kennedy Space Center Florida, Nellis AFB Neveda, and Alamosa Colorado
   c. Desoto County Florida, Nellis AFB Neveda, and Alamosa Colorado

4. Where is the largest offshore solar energy farm in the United States planning to be built?
   a. Off the coast of the Atlantic Ocean
   b. Off the coast of the Gulf Coast
   c. Off the coast of the Pacific Ocean

5. Which country has the most of the top 10 PV solar farms?
   a. Germany
   b. Spain
   c. Portugal.
   d. China

6. Which country produced the most electrical power from solar energy in 2008?
   a. Japan
   b. China
   c. Germany
   d. The United States

Basic Photovoltaic (PV) Principles and Types of Solar PV Cells (Converting Solar Energy to Electricity)

OBJECTIVES FOR THIS CHAPTERS

When you have completed this chapter, you will be able to

- Explain what valence electrons are and what the valence shell is.
- Identify an electron, proton, and nucleus of an atom and identify the polarity of each.
- Explain how two materials are combined to make P-type material.
- Explain how two materials are combined to make N-type material.
- Explain how P-type material and N-type material are combined through doping to make a silicon PV cell.
- Identify the materials on the periodic table of elements that are conductors.
- Identify the materials on the periodic table of elements that are insulators.
- Identify the materials on the periodic table of elements that are semiconductors.
- Explain how DC electric current is caused to flow when light strikes the silicon PN junction.

TERMS FOR THIS CHAPTER

Atom	Load Resistor
Atomic Number	N-Type Material
Conductance Band	Neutron
Conductor	Nucleus
Diode	PN Junction
Doping	Periodic Table Of Elements
Electrode	Proton
Electron	P-Type Material
Element	Semiconductor
Free Electron	Shell
Hole	Valence Electron
Insulator	Valence Shell
Lattice Structure	

5.0 OVERVIEW OF THIS CHAPTER

This chapter will explain how the types of solar photovoltaic (PV) cells that you will find installed around the world and in the United States are made. The information in this chapter will explain how silicon solar photovoltaic cells are manufactured and how their electrons and protons are manipulated to create P-type material and N-type material. Information about the atomic periodic table of elements is provided, and you will see how elements with different number of electrons are combined to create material used inside PV cells. This chapter will explain how basic materials used in the manufacture of photovoltaic cells can be broken down into three categories: conductors, insulators, or semiconductors. You will learn about the atomic makeup of this material and how it is exploited to take advantage of conditions that will make solar PV cells more efficient and less expensive to manufacture. You will also begin to see why specific types of materials are used in combination to make PV cells.

The information in this chapter will explain how the PV material reacts to sunlight striking it and how the PV cells create a flow of electrons that become DC current. The final part of the chapter will provide detailed diagrams of each of the types of PV cells that are commonly used and will show the different layers of materials that make them operate and convert sunlight into DC electrical power.

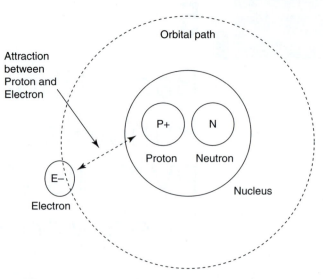

FIGURE 5-1 An atom with a positively charged proton and neutrally charged neutron in its nucleus and a negatively charged electron orbiting the nucleus. The dashed line shows the attraction between the proton and electron that keeps the electron in its orbit as it moves around the nucleus.

5.1 CONDUCTORS, INSULATORS, AND SEMICONDUCTORS

You need to understand several terms when you begin to learn about the way silicon converts sunlight to electrical energy. In this section you will learn more about the elements that are used to combine with silicon to create a solar photovoltaic cell. Since the photovoltaic cell will create electrical energy when light shines on it, you also will learn more about electricity, and which elements are conductors, insulators, or semiconductors. A *conductor* allows electrical current to flow and has less than four electrons in its valence shell, and an *insulator* prohibits the flow of electrical current and it has more than four valence electrons. *Semiconductors* are materials that have exactly four valence electrons that can be made into either conductors or insulators by combining them with other material in a process called *doping*, whereby an element with less than four valence electrons or more than four valence electrons is combined with a semiconductor material that has exactly four valence electrons.

In earlier chapters you learned that all material is made of atoms and that *atoms* have *protons* and *neutrons* in their *nucleus* and *electrons* that move around the nucleus in an orbital path. The materials that have atoms with more than one electron, such as silicon, the electrons will be located in different orbits as they move about the nucleus. The orbital path is also called a *shell* and the different shells are located a specific distance from the nucleus. You will learn in this section how the electrons in the different shells can be manipulated when a

silicon solar PV cell is made so it will convert light energy to electrical power.

An atom is the smallest part of an element that retains the chemical properties of the element. Figure 5-1 is a diagram of a simple hydrogen atom with one electron, one proton, and one neutron. A proton is the positively charged particle that is located in the nucleus in the center of the atom. The neutron is the neutrally charged particle that is also located in the nucleus. The electron is the negatively charged particle that moves about the nucleus at a specific distance in an orbit, and it is held to the nucleus in the orbit because the proton in the nucleus has a positive charge that attracts the negative charge of the electron.

There are currently 118 known substances that are called *elements*. The *periodic table of elements* is shown in Figure 5-2. It shows 103 of the elements with their abbreviations and the number of electrons and atomic weight of each element. The atomic weight is a number that does not have any units and it is the ratio of the average element mass to 1/12 of the mass of an atom of carbon. Each element has a unique number assigned to it, which is the number of electrons it has. This number is also called the *atomic number* for the element. Since no two elements have the same number of electrons, the number of electrons in each element is used as its atomic number.

Since the number of electrons in the atom of each element is different, you can get an idea of how the electrons move about the nucleus of each atom by studying one element. Silicon is one of the most widely used materials in making solar PV cells, so we will review the location of the electrons as they move about the nucleus of the silicon atom. The silicon atom has an atomic

Periodic Table of the Elements
Abundance of Elements
in Seawater near the Surface

all values are in mg/L

http://chemistry.about.com
© 2010 Todd Helmestine
About Chemistry

1A																	**8A**
1 **H** 108000 Hydrogen	**2A**											**3A**	**4A**	**5A**	**6A**	**7A**	2 **He** 7×10^{-6} Helium
3 **Li** 1.8×10^{-1} Lithium	4 **Be** 5.6×10^{-8} Beryllium											5 **B** 4.44 Boron	6 **C** 28 Carbon	7 **N** 5×10^{-1} Nitrogen	8 **O** 857000 Oxygen	9 **F** 1.3 Fluorine	10 **Ne** 1.2×10^{-4} Neon
11 **Na** 10800 Sodium	12 **Mg** 1290 Magnesium	**3B**	**4B**	**5B**	**6B**	**7B**	**8B**	**8B**	**8B**	**1B**	**2B**	13 **Al** 2×10^{-2} Aluminum	14 **Si** 2.2 Silicon	15 **P** 6×10^{-2} Phosphorus	16 **S** 905 Sulfur	17 **Cl** 19400 Chlorine	18 **Ar** 4.5×10^{-1} Argon
19 **K** 399 Potassium	20 **Ca** 412 Calcium	21 **Sc** 6×10^{-7} Scandium	22 **Ti** 1×10^{-3} Titanium	23 **V** 2.5×10^{-3} Vanadium	24 **Cr** 3×10^{-4} Chromium	25 **Mn** 2×10^{-4} Manganese	26 **Fe** 2×10^{-3} Iron	27 **Co** 2×10^{-5} Cobalt	28 **Ni** 5.6×10^{-4} Nickel	29 **Cu** 2.5×10^{-4} Copper	30 **Zn** 4.9×10^{-3} Zinc	31 **Ga** 3×10^{-5} Gallium	32 **Ge** 5×10^{-5} Germanium	33 **As** 3.7×10^{-3} Arsenic	34 **Se** 2×10^{-4} Selenium	35 **Br** 67.3 Bromine	36 **Kr** 399 Krypton
37 **Rb** 1.2×10^{-1} Rubidium	38 **Sr** 7.9 Strontium	39 **Y** 1.3×10^{-5} Yttrium	40 **Zr** 3×10^{-5} Zirconium	41 **Nv** 1×10^{-5} Niobium	42 **Mo** 1×10^{-2} Molybdenum	43 **Tc** Technetium	44 **Ru** 7×10^{-7} Ruthenium	45 **Rh** Rhodium	46 **Pd** Palladium	47 **Ag** 4×10^{-5} Silver	48 **Cd** 1.1×10^{-4} Cadmium	49 **In** 2×10^{-2} Indium	50 **Sn** 4×10^{-6} Tin	51 **Sb** 2.4×10^{-4} Antimony	52 **Te** 1.9×10^{-6} Tellurium	53 **I** 6×10^{-2} Iodine	54 **Xe** 5×10^{-5} Xenon
55 **Cs** 3×10^{-4} Caesium	56 **Ba** 1.3×10^{-2} Barium	57-71 Lanthanoids	72 **Hf** 7×10^{-8} Hafnium	73 **Ta** 2×10^{-6} Tantalum	74 **W** 1×10^{-4} Tungsten	75 **Re** 4×10^{-6} Rhenium	76 **Os** Osmium	77 **Ir** Iridium	78 **Pt** Platinum	79 **Au** 4×10^{-8} Gold	80 **Hg** 3×10^{-5} Mercury	81 **Ti** 1.9×10^{-5} Thallium	82 **Pb** 3×10^{-5} Lead	83 **Bi** 2×10^{-5} Bismuth	84 **Po** 1.5×10^{-10} Polonium	85 **At** Astatine	86 **Rn** 6×10^{-16} Radon
87 **Fr** Francium	88 **Ra** 8.9×10^{-11} Radium	89-103 Actinoids	104 **Rf** Rutherfordium	105 **Db** Dubnium	106 **Sg** Seaborgium	107 **Bh** Bohrium	108 **Hs** Hassium	109 **Mt** Meitnerium	110 **Ds** Darmstadtium	111 **Rg** Roentgenium	112 **Cp** Copernicium	113 **Uut** Ununtrium	114 **Uuq** Ununquadium	115 **Uup** Ununpentium	116 **Uuh** Ununhexium	117 **Uus** Ununseptium	118 **Uuo** Ununoctium

Lanthanides

57 **La** 3.4×10^{-4} Lanthanum	58 **Ce** 1.2×10^{-4} Cerium	59 **Pr** 6.4×10^{-7} Praseodymium	60 **Nd** 2.8×10^{-6} Neodymium	61 **Pm** Promethium	62 **Sm** 4.5×10^{-7} Samarium	63 **Eu** 1.3×10^{-7} Europium	64 **Gd** 7×10^{-7} Gadolinium	65 **Tb** 1.4×10^{-7} Terbium	66 **Dy** 9.1×10^{-7} Dysprosium	67 **Ho** 2.2×10^{-7} Holmium	68 **Er** 8.7×10^{-7} Erbium	69 **Tm** 1.7×10^{-7} Thulium	70 **Yb** 8.2×10^{-7} Ytterbium	71 **Lu** 1.5×10^{-7} Lutetium

Actinides

89 **Ac** Actinium	90 **Th** 1×10^{-6} Thorium	91 **Pa** 5×10^{-11} Protactinium	92 **U** 3.2×10^{-3} Uranium	93 **Np** Neptunium	94 **Pu** Plutonium	95 **Am** Americium	96 **Cm** Curium	97 **Bk** Berkelium	98 **Cf** Calfornium	99 **Es** Einsteinium	100 **Fm** Fermium	101 **Md** Mendelevium	102 **No** Nobelium	103 **Lr** Lawrencium

Legend: $<10^{-12}$ | $10^{-12}-10^{-9}$ | $10^{-9}-10^{-6}$ | $10^{-6}-10^{-3}$ | $10^{-3}-1$ | $1-10^{3}$ | $>10^{3}$

FIGURE 5-2 Periodic table of elements.

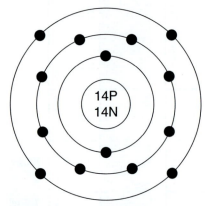

FIGURE 5-3 A simplified diagram of the 14 electrons that are moving about the nucleus of the silicon atom in three distinct orbits that are also called shells.

number of 14, which means it has 14 electrons moving about its nucleus in three separate *orbits*. Figure 5-3 shows the layout of the electrons in the silicon atom in a simplified two-dimensional diagram. Two electrons are located in the orbit or shell that is closest to the nucleus. The second orbit or shell has eight electrons in it, and the outermost orbit or shell has four electrons. The outermost shell is called the *valence shell* and the electrons in that shell are called *valence electrons*. Figure 5-3 shows silicon has four electrons in its outer shell so it is a semiconductor material.

Figure 5-4 shows the silicon atom with its electrons orbiting in a variety of directions. In reality this is exactly how the electrons orbit the nucleus so they never bump into one anther. When we are discussing the electrons in

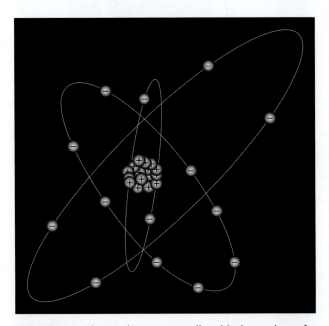

FIGURE 5-4 The 14 electrons actually orbit the nucleus of the silicon atom in a variety of directions as shown in this model.

each shell, it is easier to use the two-dimensional model to count the electrons in each orbit (shell).

Some elements tend to be more stable than others, which is important when solar PV cells are made. In the simplest terms the atoms of the most stable elements in the periodic table have eight valence electrons, which are found as four pairs. The electrons in each pair make the atom more stable because one electron spins clockwise and the other spins counterclockwise. When an atom has four pairs of electrons in its outermost shell, it combines well with other similar atoms.

When elements are combined to make solar PV cells, the material that is used may have eight electrons in the second or third shell depending on the total number of electrons. For example silicon has four electrons in its third shell; it can be combined with other atoms to add additional electrons to fill the third shell, or it can be combined with an atom that will take away electrons, which would leave a full second shell of eight electrons.

When scientists are combining elements with silicon to make a solar PV cell and the element does not have exactly eight electrons in its outer shell, the new material can add electrons or give up electrons so the new combined material will get exactly eight electrons in its outer (valence) shell. Doping is this process of combining two materials to exchange electrons.

During the doping process, materials such as silicon are selected to give up electrons or take on electrons from the material they are combined with. The new combined material can be described by the amount of energy that is required to get to the eight electrons in a full shell: whether it will take less energy to add electrons to get a full shell (eight) or to give up these electrons to achieve the stable configuration of a full shell with eight electrons. Not all new combined material will have a full shell of eight electrons, but it is important to remember that the activity each atom exhibits when it is combined will always be in reference to giving up or taking on electrons to get the eight electrons to make its shell full. This concept will help explain some of the activity that goes on inside material that has been combined through doping.

Another point about each shell or orbit is that as the electrons move about the nucleus in each orbit, the movement represents the amount of energy needed to free electrons from that shell. The electrons that are in the first shell that is closest to the nucleus and the positive proton will require the most energy to free electrons because the attraction between the negative electron and the positive proton is very strong. The electrons in the second shell (orbit) require less energy to free an electron than those in the first shell. The electrons in the outermost shell require the smallest amount of energy to free them from their orbit, because they are so far from the nucleus and the attraction of the proton is very small. This is why the electrons in the valence shell are so important, because they can be freed from their shell with the smallest amount of energy expended. If you study the elements in periodic table in more depth, you would find it is possible to calculate the amount of energy needed to free an electron from any shell. In this chapter we only need to know that the electrons in the outermost shell, the valence shell, require the smallest amount of energy to allow them to break free of the shell.

When we've combined material (other elements) with silicon during the doping process, the resulting new material will no longer be a pure semiconductor with exactly four electrons in its outermost shell. The new material will add electrons to or remove electrons from the silicon atom. When electrons are added to the silicon atom, we call the new material *P-type material;* when electrons are subtracted from the silicon atom, we will call the new material *N-type material.* We will also be able to describe the P-type and N-type material in reference to whether it acts more like an electrical conductor or an electrical insulator. A conductor is a material that allows electrons (electrical current) to flow easily through it, and an insulator does not allow electrons (electrical current) to flow through it. An example of a conductor is copper that is used for electrical wiring. An example of an insulator material is rubber or plastic. The atomic structure of a conductor makes it easier for electrons to flow through it, and the atomic structure of an insulator makes it nearly impossible for any electrons to flow through it.

Another way to show this is to look at the number of electrons in the outermost shell (valence shell) of a material and describe the activity of the electrons in the material when a voltage potential is applied to it. If the material gives up electrons freely and easily when an electrical voltage potential is applied to it, it is a good conductor. If the material takes on free electrons easily, it will pull them out of the current flow and this material is an insulator.

5.2 CONDUCTORS, INSULATORS, AND SEMICONDUCTORS IN THE PERIODIC TABLE OF ELEMENTS

Figure 5-5 shows seven columns from the periodic table of elements for family 1B, 2B, 3A, 4A, 5A, 6A, and 7A. The elements in family 1B all have one valence electron, and the elements in family 2B all have two valence electrons. The elements from family 3A all have three valence electrons. The elements in families 1B, 2B, and 3A are all considered electrical conductive elements in that these materials will easily conduct electricity. You can see that copper (Cu) is element number 29 and it is found in column 1B, and aluminum (Al) is element number 13 and is found in column 3A. Both copper and aluminum are used to make wire conductors.

The elements in the column for family 4A are all considered to be semiconductor materials since they all have exactly four valence electrons. You can see that silicon (Si) is element number 14 and germanium (Ge) is element number 32. Both of these elements have been used in the manufacture of semiconductor electronic components such as transistors and diodes that have

1B	2B	3A	4A	5A	6A	7A
		5 **B** 4.44 Boron	6 **C** 28 Carbon	7 **N** 5×10^{-1} Nitrogen	8 **O** 857000 Oxygen	9 **F** 1.3 Fluorine
		13 **Al** 2×10^{-3} Aluminum	14 **Si** 2.2 Silicon	15 **P** 6×10^{-2} Phosphorus	16 **S** 905 Sulfur	17 **Cl** 19400 Chlorine
29 **Cu** 2.5×10^{-4} Copper	30 **Zn** 4.9×10^{-3} Zinc	31 **Ga** 3×10^{-5} Gallium	32 **Ge** 5×10^{-5} Germanium	33 **As** 3.7×10^{-3} Arsenic	34 **Se** 2×10^{-4} Selenium	35 **Br** 67.3 Bromine
47 **Ag** 4×10^{-5} Silver	48 **Cd** 1.1×10^{-4} Cadmium	49 **In** 2×10^{-2} Indium	50 **Sn** 4×10^{-5} Tin	51 **Sb** 2.4×10^{-4} Antimony	52 **Te** 1.9×10^{-5} Tellurium	53 **I** 6×10^{-2} Iodine
79 **Au** 4×10^{-6} Gold	80 **Hg** 3×10^{-6} Mercury	81 **Ti** 1.9×10^{-5} Thallium	82 **Pb** 3×10^{-5} Lead	83 **Bi** 2×10^{-5} Bismuth	84 **Po** 1.5×10^{-15} Polonium	85 **At** Astatine

FIGURE 5-5 The elements from families 1B, 2B, 3A, 4A, 5A, 6A, and 7A from the periodic table of elements.

been used in electronic circuits for more than 50 years. The elements in column 4A all have exactly four electrons in their valence shells.

The elements in columns 5, 6, and 7 are all considered to be insulators. The elements in column 5A all have five valence electrons in their valence shells, the elements in column 6A all have six valence electrons, and the elements in column 7A all have seven valence electrons in their valence shell. All of the elements in columns 5A, 6A, and 7A are considered to be insulators and are used in materials that are electrical insulators.

5.3 SIMPLIFIED STRUCTURE OF A CONDUCTOR

Figure 5-6 shows a simplified atomic structure of an element that is considered to be a conductor. The element shown in this figure has one valence electron. Any element whose atoms have one, two, or three valence electrons is a conductor. Since all atoms will try to gain or lose sufficient electrons to achieve a configuration of eight electrons (four pairs) in their outermost (valence) shell, it takes less energy for conductors to give up these electrons (one, two, or three) so that the valence shell will become empty and the next innermost shell will be full with four pairs (eight electrons). At this point the atom becomes more stable without the electrons in the outer shell, leaving behind the electrons in the next innermost shell, which is full. For example if the element has one, two, or three electrons in its third shell, and these electrons are released when an electrical potential voltage is applied, the material will then have two shells with the second shell being full with four pairs of electrons making a total of eight valence electrons. When the electrons leave, the next outermost shell becomes the

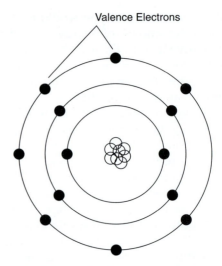

FIGURE 5-7   Simplified structure of an insulator.

new valence shell with eight electrons. The one, two, or three electrons that are given up are free to move as current flow.

5.4 SIMPLIFIED STRUCTURE OF AN INSULATOR

Figure 5-7 shows the simplified atomic structure of an insulator. Any element that has five, six, or seven valence electrons is an insulator. In this example the atom has seven valence electrons. This structure makes it easy for insulators to take on one extra electron to get eight valence electrons. When electrons are captured by an element to fill the valence shell, they are usually the electrons that would normally be free to flow as current. When the electrons are captured and taken into the valence shell, and the shell becomes full, the material becomes stable and the electrons are not freely given up, which means the electrons are taken out of current flow. This is basically how insulators prevent electrons from flowing.

5.5 SIMPLIFIED STRUCTURE OF A SEMICONDUCTOR

Semiconductors are materials whose atoms have exactly four valence electrons. They can take on one, two or three new valence electrons when they are combined with a new material so that they act more like an insulator, or they can release one, two or three valence electrons when they are combined with another material so that the new material acts more like a conductor. Figure 5-8 shows the simplified atomic structure of a semiconductor material with exactly four valence electrons.

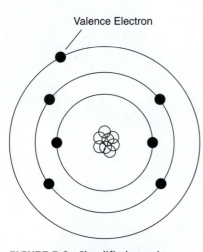

FIGURE 5-6   Simplified atomic structure of a conductor.

Valence Electron

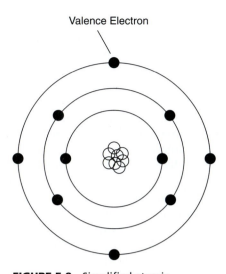

FIGURE 5-8 Simplified atomic structure of a semiconductor material.

5.6 COMBINING SILICON ATOMS

When solar PV material such as a silicon PV cell is manufactured, large numbers of silicon atoms are placed side by side during the doping process. The resulting structure becomes very stable at this point because each silicon atom shares its electrons with the silicon atom next to it. Figure 5-9 shows an example of what happens when a large number of silicon atoms are placed next to each other during the doping process. In this diagram only the four valence electrons for each silicon atom are shown. When the silicon atoms are placed close to each other, they will share an electron with each of the silicon atoms that are close to it. In this figure five silicon atoms are shown in close proximity. If you look closely at the center atom, you can see that it appears to have four pairs of electrons instead of four single electrons, because each of the atoms that is near it is sharing one of its atoms. When this process is duplicated with billions of silicon atoms, the resulting material becomes extremely stable since each silicon atom that has four valence electrons

now appears to have eight valence electrons. When atoms share electrons, the material takes on what is called a *lattice structure*. The lattice structure is considered to be the strongest atomic structure and the most stable structure for an atom.

The next area we will study is combining silicon atoms with atoms of other materials to make P-type and N-type material.

5.7 COMBINING PHOSPHOROUS OR ARSENIC WITH SILICON TO MAKE N-TYPE MATERIAL

When silicon is used to make a solar photovoltaic cell, it is combined with other materials to make N-type material or P-type material. When the silicon N-type material and the silicon P-type material are combined, the resulting new product is called a solar photovoltaic (PV) cell, and it has the ability to make sunlight cause electrons to flow in the material, creating an electrical current flow.

N-type silicon material is created by doping (mixing the silicon with its four valence electrons) with elements that contain one more valence electrons than silicon does. Two examples of elements that have five valence electrons are phosphorus and arsenic, and they are found on the periodic table under the column 5A (family 5 elements).

Since only four electrons are required to bond with the four adjacent silicon atoms, the fifth valence electron is available for conduction.

Phosphorus (P) and arsenic (As) have five valence electrons. When they combine with silicon atoms, which have four valence electrons, they create a new N-type material. It is called N-type material because it has an extra electron which has a negative charge. Figure 5-10 is a diagram of four silicon (Si) atoms combined with one atom of arsenic (As). This combination creates a strong lattice structure. Each silicon atom donates one of its valence electrons to pair up with each

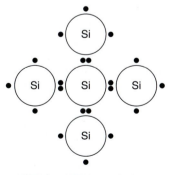

FIGURE 5-9 Atoms of silicon semiconductor material combine to create a lattice structure.

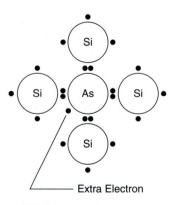

Extra Electron

FIGURE 5-10 N-type material formed by combining four silicon atoms with a single arsenic atom.

of the valence electrons of the arsenic atom, just as if it were another silicon atom. Since the arsenic atom has five valence electrons, one of the electrons will not be paired up and will become displaced from the atom. This electron is called a *free electron* and it can go into conduction as electrical current with minimal energy. Since this new material has a free electron and electrons have a negative charge, the new material is N-type material.

5.8 COMBINING BORON AND SILICON TO MAKE P-TYPE MATERIAL

P-type silicon is created by doping (mixing) it with compounds containing one less valence electron than silicon has. The elements that have three valence electrons come from the column in the periodic table that shows family 3 elements (see Figure 5.5). Typical elements from family 3 that are used in this doping process include boron (B), gallium (Ga), and indium (In).

When silicon, with four valence electrons, is doped with atoms of elements such as boron that have one less valence electron (three valence electrons), only three electrons from boron are available for bonding with four adjacent silicon atoms. Since only three electrons are available for bonding, the fourth electron is left without a paired electron, and this space where the electron typically would be is called a hole. Figure 5-11 is a diagram of four silicon atoms combined with one atom of boron. *Hole* is the name given to the space where the fourth electron would have paired up with the fourth electron in the silicon; it is identified in the diagram where the bottommost silicon atom shares its electron with boron. Since boron has only three valence electrons, there is no electron left to share with the silicon atom at the bottom, and the hole is formed. The important point about the hole is that it can attract any electron from any nearby atoms, especially electrons that may be flowing as part of current. When this occurs, the electron is removed from the current flow

and gathered by the hole, which basically begins to prevent the flow of electrical current. In this way the material will act like an insulator and it stops the electron flow. It is also important to understand that when one hole is filled, it will create another new hole where that electron was pulled from.

The combination of silicon and boron creates a new material called P-type material. The material is called P-type material because it has a hole that represents a positive charge. The boron atom and the silicon atoms create a strong lattice structure that is similar to that formed when atoms of silicon are combined. Since the hole does not have an electron in it, it is considered to have a positive charge since it is not occupied by a negatively charged electron. When billions of atoms of silicon are combined with millions of atoms of boron, they continue to combine at a rate of four silicon atoms to one boron atom. This combination creates a new material that has a large number of holes with a positive charge, so the new material is called P-type material.

5.9 CREATING A PN JUNCTION

One piece of P-type silicon material can be combined with one piece of N-type silicon material to make a silicon *PN junction*. Figure 5-12 shows a typical PN junction. The most commonly known solar cell is configured from silicon PN junctions. We will explain a simplified silicon PN junction as placing one single layer of N-type silicon material into direct contact with a single layer of P-type silicon material. However, it is important to understand that the PN junctions of silicon solar cells are not made in this way, but rather by diffusing an N-type dopant into one side of a P-type wafer (or vice versa). N-type dopant is N-type material that is created specifically to be added to silicon for making PV cells. A wafer is a slice of material that is cut as thin as possible so it can be used in the PV manufacturing process. Diffusing means that molecules of P-type material that have a higher concentration of holes are mixed with molecules of N-type material that have a higher concentration of electrons,

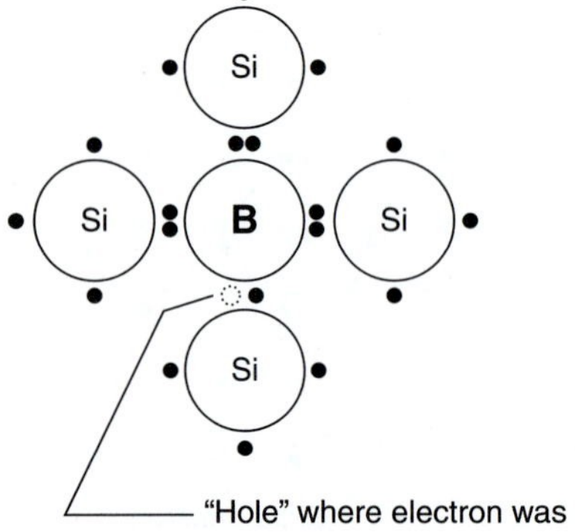

FIGURE 5-11 P-type material formed by combining four silicon atoms with a single boron atom.

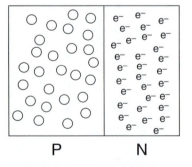

FIGURE 5-12 An example of a piece of P-type material joined to a piece of N-type material.

which results in the gradual mixing of the two materials into specific PN junctions.

When the piece of P-type silicon is placed in contact with a piece of N-type silicon, the diffusion of electrons occurs from the region of high electron concentration (the N-type side of the junction) into the region of low electron concentration (P-type side of the junction). When the electrons diffuse across the PN junction, they combine with holes on the P-type side. The diffusion of electrons and holes does not continue indefinitely because charges build up on either side of the junction and create an electrical field. The region where electrons and holes have diffused across the junction is called the depletion region because it no longer contains any free electrons or free holes as carriers.

5.10 CONNECTING THE PN JUNCTION TO AN EXTERNAL LOAD

Electrical contacts called *electrodes* are connected to both the N-type and P-type sides of the solar cell. When these electrodes are connected to an external load called a *load resistor*, the electrons that are created on the N-type material and have been "collected" by the junction are swept onto the N-type material side, and they may travel through the load resistor as current and continue through the wire until they reach the P-type material where they will complete the circuit.

Figure 5-13 is an example of this equivalent circuit, which shows the PN junction as a source of load current and a diode. A diode is an electronic component that allows current to flow in only one direction in a circuit. The direction of current flow is against the direction the arrow is pointing. The source of the load current is a circle with an arrow pointing upward, and next to it is the diode that is symbolized by an open arrowhead that is pointed downward to a line. The next component connected in parallel with the current source and the diode is a resistor that is identified as R_{SH} and it represents the internal parallel or shunt resistance of the PN junction. The current that flows internally through the diode is identified as I_{SH}. The last resistor in the circuit that is identified as the internal series is identified as R_S. The positive and

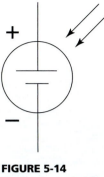

FIGURE 5-14
Electronic symbol for the photovoltaic cell.

negative terminals for the equivalent circuit are also identified in the diagram.

The electrical symbol for the photovoltaic cell is shown in Figure 5-14. It is a circle with two arrows pointing into the circuit that represent light going into the photovoltaic cell. The top of the photovoltaic cell is identified with a plus sign (+) that represents the positive terminal of the component, and a minus sign (-) that represents the negative terminal of the cell.

5.11 BASIC EXPLANATION OF HOW A SILICON PV CELL PRODUCES ELECTRICAL POWER WHEN LIGHT STRIKES IT

Part of the energy from the sun reaches the earth in the form of photons in electromagnetic waves, which we see as visible light waves. When visible light reaches the surface of the PN silicon junction that makes up the silicon solar PV cell, several things happen. The lower-energy photons in the visible sunlight generally will pass directly through the silicon PV cell. Other photons will be reflected off the surface of the PV cell, and the remaining photons will be absorbed by the silicon PV cell. When the photon is absorbed, its energy is given to an electron in the valence band, and it provides enough energy to excite the free electrons in the N-type material into the conductance band where it is able to move around inside the semiconductor. The *conductance band* is a place inside the PN junction where electrons move when they are freed up from their shells. Once electrons make it to the conduction band from their shells, they are free to move as electrical current flow. Figure 5-15 shows the photons striking the side of the PV cell that has the N-type material and the electron flow is out of the N-type material to the place where electrical power is used (at the load) or on the grid or to the battery and then the flow returns back to the P-type material to complete the circuit. The electrical current from the PV cell is DC (direct current) electricity.

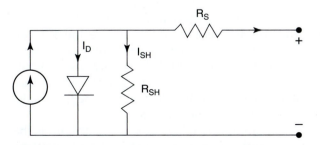

FIGURE 5-13 Photovoltaic cell equivalent circuit.

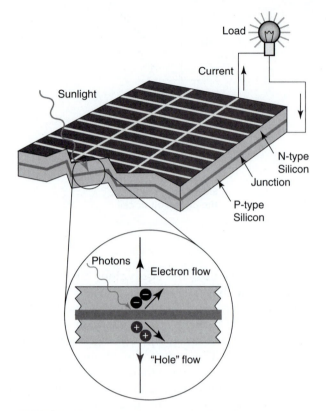

FIGURE 5-15 Current flow from the PV cell when sunlight strikes it. (RESLAB)

5.12 GALLIUM ARSENIDE (GaAs) PHOTOVOLTAIC CELLS

Another type of photovoltaic cell is made from gallium and arsenide. If you refer back to the periodic table found in Figures 5-2 and 5-5, you can see gallium (Ga) is element number 31 and is found in the column for family IIIA (3A). Arsenide (As) is element number 33 and is found in the column for family VA (5A). Gallium arsenide is used for the production of high-efficiency solar cells. It is often utilized in concentrated PV systems and space applications. Its efficiency is up to 25% 28% at concentrated solar radiation. Special types of cells have efficiencies of more than 30% when they are subjected to concentrated sunlight that is hundreds of times the intensity of standard sunlight.

Gallium arsenide material is much better than silicon material to make high-efficiency solar cells.

The energy gap between the gallium and arsenide material is ideal for single junction solar cells and is much better than the energy gap found in silicon PN material. The energy gap is representative of the amount of energy required to get electrons to move into conduction. The smaller the energy gap, the more efficient the material is. Also the gallium arsenide PV cells are not sensitive to heat buildup on the cells, as silicon PV cells are. When solar energy strikes a silicon PV cell, not only

does the sunlight make it to the PV cell, but the heat from the sunlight also gets to the silicon PV cell. Heat causes the cell to have lower efficiency since it reduces the voltage output of the cell, and over time the heat builds up in the cell and causes it to begin to degrade in its ability to convert light into electrical energy through the silicon PN material.

It is also important to understand that the amount of gallium arsenide needed for a PV cell is far less than that needed to produce the same amount of electrical power as the silicon material. For example a gallium arsenide cell is only about 1–2 microns thick, whereas a comparable silicon PV cell is about 100 microns thick. This means that 100 times as much silicon is needed to produce the same amount of electrical power as the amount of gallium arsenide. Since the gallium arsenide is more expensive, and less available, it is mainly used for space applications where solar PV panels are used to produce electrical power for space craft such as the international space station (ISS).

Another difference between silicon PV cells and gallium arsenide cells is the way the gallium arsenide PN junction is formed. With silicon PV cells, the PN junction is formed by treating the top of a P-type silicon wafer by adding N-type silicon material. When gallium arsenide PV cells are manufactured, all the active parts are made by growing sequences of thin, single-crystal layers on a single-crystal substrate. As each layer is grown, it is doped in different ways to form the PN junction and to control other aspects of cell performance.

5.13 CADMIUM TELLURIDE (CdTe) PHOTOVOLTAIC CELLS

Another type of photovoltaic cell material is made from cadmium and telluride (CdTe). If you refer back to the periodic table shown in Figure 5-2, you can see that cadmium is element number 48, and it can be found in the column IIB (2B), which means that its outer shell has two electrons. Telluride comes from the element tellurium, which is element 52 and can be found in the column for family VIA (6A), which means it has six electrons in its valence shell. Cadmium and telluride are placed on a thin-film material by a process called deposition or sputtering. The thin-film material is useful in that it can conform to a variety of applications to match the curvature of different products where the PV cell is installed. These processes are a promising low-cost solution for photovoltaic applications. The manufacturing process is in its infancy and many people believe that its advantages of low cost and flexibility of the final material will be realized in the near future.

The main disadvantage of using cadmium is that it is a poisonous material. It must be tightly controlled when it is used in the production process so that lab technicians are not exposed to this poisonous material. Once the PV cell is made, the cadmium is more stable and handling the solar cells is much like handling a nickel-cadmium (NiCad) battery and does not pose a threat to humans. The efficiency of the cadmium telluride PV cells is approximately 8%, but it is competitive because of its lower manufacturing costs. This means that almost double the amount of CdTe solar cells must be installed to get the same amount of DC electrical power from the PV cells as compared to similar silicon PV cells. Since the CdTe is much less expensive to manufacture, the final cost per watt of DC electrical power produced makes the CdTe slightly less expensive than the equivalent silicon PV installation.

Figure 5-16 shows the material used to make the CdTe photovoltaic cell. The material is deposited on a back contact material that is shown at the top of the stack. The P-type material in this solar PV cell is the CdTe material. The next layer of material is the N-type material that is made of CdS, a mixture of cadmium and sulfur. The bottom layer is made of tin (Sn) oxide (O_2) and is placed directly on a glass substrate. The glass substrate allows light to pass through it to reach the CdTe photovoltaic material, and the light causes the electrons to break free for electrical current flow.

5.14 COPPER INDIUM GALLIUM (DI)SELENIDE (CIGS) PHOTOVOLTAIC CELLS

Another type of PV cell is made by combining copper, indium, gallium, and selenide. Copper is element number 29 and is found in family 1B; indium is element number 49 and is found in family 3A; gallium is element number 31 and is found in family 3A; and selenide comes from the element selenium, which is element number 34 and is found in family 6A. The combined form of these materials makes a photovoltaic cell that is called $CuInGaSe_2$. Figure 5-17 is a diagram of this type of PV cell. The top layer of the PV cell is made of nickel (Ni) and aluminum (Al). The next layer is made from zinc (Zn) oxide (O) and aluminum (Al), and the next level is made of zinc oxide and iodine (I). The semiconductor layer lies below in the next layer and is made from N-type material cadmium (Cd) sulfur (S). The layer with the P-type material is made of copper (Cu), indium (In), gallium (Ga), and selenide (Se_2). The final two layers are molybdenum (Mo) and glass. Glass is the outermost layer. The positive electrode is connected to the molybdenum layer, and the negative electrode is connected to the top layer of nickel and aluminum (Ni/Al) material.

This type of PV cell is applied on a thin-film material and has an efficiency of up to 17%. The material could become one of the new PV cells, but the complexity of the cell is causing some manufacturing problems that increase the costs of production.

5.15 MASS-PRODUCING SILICON FOR PV CELLS

Slicon PV cells are mass-produced by creating silicon ingots. Figure 5-18 shows two of the types and sizes of ingots that are made. The ingots are produced by manufacturing processes that dope the silicon P-type material and the silicon N-type material. After the ingots are made, a precision computer numerical control (CNC) machining system cuts ingots into thin slices called silicon wafers. Figure 5-19 shows the thin wafers that have been cut from the silicon ingot. These wafers are called cells and the

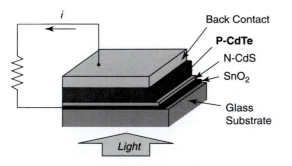

FIGURE 5-16 The makeup of the CdTe solar PV cell.

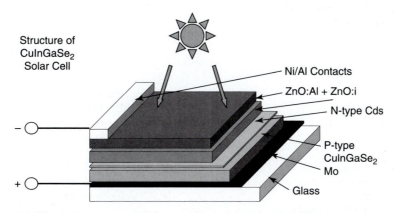

FIGURE 5-17 A copper, indium, gallium, selenide solar cell.

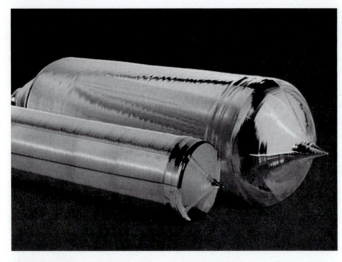

FIGURE 5-18   Silicon ingots   (Courtesy Silfex)

FIGURE 5-19   Silicon wafers cut from an ingot.   (Courtesy Silfex)

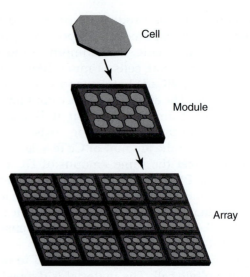

FIGURE 5-20   Combining a single silicon solar PV cell to make a PV module and then a solar PV array.

silicon cells are combined to make a solar module. Figure 5-20 shows how the individual wafers are made into modules and the modules are combined to make a silicon solar PV array. When silicon is made into arrays, the ingots are mass-produced to provide enough wafers for the large numbers needed to make the solar module and then the solar array. Currently there is substantial waste in slicing the wafers and newer CNC cutting processes are being researched. Any material that is wasted during the cutting process can be reused.

Questions

1. Explain what valence electrons are and what the valence shell is.
2. Identify an electron, proton, and nucleus of an atom and the polarity of each.
3. Explain how two materials are combined to make P-type material.
4. Explain how two materials are combined to make N-type material.
5. Explain how P-type material and N-type material are combined through doping to make a silicon PV cell.
6. Identify the elements on the periodic table that are conductors.
7. Identify the elements on the periodic table that are insulators.
8. Identify the elements on the periodic table that are semiconductors.
9. Explain how DC electric current is caused to flow when light strikes the silicon PN junction.
10. Make a sketch of a silicon PV cell and show the layers and how light is absorbed in one of the layers and causes DC current to flow.

Multiple Choice

1. An electron has a _____
   a. Positive charge.
   b. Negative charge.
   c. Neutral charge.
2. An electron _____
   a. rotates around the nucleus in an orbit.
   b. is part of the nucleus and protons rotate around it.
   c. is the positive part of the atom.
3. P-type silicon material is made with material from _____
   a. Family 3 elements.
   b. Family 4 elements.
   c. Family 5 elements.
4. N-type silicon material is made with material from _____
   a. Family 3 elements.
   b. Family 4 elements.
   c. Family 5 elements.
5. Semiconductor material has _____
   a. 3 valence electrons.
   b. 4 valence electrons.
   c. 5 valence electrons.

6

Construction and Manufacturing of Solar PV Panels

OBJECTIVES FOR THIS CHAPTER

When you have completed this chapter, you will be able to

- Explain the steps involved in making a silicon solar panel.
- Explain the steps involved in making a thin-film solar panel.
- Identify the information in a solar panel data sheet.
- Identify typical panel test standards.

TERMS FOR THIS CHAPTER

Back Film Material
Cable Connector
Ethylene Vinyl Acetate (EVA)
   Material Frame
Junction Box

Solar PV Panel
Strings
Technical Data Sheet
Tempered Glass

6.0 OVERVIEW OF THIS CHAPTER

This chapter will explain how the different types of solar photovoltaic (PV) cells are manufactured into photovoltaic panels. In the last chapter you learned how different types of PV cells were manufactured from solid-state materials. In this chapter you are going to see how different cells are grouped together in strings and the strings are used to manufacture complete solar panels that are ready to be mounted in the field. This chapter will explain the steps in manufacturing several of the most typical types of solar panels. Since each of these panels uses a different type of material and is designed for different applications, the steps to manufacture a completed panels will be slightly different.

This chapter will also provide detailed data sheets for typical solar panels and explain how the data for these sheets are interpreted and used when installing a panel. Chapter 11 will go into more detail about installing the panels on rooftops and on ground-mounted systems. This chapter will provide detailed diagrams of how the panels are assembled, and then how they are installed and wired together.

6.1 PANEL TEST STANDARDS

The International Electrotechnical Committee (IEC) International Standard 61215 for photovoltaic (PV) modules (panels) relates to crystalline silicon terrestrial PV modules. Design qualifications and approval tests are used to standardize the manufacturing standards for many solar PV panels.

TABLE 6-1	Generic standards tests for PV modules.

Visual inspection

Maximum power determination

Insulation test

Measurement of temperature coefficients

Measurement of nominal operating cell temperature (NOCT)

Performance at STC and NOCT

Outdoor exposure test

Hot-spot endurance test

UV preconditioning test

Thermal cycling test

Humidity-freeze test

Damp-heat test

Robustness of terminations test

Wet leakage current test

Mechanical load test

Hail test

Bypass diode thermal test

Typical generic categories of tests for solar panels are listed in Table 6-1. The IEC tables for tests are available through IEC standards. IEC standards can be purchased at IEC's website at http://www.iec.ch/.

In the IEC protocol each of the tests is spelled out in the complete standard. The first set of tests is for the exterior of the solar panel. Another test determines how well the panel withstands ultraviolet exposure. Another set of tests determines how well the panel can withstand temperature extremes; humidity; and moisture from rain, snow, and ice under freezing conditions. Damp heat tests determine how well the panel will hold up as it is heating up with moisture. Yet another test checks the termination connectors and determines how well they hold up under various conditions, including checking for any current loss due to moisture integration into the terminal connection. A mechanical load test is used to determine how much dead weight the glass on the top of the panel can withstand before it breaks. Another test subjects each panel to hailstones of various sizes to determine at what point the PV panel becomes damaged, and eventually what size hail will cause the glass to break or shatter. The final test is a bypass diode thermal test; the job of this diode is to protect the cell and bypass current when the output voltage of various solar panels is not the same. Solar panels must meet a variety of other tests and standards in order to be sold in the American market and in certain states. You can check the Internet for your area to see what standards the panels must meet.

6.2 MAKING A RIGID FRAME SOLAR PANEL

This section will explain how a rigid frame solar panel is manufactured. Figure 6-1 shows a typical panel that is complete and ready to install. Figure 6-2 shows the sections of the same panel labeled "a" through "h." This section will explain how the different layers are put together to manufacture solar PV panels.

6.2.1 Tempered Glass

The top section of the solar panel shown in Figure 6-2 consists of a sheet of *tempered glass* (labeled "a") that covers the entire solar panel and all the material is underneath of it. The tempered glass is designed to allow light to pass through it but to keep out rain, snow, sleet, dirt, and anything else that might harm the system. The tempered glass is designed to be rugged enough to withstand some damage from hail or other debris that might strike it, but this glass can be fractured if the object is large enough. If the tempered glass cracks or fractures, it would allow moisture to get inside the complete panel and cause damage when the moisture condenses. The outside surface of the tempered glass will eventually gather dirt and dust over time and must be cleaned, to keep the PV panel operating at its rated conditions. In most cases rainfall will occur often enough to clean the tempered glass on the panels. In regions of the country where rainfall is limited, maintenance crews may have to perform glass-cleaning task periodically to clean the dirt and grime off the tempered glass surface.

The tempered glass on the newest panels is cerium free, which allows the glass to have high transmittance.

FIGURE 6-1 A complete solar PV panel that is ready for installation. (Courtesy NREL and U.S. Department of Energy.)

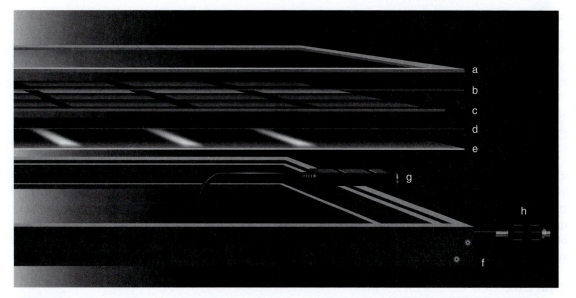

FIGURE 6-2   The segments of a solar panel and how it is manufactured: (a) tempered glass, (b) ethylene vinyl acetate (EVA) film, (c) solar cells (strings), (d) EVA film, (e) back film, (f) frame, (g) junction box, and (h) connection.

A high transmittance value is important because it allows more sunlight to reach the solar cells. This glass enables the solar panel to have a higher efficiency rating. Traditionally cerium has been added to glass to help it absorb more ultraviolet rays. This is a desirable feature for glass that is made for windows, but when cerium is added to glass that is used in solar panels in high concentrations, it tends to break down and discolor, blocking some of the solar energy from passing through the glass to the solar PV cells. An antireflective coating is placed on the outer surface of the glass, and this coating shuts out solar rays that reflect off solar cells, which makes the cell more efficient. When too much sunlight reflects off the surface of the glass for the solar panel, it reduces the amount of solar energy that is absorbed and turned into electrical energy.

6.2.2 Ethylene Vinyl Acetate Material

The next material to be added to the solar panel consists of two sheets of *ethylene vinyl acetate* (EVA) *material.* The EVA material is a sheet of clear vinyl acetate that will become laminated to the solar cell strings on each side to protect them from moisture or other harmful elements. In the diagram of the solar panel in Figure 6-2 the EVA material is identified with the letters "b" and "d," and the EVA seals the solar strings between the two sheets. The solar cell strings are covered with one sheet from below and one sheet from above, which are laminated during the manufacturing process.

6.2.3 Solar Cells (Strings)

In Chapter 5 you learned that individual solar PV cells are made by combining P-type and N-type semiconductor materials into individual cells. These cells are put together in multiple groups called *strings* that are made into larger sheets before they are assembled into the solar panels. The sheet is identified as letter "c" in Figure 6-2. Each string has a specific voltage rating that the panel will produce when solar light strikes the cell. The strings are cut into sheets that are the same size as the tempered glass and they will fit into the solar panel frame. At this point in the manufacturing process the sheets that have the strings also have positive and negative terminals. Later in the manufacturing process the cables that have connectors on the ends will be connected to these terminals. The sheet of solar cell strings is laid on one of the pieces of the EVA material, and then another sheet of EVA material is laid on top of it. Once the solar cell strings are sandwiched between the two pieces of EVA material, the EVA material is laminated to provide a protective skin over the cells.

6.2.4 Back Film

The next part of the manufacturing process involves putting the laminated sheet of solar cells on a back film. The back film is a material that is dark-colored is known as polyethylene terephthalate (PET) film. The back film is identified as letter "e" and it provides support for the material resting on it. The back film is the backside of the solar panel and is exposed to anything that may scratch or tear it during the installation process. It is very important to handle the solar panels with as much care as possible so the back film is not damaged during the installation process.

6.2.5 Frame

The *frame* for the solar panel is made from high-strength aluminum. It is identified by the letter "f" in Figure 6-2.

The aluminum frame encases the entire group of components that make up the solar panel. Aluminum is used because it provides strength yet is lightweight. The frame has all of the mounting holes to attach brackets and other hardware to keep the solar panel mounted in place. When all of the components for the solar panel have been assembled and laminated, the aluminum frame is wrapped around the solar panel to cover the four edges. The frame may be one long piece, or it may be four separate pieces, which would be one for each edge. The frame may be one long piece of the panel. If the frame is made from one long piece, it will be folded around the cell and have only one final location where the corner segments are joined together. If the frame is made of four separate pieces, each corner must be joined together either by welding or some other method of securing it.

6.2.6 Junction Box

The *junction box* is identified in Figure 6-2 as letter "g," and it is where the electrical connections are made between the positive and negative terminals of the solar PV strings and the wiring that allows the solar panels to be connected together. The positive and negative terminals from the PV cells are bent at 90° so they are pointing upward from the surface of the solar PV strings sheet. The junction box is placed directly over these terminals and adhesives and sealants are used to mount the box to the sheet so that it will not be affected by moisture. The terminals can be connected inside the junction box with a solder connection or a solder-less connection that makes a permanent connection between the PV strings and the cable. A diode is built right into the junction box and it is connected between the positive and negative terminals.

This diode is a bypass diode and it allows current to bypass the cell if it becomes shaded or produces lower power for any reason. If the cell is not bypassed it would otherwise cut off the current flow completely for that cell and any cells connected in series with it. The bypass diode complies with the IEC61215 second edition bypass diode thermal test.

After the cable connections are made, sometimes the entire junction box is filled with potting material, which is a sealant material that ensures that the water cannot permeate the junction box. The potting material fills the entire volume of the junction box, so there is no room for any moisture to seep into it. Figure 6-3 is a sketch of a typical junction box. It has a metal barrier cover, a resin cover, and finally a metal barrier cover. This makes a permanent connection between the cable and the PV strings.

6.2.7 Connector

The MC4 connector is shown as letter "h" in Figure 6-2. A drawing of a typical *cable connector* is shown in Figure 6-4. One of the connector ends is male and the other is female, which allows them to be plugged into each other. Basically the cable connected to the positive terminal of the strings of PV cells will be the male end and the female end is connected to the negative terminal. When the male plug end is inserted into the female plug end, it will connect the positive terminal of one solar panel with the negative terminal of the next panel, which means they are connected in series. Splice connectors are available to connect multiple solar panel cables on an installation. Figure 6-5 shows how panels are connected together in series and parallel with the connectors.

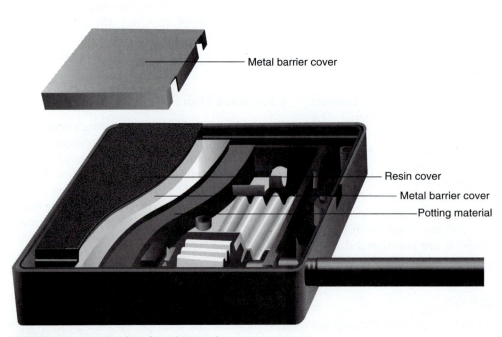

FIGURE 6-3 Junction box for solar panel.

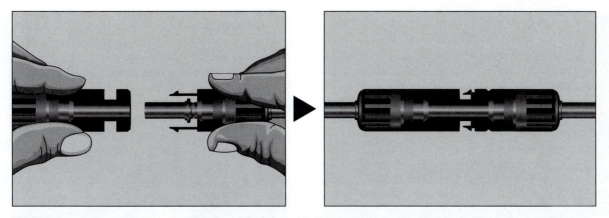

FIGURE 6-4   Solar cable connectors on the end of the cable that connects to the PV strings. The male end of the cable is connected to the positive terminal of the PV string.

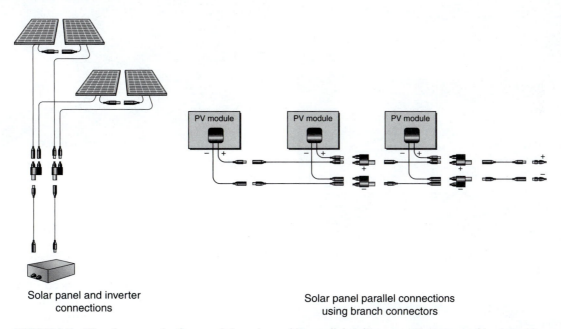

Solar panel and inverter
connections

Solar panel parallel connections
using branch connectors

FIGURE 6-5   Ways to connect solar panels in series and in parallel. Splice connectors are used to connect the panels in series and parallel.

6.3 MAKING SOLAR PANELS FROM POLYCRYSTALLINE CELLS

The process of making a solar panel from polycrystalline cells is a fairly simple multistep process. Some companies have several additional steps in their assembly process to ensure that the panels maintain the most effective conversion of light to electricity. They may also add some products into the assembly to ensure that each panel will have the longest life possible. This section will explain the steps in automated manufacturing processes that are followed to make high-quality solar panels using polycrystalline PV cells. Some companies use robots and other automated equipment to assemble the parts into a completed solar panel. In some smaller companies more of the assembly work is completed by hand. Figures 6-6 and 6-7 show solar panels in various stages of assembly.

6.3.1 GLASS COVERING FOR THE PANEL

The first step in creating a solar panel is to select the glass covering for the solar modules. This glass is specifically designed for the outermost layer of the solar panels and to let the fullest amount of light pass through, as well as to be resistant to breaking down under the constant presence of UV rays from sunlight. An important test is to stress test the glass to ensure that it will hold up as the outermost covering of the solar panel. In the most automated assembly process, robots are used to pull each sheet of glass from a stack and stress test it. After the robot has stress tested the glass to ensure that it meets the standards, it presents the glass to a cleaning station where the glass is cleaned with water treated to prevent bacteria growth between the sheets of materials inside the completed PV panel. Some companies apply a liquid UV

FIGURE 6-6 Solar panels in various states of assembly. (Courtesy Schott Solar PV Inc.)

FIGURE 6-7 Solar panels during the assembly process. (Courtesy Schott Solar PV Inc.)

prevention, whereas others panels use other materials, for UV protection. UV prevention is important to extend the life of the panel so that the sunlight does not destroy the panel over time. Figure 6-8 shows solar panels in the early stages of the assembly process.

FIGURE 6-8 Solar panel in the early stages of production. (Courtesy First Solar, Inc.)

6.3.2 Laying the Intermediate Film

The next step in the process of making a solar module is laying the intermediate film. The intermediate film is a translucent film. It can be placed on the glass plate by a robot if the process is automated or it can be placed on the glass manually. In the automated process the film is cut to size and it is handled by a robot fixture with suction cups. The film generally comes from a large roll, and the robot and automated equipment pull the continuous sheet of film off the roll and cut it to size. In the manual process the technician pulls the film off the roll and a slitter blade cuts it to the exact dimension. The technician then precisely locates the film so it covers the glass. The film ensures that a maximum degree of cross-linking between the PV cells and the glass plate occurs. This makes it possible to get the most efficient transfer of light energy into the PV cells.

6.3.3 Combining PV Cells into Stringers

In this step a thin sheet that has PV cells in groups called the stringer are strategically placed together into sheets. The stringer consists of strings of PV cells that are interconnected so that electrical current can pass through them easily when light strikes the cells and creates the current. All of the material used in assembling the solar

FIGURE 6-9 All materials that are used in assembling the solar panels are thoroughly inspected prior to assembly.   (Courtesy Schott Solar PV Inc.)

FIGURE 6-11 Technician handles solar cells during assembly process.   (Courtesy Schott Solar PV Inc.)

panels is thoroughly inspected prior to use. Figure 6-9 shows a technician inspecting the solar PV material.

The PV strings that are used in the solar panels are the same photovoltaic cells that were discussed in the previous chapter. The size and shape of the sheets of stringers will match the frame and glass of the PV panel. In this process the solar cells are manufactured to very specific standards for the modules. The quality of the stringer sheet is controlled so that it will operate at stated standards throughout its life. If the solar panel manufacturing process is automated, in the next step the robots will use vacuum suction cups to pick up the stringers and place them in the correct location. If the process is manual, technicians select the stringers and place them on the solar panels. Figure 6-10 shows polysilicon cells ready for the assembly process, and Figure 6-11 shows a technician handling photo cells during the assembly process.

6.3.4 Lay Up

The next step in the assembly process is called the lay up. This step involves taking the cells that have been soldered into strings and placing them precisely on the intermediate film and glass plates that have been previously positioned. If this process is fully automated, each string is positioned correctly by a robot or automated positioning system. If the panel is being manufactured manually, the technician will need to ensure that the placement of the stringers on the intermediate film and glass plate is correct.

6.3.5 Bussing

The next step in the process is called bussing. Bussing is the process of soldering the electrical connections to the PV strings to provide the external connections for the solar panels. The electrical connection on the solar panel is the point at which the panel is most likely to fail after years of experience. The better the bussing process is, the better the electrical connections will hold up over the years. In the most automated manufacturing processes bussing is fully automated in a chamber equipped with the latest laser technology, which ensures a solid soldering connection. In this automated system all the connections that need to be soldered are soldered with a laser process, which ensures a high degree of accuracy and a quality soldered connection.

6.3.6 Second Intermediate Film

During the next step a second piece of intermediate film is laid on the backside of the solar cell. The second layer of film is used to provide a particularly high degree of cross-linking. If the process uses manual operations, a technician lays the second layer of film on the backside of the solar cells. If the process is fully automated, a robot places the second piece of intermediate film. The second

FIGURE 6-10 Polysilicon cells ready for assembly.   (Courtesy Schott Solar PV Inc.)

piece of intermediate film is similar to the first in that it is a translucent film. Once the second layer of film is in place, the material is ready to be laminated.

6.3.7 Lamination

During this step the completed modules are laminated. During lamination the cells, films, and glass plate get fused into a solid unit within a vacuum. During the vacuum process the assembly is moved to a station where the film creates a seal around components, and all the air is evacuated through a vacuum process. When the two sheets are laminated, the solar cells, intermediate films, and glass plate get fused into a solid unit and all unwanted air is removed with a vacuum process. This process ensures that the cells are permanently protected from moisture and dirt and will withstand severe conditions throughout their lifetimes. Several laminators are used at the same time to ensure the highest production rate.

6.3.8 Inspection Point

The next step in the process after the sheets have been laminated is to completely inspect each sheet individually to ensure that it meets quality control standards. Quality control technicians analyze and examine the final laminated sheets and look for any defects in the lamination process or in the assembly process to this point. Any panels with defects are removed from the manufacturing process, ensuring that only the highest-quality panels make it through the final steps of assembly. Figure 6-12 shows a quality control technician inspecting a panel.

6.3.9 Junction Box Installation

After the panel has passed a rigorous inspection process, the panels are moved to a station where the junction box is installed. The junction box is placed on the panel; this is

FIGURE 6-12 Technician inspecting solar panel. (Courtesy First Solar Inc.)

where electrical connections are made to the positive and negative terminals of the PV solar cell stringer sheet. If this process is fully automated, a robot places the terminal box so that it is precisely located as it is placed down on the electrical terminals of the stringer sheet. A robot picks up a junction box and moves it to a station where sealant is applied only to the base of the box so that only the base is glued to the module. In the fully automated process the robot moves the box under the sealant dispenser so that the sealant is applied evenly to the base of the box. The sealant will ensure that the junction box fits tightly to the surface and does not allow any moisture or water to get into the box or to the electrical terminals. The robot places the box precisely over the electrical terminals and applies enough pressure to ensure that the sealant is evenly displaced and a seal is created between the box and the module. The sealant performs a second function in that it is also a glue to hold the box to the module permanently. The electronics are located inside the box and the cover is installed later. The contacts are bent precisely in a fully automated process that is monitored by a high-performance video camera. The base of the junction box is attached to the backside of the module. During this process the contacts can be welded or soldered. The most innovative processes use laser welders to permanently secure electrical connections. If the process is completed manually, a technician welds or solders the terminals of the electrical box to the electrical terminals of the solar panel.

6.3.10 Module Frame

After the junction box has been attached, the assembly is nearly complete. The last step in the assembly process is to mount aluminum frames on all four sides of the solar panel. Before the framing is started, the overlaying films are removed. If this step is completed by robots, usually two or more robots work together to encase each solar module within its aluminum frame. The frames are pressed onto the edges of the solar panel and are permanently mounted to ensure that the frame stays in place throughout the life of the panel. The aluminum frame has threaded holes and other holes for mounting hardware to be connected when the solar panel is being installed. The mounting hardware will be slightly different for a roof installation versus a pole installation or ground installation.

6.3.11 Flasher

A crucial manufacturing station is the flasher that is used to shine simulated sunlight on all modules. The actual performance figures are then recorded and documented. This is followed by safety and high-voltage tests. In this process each solar panel is tested with a predetermined amount of light, and the amount of voltage it produces is measured. This test ensures that the completed assembled solar panel produces the amount of electrical DC power that is specified when it receives light. The test is conducted by flashing a specific amount of light directly onto the solar panel for a short amount of time and measuring the amount of electrical DC current that the panel produces. The amount of electrical current the panel produces is compared to the rating, and it must meet or exceed the amount of current in the rating to be allowed to continue to the process for shipping.

6.3.12 Shipping Preparation

During the final step each module receives a label with its serial number, power class, and other important data. Figure 6-13 shows the panel getting the final inspection, and Figure 6-14 shows a technician placing the final stickers on the solar panel. The sticker has the product information and panel serial number on it. The serial number can be used to find relevant data and process parameters of a module. The modules are presorted by power class. A special logistics system reduces the cost of

transportation and storage in the production facility. Top-quality modules stay in storage only a short time before they are shipped out to solar power installation sites across the world.

The panels are stacked in piles so that they can easily be shipped in groups. By stacking the panels on top of one another, they can safely be shipped long distances to the site where they will be installed. Being able to ship multiple panels in one box reduces the cost of shipping. It is important that the panels are placed into the cardboard shipping containers in such a way that they are not damaged during the shipping process.

6.4 CREATING SOLAR PV PANELS USING THIN-FILM AMORPHOUS SILICON

Another type of solar PV panel is a thin-film amorphous silicon panel. This type of solar cell material is not typically found in panels with metal frames as are the previous panel discussed in this chapter. The thin-film material shown in Figure 6-15 is manufactured in 13-inch wide material and is sold in rolls that are 2,400 feet long. Since the material is so thin, it can be installed on surfaces such as roofing material or siding.

The material is manufactured in a long roll, which significantly reduces the manufacturing costs. The solar panel is manufactured as a monolithically integrated material that is mounted on a plastic roll. The plastic material is a durable polymer (polyimide) substrate that results in a thinner panel and lighter weight. The substrate material is typically 1 mil (0.025 mm) thick.

Figure 6-16 shows a cross-sectional close-up of the thin-film PV material. The top layer is a transparent conductor that allows light to move through it. The next layer is made of the P-type material and the

FIGURE 6-13 Final inspection of finished solar panels. (Courtesy Schott Solar Inc.)

FIGURE 6-14 Technician puts final stickers on finished panel. (Courtesy Schott Solar Inc.)

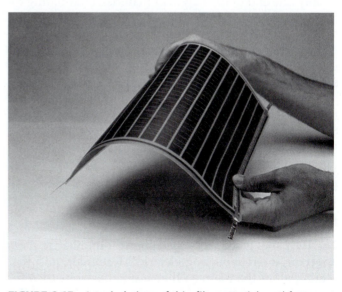

FIGURE 6-15 A typical piece of thin-film material used for a solar panel. (Courtesy NREL and U.S. Department of Energy.)

Cross section close-up

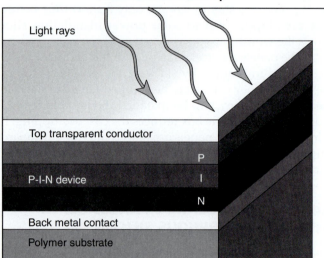

FIGURE 6-16   Cross section close-up of thin-film PV material.

N-type material that are separated by an insulator. The last two layers are a back metal contact that gives support to the thin film. The final layer at the bottom is called the polymer substrate and it provides support at the bottom as well as ensures that the material cannot be penetrated by moisture. The last layer can be made from fabric metal or fiberglass and it can be custom designed for any customer.

The solar cell material is thin-film amorphous silicon that acts as the absorber layer for the solar panel. This type of solar panel uses approximately 1% of the amount of silicon material as the traditional solar panel. Since this type of panel uses silicon P-type and N-type materials, it is cadmium free. The solar material is totally encapsulated in long-lasting film that allows sunlight to shine through it.

Connections for this type of panel need to be made at only one end of the length of material, eliminating the need for cables and connectors every 3 to 4 feet between smaller shorter panels. Since the solar generating material is mounted on one long sheet, it produces electrical power in a similar fashion to a number of traditional solar panels that are wired in parallel.

The thin-film panels can be mounted on a more ridged backing material if the application needs more substance. It can also be specially made into specific building materials such as roofing materials or siding materials.

6.5 GENERAL TECHNOLOGY USED TO BUILD A DUPONT SOLAR PHOTOVOLTAIC PANEL

The PV panel is made by assembling a number of parts, applying interconnecting electrical terminals, and then sealing all of the parts so they are impervious to water infiltration from rain, snow, frost, condensation, or other weather conditions.

One of the basic components to every solar panel are the PV cells that you learned about in the previous chapter. The PV cells can be manufactured from silicon, cadmium, or other semiconductor materials, and they are manufactured as monocrystalline, polycrystalline, and amorphous solar modules. Once the individual PV solar cells are manufactured, they are connected into strings. Most commercial crystalline modules consist of 36 or 72 cells. Solar cells are connected electrically and are typically interconnected with thin contacts on the upper side of the semiconductor material, which can be seen as a metal net on the solar cells. This network of electrical connections ensures that each cell is electrically connected to the others so voltage and current can build as photons strike the PV cells. The wiring interconnection net on the solar cells must be as thin as possible and the finished solar cell layer is placed (sandwiched) between a Tedlar® plate on the bottom and tempered glass on the top. Figure 6-17 shows an example of Tedlar® material, which is made by DuPont. Tedlar® is a flexible polyvinyl fluoride film that is used as a back sheet.

The plate glass on the front side material (superstrate) is usually low-iron, tempered glass. Some special modules, however, use other types of front side materials such as DuPont™ Tefzel® or nontempered glass.

The solar cells have a sheet of ethylene vinyl acetate (EVA) plastic film placed on top of them between the solar cell sheet and the glass, immediately below the solar cell sheet, on the bottom. The EVA plastic sheets are sealed so the solar material is laminated between them. Figure 6-18 shows a typical sheet of EVA material. Many other materials could be used for encapsulation instead of the EVA, such as PVB (polyvinyl-butiral), which is also used in automotive windshields to prevent them from shattering when they are broken. PVP is typically used as encapsulation material in transparent modules and EVA is used for encapsulation of cells in standard modules.

FIGURE 6-17   Tedlar® polyvinyl fluoride films are in high demand for use in durable, weather-resistant back sheets for PV modules.   (Courtesy DuPont.)

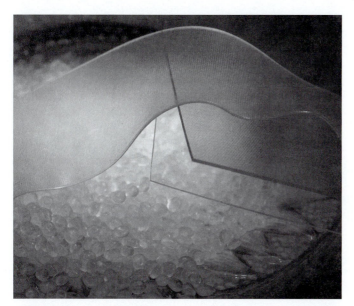

FIGURE 6-18   EVA flexible and rigid sheet products.   (Courtesy DuPont.)

When all of the parts are fully assembled, the final step of the production process is to enclose all the components in a frame. Typically the frame is aluminum, plastic, or stainless steel. The frame must be sturdy enough to hold all the PV panel components safely on the roof or pole so that they are not damaged or moved by the wind. The frame must also be lightweight and weatherproof so that it does not rust or deteriorate in the weather or sunlight.

Figure 6-19 shows two types of solar panels from the DuPont company. The diagram on the left shows all of the material layered together to make a tradition silicon crystalline solar panel and the diagram on the right shows the material to make a thin-film solar panel. DuPont uses all of its own material in making its solar panels. The layers are identified with the letters "A" through "F." The top layer is called the front sheet material and is identified by letter "A." This material is a clear sheet called Teflon film, and it acts like the pane of glass that is used in some solar panels. "B" is the EVA film

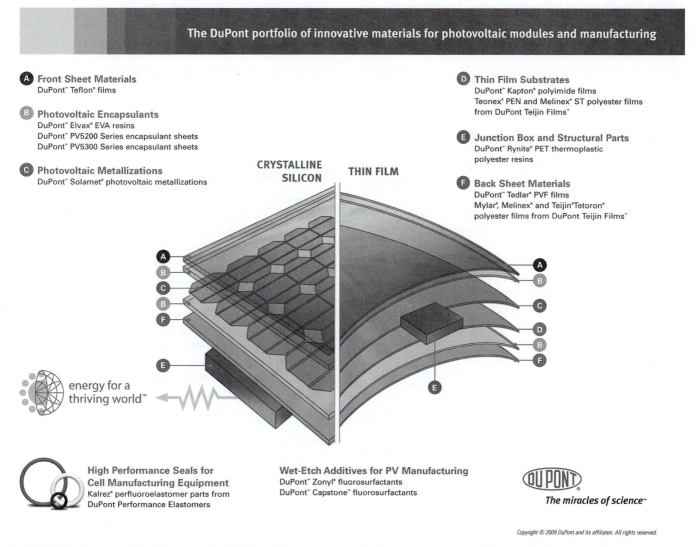

The DuPont portfolio of innovative materials for photovoltaic modules and manufacturing

A Front Sheet Materials
DuPont™ Teflon® films

B Photovoltaic Encapsulants
DuPont™ Elvax® EVA resins
DuPont™ PV5200 Series encapsulant sheets
DuPont™ PV5300 Series encapsulant sheets

C Photovoltaic Metallizations
DuPont™ Solamet® photovoltaic metallizations

D Thin Film Substrates
DuPont™ Kapton® polyimide films
Teonex® PEN and Melinex® ST polyester films
from DuPont Teijin Films™

E Junction Box and Structural Parts
DuPont™ Rynite® PET thermoplastic
polyester resins

F Back Sheet Materials
DuPont™ Tedlar® PVF films
Mylar®, Melinex® and Teijin®Tetoron®
polyester films from DuPont Teijin Films™

CRYSTALLINE SILICON   THIN FILM

energy for a thriving world™

High Performance Seals for Cell Manufacturing Equipment
Kalrez® perfluoroelastomer parts from
DuPont Performance Elastomers

Wet-Etch Additives for PV Manufacturing
DuPont™ Zonyl® fluorosurfactants
DuPont™ Capstone™ fluorosurfactants

DUPONT
The miracles of science™

FIGURE 6-19   The formation of a crystalline silicon solar panel and a thin-film solar panel manufactured by DuPont.   (Courtesy DuPont.)

material that is used on top and bottom of the solar cells to encapsulate them so that moisture and other types of water do not get into them. You can see on the right side of the figure that two sheets of the EVA film are used and both are identified by the letter "B."

"C" shows the PV cells on a sheet called PV metallizations. The DuPont solar PV cells are manufactured from crystalline silicon in a single sheet called Solamet.

"D" shows the film that sits below the sheet of PV material. In the DuPont ^{TM} solar panel, this film is called Kapton®, and it is a polyimide film. The film could also be a product called Teonex® "PEN and Melinex®" ST - polyester films from DuPont Teijin Films ^{TM}. One sheet of one of these types of films is pressed together with the top sheet of EVA material to encapsulate the sheet of solar PV material so that it does not get damaged from moisture.

The junction box has electrical terminals that are connected to the electrical terminals of the solar PV sheet. The junction box is attached securely to the sheet. The junction box is shown with the letter "E" and has the electrical cable that is used to connect solar panels in series or parallel. The electrical cables from the solar panels are connected to the positive and negative terminals of cell so that the DC electrical power can flow from the solar panels to the electrical circuit. This may be a bank of batteries, or it may be the input terminals of an electronic inverter, which will convert the DC voltage to AC voltage. The junction box on this type of solar cell is made of Rynite® PET plastic. PET plastic is made from *polyethylene terephthalate* and is used because of its strength and its thermostability. PET plastic is also recyclable when the solar panel is recycled.

In the traditional crystalline silicon panel the junction box is mounted at the very bottom of the panel, as shown on the left diagram in Figure 6-19. The box is mounted on the bottom of the solar panel, but the electrical connections are made between the positive and negative terminals on the PV sheet and the electrical terminals in the junction box. In the thin-film solar panel the junction box is connected directly to the solar PV sheet, and the electrical connection is made to the positive and negative terminals on the sheet.

The final layer, called the back layer or the back sheet, is made by using DuPont ^{TM} "Tedlar" PVF films such as Mylar®, Melinex®, and Teijin® Tetoron® polyester films. These products are all made by DuPont's Teijin Films ^{TM} division. The back sheet provides a protective covering on the bottom section of the solar panel. On the thin-film type panel the back sheet also provides some support for the remaining parts and layers of the panel.

Thin-film materials are manufactured so that they are more pliable than rigid solar panels. Figure 6-20 is an example of this type of thin solar material that can be shaped to virtually any form on a roof or other

FIGURE 6-20   An example of thin-film solar PV material. This material can be applied directly to a roof or siding during the construction process of a building or home. (Courtesy DuPont.)

mounting surface. One of the strengths of this material is that it can be built right into a rooftop during the construction process. It can also be laid out in long sheets, which means that a cable connection does not need to be made every 3 or 4 feet as with rigid solar panels. The manufacturing process for the thin-film sheets is also simpler than that for the multilayered process of making rigid solar panels, so the thin-film will be less expensive to produce. However, the efficiency of the thin-film material is not as good as that of the rigid type solar panel.

6.6 TECHNICAL DATA SHEETS

After a solar panel has been manufactured, a data label is applied to the back of the module and a *technical data sheet* is produced and is packed with the panel and sent to the end user. The data sheet shows all of the detailed parameters about that solar panel. The most important module parameters include the peak power, the rated voltage and rated current, a short circuit current, an open circuit voltage, and a nominal voltage, which are all established based on a standard test condition of 1,000 watts per square meter (w/m^2) or irradiance (light shining on the panel), with an air mass of 1.5 and a cell temperature of 77°F (25°C). This test standard is used for every model of solar PV panel that is manufactured.

Figure 6-21 shows a typical data sheet. This data sheet is for a SunPower 215 solar panel and shows the following:

Peak Power—P_{max} 215 W

Rated Voltage—V_{mp} 39.8 V

Rated Current I_{mp}—5.4 A

Open Circuit Voltage—V_{oc} 48.3 V

Short Circuit Current—I_{sc} 5.8 A

Maximum System Voltage—1000 V/600 V

SUNPOWER·

215 SOLAR PANEL
EXCEPTIONAL EFFICIENCY AND PERFORMANCE

Electrical Data

Measured at Standard Test Conditions (STC): irradiance of 1000/m², air mass 1.5g, and cell temperature 25° C

Peak Power (+/-5%)	Pmax	215 W
Rated Voltage	Vmp	39.8 V
Rated Current	Imp	5.40 A
Open Circuit Voltage	Voc	48.3 V
Short Circuit Current	Isc	5.80 A
Maximum System Voltage	IEC, UL	1000 V, 600 V
Temperature Coefficients		
	Power	–0.38% /°C
	Voltage (Voc)	–136.8 mV/°C
	Current (Isc)	3.5 mA/°C
Series Fuse Rating		15 A
Peak Power per Unit Area		173 W/m², 16.1 W/ft²
CEC PTC Rating		198.5 W

Mechanical Data

Solar Cells	72 SunPower all-back contact monocrystalline
Front Glass	3.2 mm (1/8 in) tempered
Junction Box	IP-65 rated with 3 bypass diodes
Output Cables	900 mm length cable / Multi-Contact connectors
Frame	Anodized aluminum alloy type 6063
Weight	15 kg, 33 lbs

IV Curve

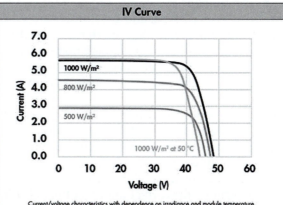

Current/voltage characteristics with dependence on irradiance and module temperature.

Tested Operating Conditions

Temperature	– 40° C to +85° C (–40° F to +185° F)
Max load	50 psf (2400 Pascals) front and back
Impact Resistance	Hail – 25 mm (1 in) at 23 m/s (52 mph)

Warranty and Certifications

Warranty	25 year limited power warranty
	10 year limited product warranty
Certifications	IEC 61215 , Safety tested IEC 61730; UL listed (UL 1703), Class C Fire Rating

Dimensions

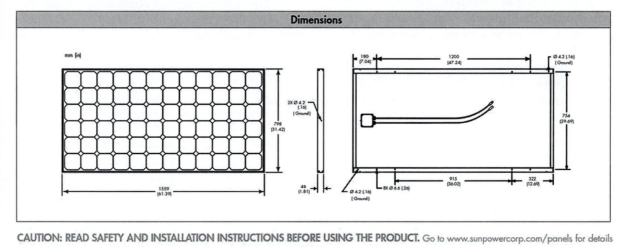

CAUTION: READ SAFETY AND INSTALLATION INSTRUCTIONS BEFORE USING THE PRODUCT. Go to www.sunpowercorp.com/panels for details

About SunPower

SunPower designs, manufactures and delivers high-performance solar electric technology worldwide. Our high-efficiency solar cells generate up to 50 percent more power than conventional solar cells. Our high-performance solar panels, roof tiles and trackers deliver significantly more energy than competing systems.

Document #001-42024 Rev **

♻ Printed on recycled paper

www.sunpowercorp.com

FIGURE 6-21   Data sheet for a SunPower 215 solar panel.   (Courtesy SunPower.)

6.7 OTHER TYPES OF DATA SHEETS FOR A SOLAR PANEL

Along with manufacturing information each solar panel must have a data sheet. Table 6-2 shows a typical data sheet for two similar polycrystalline silicon panels that are manufactured by Mitsubishi. The first item in the table is the model number and name. The next two rows of information are the cell type, and the number of cells. This indicates the number of PV cells that are used to make the panel and the amount of wattage that each can provide. You can see in this data sheet that the wattage ranges from 130 W for the largest panel, and 120 W for the smaller panel.

The next four rows of data provide information about the voltage and current for the system. This includes the open circuit voltage (V_{oc}), the short circuit current (I_{sc}), the maximum power voltage (V_{mp}), and the maximum power current (I_{mp}). This information helps the designers determine the correct number of panels for a residence or commercial application. By calculating the amount of voltage each panel can provide, the designer can determine how many panels to put in series to get the maximum voltage. Additional panels can be put in parallel to increase the amount of current the panels will produce.

The next row of information provides the normal operating cell temperature in degrees centigrade. This is the normal temperature when the cell is producing maximum voltage. If the temperature gets higher than this level, then designers understand that the cell will produce less voltage and become less efficient.

The next row of information indicates the maximum voltage this PV panel can be exposed to. These panels' maximum voltage is 1,000 V DC. It is possible for voltage to back feed from the other sources into these panels, so it is important to understand what the maximum voltage any one panel can sustain. In the United States the voltage after environmental corrections cannot exceed 600 V, and 1,000 V is the limit for other parts of the world.

Some data sources for PV panels will provide information that indicates that each string of the panels should have a series fuse rating and the size of the fuse will be specified. The fuse is provided in the circuit in series with each panel or groups of panels to prevent them from having excessive current back fed from a short circuit between two or more strings. It is important to understand that excessive current will melt wiring and possibly cause a fire, damage the terminal connections, and cause permanent damage to a solar panel. The fuse should open anytime amperage exceeds the 15 A rating, whether the current is normal current that is fed backwards through the cell, to protect the solar panel from any damage due to excessive current. Solar modules are current limited and cannot exceed the amperage shown in their data sheets; the fuse is only for a short circuit condition when parallel string amperages could combine and back feed into the string.

The next two rows of information provide the dimensions for each panel and the weight for each panel. Notice that the weight is provided in pounds so that you can quickly determine the maximum weight a set of panels will have on a roof or other mounting brackets.

The information for some solar PV panels will also indicates the type of output terminal connectors that are provided with each panel. It is important to match the

TABLE 6-2	Standard data sheet for typical solar panels.	
Panel names	**Solar panel 1**	**Solar panel 2**
Type of cells	Polycrystalline	Polycrystalline
Number of cells	36 cells in series	36 cells in series
Fuse rating	15 A (300 V)	15 A (300 V)
Short circuit current (Isc)	8.4 A	7.2 A
Open circuit voltage (V_{OC})	22.1 V	21.6 V
Power rating (P_{max})	130 W	120 W
Tolerance for power rating	+ 10% −5%	+ 10% −5%
Maximum power voltage (Vmp)	17.3 V	17.1 V
Maximum power current (I_{MP})	7.51 A	7.32 A
Normal operating cell temperature	47 degree C	47 degree C
Maximum system voltage	1000 V DC	1000 V DC
Module efficiency	12.4%	11.6%
Dimensions	60 × 24 × 1.9 inches	60 × 24 × 1.9 inches
Weight	28.2 lbs	28.2 lbs
Packaging	2 panels per carton	2 panels per carton

terminal connector with other similar terminal connectors so that the installation will go smoothly and the panel wiring can be completed by simply snapping the terminal connections from one panel to the next. This information is also useful for the final connections at the combiner box or other terminal location, where the wiring from the panels must be terminated.

Some literature provided with the panels will also provide information that explains how many panels are in each carton, and what certificate process the panel was tested to. One type of test that panels are tested to is IEC standard 61215 edition 2, and the amount of static load that the panel is tested to is 2,400 Pa (Pascals). Other manufacturers may include additional information but this minimal information must be provided with each panel that is shipped.

Questions

1. Explain the steps involved in making a silicon solar panel.
2. Explain the differences between a thin-film solar panel and a rigid frame solar panel.
3. Explain the steps involved in making a thin-film solar panel.
4. Identify the information in a solar panel data sheet.
5. Identify typical panel test standards.

Multiple Choice

1. What is the material on the top of the rigid frame solar panel?
   a. Back film material.
   b. Tempered glass.
   c. EVA material.
   d. String of solar cells.
2. The material used in a rigid frame solar panel that is used to seal the solar panel to keep water out is _____
   a. back film material.
   b. tempered glass.
   c. EVA material.
   d. string of solar cells.
3. What information is not found on a typical data sheet?
   a. Peak Power—P_{max} 215 W.
   b. Rated Voltage—V_{mp} 39.8 V.
   c. Rated Current I_{mp}—5.4 A.
   d. Cost of the solar panel.
4. The material on the bottom of the rigid frame solar panel is _____
   a. back film material.
   b. tempered glass.
   c. EVA material.
   d. string of solar cells.
5. What is the function of the junction box on a rigid frame solar panel?
   a. To provide an electrical connection between the solar cell and the connecting cables.
   b. To provide a point where the solar cell is laminated.
   c. To provide a point in the solar cell where the layers of glass are glued together.

Electricity for Solar Energy: Basic Electrical Principles Used for Solar PV Systems

OBJECTIVES FOR THIS CHAPTER

When you have completed this chapter, you will be able to

- Describe voltage.
- Explain current.
- Describe resistance.
- Understand Ohm's law and how it is used to calculate volts, ohms, and amperes.
- Explain the function of an electrical wiring diagram.
- Explain the function of an electrical ladder diagram.
- Explain why the voltage at each branch of a parallel circuit is the same as the supply voltage.
- Explain the symbols M, k, m, and μ and provide an example of how each is used.
- Discuss the advantage of connecting switches in series with a load.
- Explain infinite resistance.
- List two things to be aware of when making resistance measurements with an ohm meter.
- Explain why the voltmeter and ammeter should be at the highest setting for the first measurement of an unknown voltage or current.

TERMS FOR THIS CHAPTER

Coil	Overload
Contactor	Parallel Circuit
Current	Power Circuit
Fuse	Relay
Fuse Disconnect	Series Circuit
Milliampere	Series-Parallel Circuit
Normally Closed	Switch
Normally Open	Transformer
Open Circuit	Voltage
Open Fuse	Voltage Drop
Open Switch	

7.0 OVERVIEW OF THIS CHAPTER

As a technician, you must thoroughly understand electricity so that you are able to troubleshoot the electrical components of systems, such as solar panels, switches, and controls. This chapter will provide you with the basics of electricity, which will be the building blocks for more complex circuits that you will encounter in the field. You do not need any previous knowledge of electricity to understand this material. The simplest form of electricity to understand is direct current (DC) electricity, so many of the examples in this chapter will use DC electricity. It is also important to understand that the majority of electrical components and circuits in residential and commercial systems have equipment that uses alternating current (AC) electricity, so some of the examples in this chapter will explain the differences between AC and DC electricity. This chapter will begin by defining basic terminology such as voltage, current, and resistance and by showing simple relationships between them.

7.1 BASIC ELECTRICITY AND A SIMPLE ELECTRICAL CIRCUIT

It is easier to understand the basic terms of electricity if they are connected with a working circuit with which you can identify. As an example, we will use the motor that moves a furnace fan for a solar heating application at a residence or commercial location. The motor must have a source of voltage and current. *Current* is a flow of electrons, and *voltage* is the force that makes the electrons flow. This is just an introduction; much more detail about the terms is provided later in this chapter.

This circuit is shown in Figure 7-1. It shows the source of voltage is 110 V, which is identified by L1 and N. The switch is a manually actuated switch for this circuit, so the switch provides a way to turn the motor on and off. When the switch is moved to the on position, or closed, voltage and current move through the switch and wires to the motor. When voltage and current reach the motor, its shaft begins to turn. The wires in the circuit are conductors and provide a path for voltage and current, as well as a small amount of opposition to the current flow. This opposition is called *resistance*.

After the motor has run for a while, the switch is opened, and the voltage and current are stopped at the switch and the motor is turned off. When the switch is closed again, the voltage and current again reach the motor.

The source of energy that makes the shaft in the motor turn is called electricity. *Electricity* is defined as a flow of electrons. You should remember from previous discussions that the definition of *current* is also a flow of electrons. In many cases the terms *electricity* and *current* are used interchangeably. Electrons are the negative parts of an atom, and atoms are the basic building blocks of all matter. All matter can be broken into one of 118 materials known as elements. Two of the elements that are commonly used in electrical components and circuits are copper, which is used for wire, and iron, which is used to make many of the parts, such as motors.

To understand electricity at this point, you need only to understand that the atom has three parts: electrons, which have a negative charge; protons, which have a positive charge; and neutrons, which have no charge (also called a neutral charge). Figure 7-2 is a diagram of the simplest atom, which has one electron, one proton, and one neutron. The proton and neutron are located in the center of the atom, called the nucleus. The electron moves around the nucleus in a path called an orbit, in much the same way the earth revolves around the sun.

Because the electron is negatively charged and the proton is positively charged, an attraction between these two charges holds the electron in orbit around the nucleus until sufficient energy, such as heat, light, or magnetism, breaks it loose so that it is free to flow. Any time an electron breaks out of its orbit around the proton and is free to flow, an electrical current is produced. An electrical current can occur only when the electron is free from the atom and can flow. The energy to create electricity can come from burning coal, nuclear power reactions, or hydroelectric dams.

The atom shown in Figure 7-2 is a hydrogen atom; it has only one electron. In the circuit presented in Figure 7-1, the copper wire is used to provide a path for the electricity to reach the motor. Because copper is the element most often used for the conductors in circuits, it is important to study the copper atom. The copper atom is different from

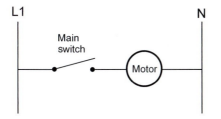

FIGURE 7-1   Electrical diagram of a manually operated switch controlling a motor.

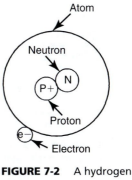

FIGURE 7-2   A hydrogen atom.

Protons and
Neutrons in
Nucleus

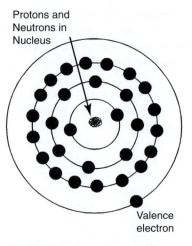

Valence
electron

FIGURE 7-3 A copper atom has 29 electrons. This atom has one electron in its valence shell that can break free to create current flow.

electrons. For example, gold—which is one of the best conductors—has only one valence electron and silver—which is also a good conductor—has three valence electrons.

In the motor circuit presented earlier, a large amount of electrical energy (voltage) is applied to the copper atoms in the copper wire. The voltage provides a force that causes a valence electron in each copper atom to break loose and begin to move freely. The free electrons actually move from one atom to the next. This movement is called electron flow; more precisely, it is called electrical current. Because the copper wire has millions of copper atoms, it will be easier to understand if we study the electron flow among three of them. Figure 7-4 shows the electrons breaking free from three separate copper atoms.

When the electron breaks loose from one atom, it leaves a space, called a hole, where the electron was located. This hole will attract a free electron from the next nearest atom. As voltage causes the electrons to move, each electron moves only as far as the hole that was created in the adjacent atom. This means that electrons move through the conductor by moving from hole to hole in each atom. The movement will begin to look like cars that are bumper to bumper on a freeway.

the hydrogen atom in that it has 29 electrons, whereas the hydrogen atom has only 1. The copper wire that provides a path for the electricity for the motor contains millions of copper atoms.

Figure 7-3 shows the copper atom with its 29 electrons, which move about the nucleus in a series of orbits. Because the atom has so many electrons, all of them will not fit in the same orbit. Instead, the copper atom has four orbits. These orbits are referred to as shells. There are also 29 protons in the nucleus of the copper atom, so the positive charges of the protons will attract the negative charges of the electrons and keep the electrons orbiting in their proper shells. The attraction for the electrons in the three shells closest to the nucleus is so strong that these electrons cannot break free to become current flow. The only electrons that can break free from the atom are the electrons in the outer shell, called the valence shell, and the electrons in the valence shell are called valence electrons. Notice that the copper atom only has one electron in the valence shell. All good conductors have a low number of valence

7.1.1 Example of Voltage in a Circuit

As discussed voltage is the force that moves electrons through a circuit. Figure 7-5 shows the effects of voltage in a circuit in which two identical light bulbs are connected to separate voltage supplies. Notice that the bulb connected to the 12-V battery is much brighter than the bulb that is connected to the 6-V battery because 12 volts can exert more force on the electrons in the circuit than 6 volts, and higher voltage will also cause more electrons to flow.

In the diagram of the motor in Figure 7-1 the amount of voltage is 110 V. The energy that produces this force comes from a generator. The shaft of the generator is turned by a source of energy such as steam. Steam is created from heating water to its boiling point by burning coal or from a nuclear reaction. The energy to turn the generator shaft can also come from water moving past a

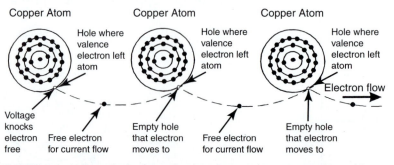

FIGURE 7-4 Voltage is the force that knocks one electron out of its orbit. In this diagram the valence electron in the atom on the far right breaks free from its orbit and leaves a hole. The electrons in the atoms to the left of the first atom move from hole to hole to create current flow.

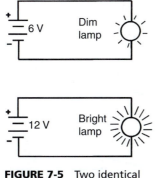

FIGURE 7-5 Two identical light bulbs with different voltages applied to them. The light bulb with 12 V is brighter than the one with 6 V.

dam at a hydroelectric facility. In the light bulb example, the voltage comes from a battery. The voltage in a battery was previously stored in a chemical form when the battery was originally charged. The larger the amount of voltage in a circuit, the larger the force that can be exerted on the electrons, which makes more of them flow. The scientific name for voltage is electromotive force (EMF).

7.1.2 Example of Current in a Circuit

The actual number of electrons that flow at any time in a circuit can be counted, and their number is very large. The unit of current flow is the *ampere* (A). The word *ampere* is usually shortened to "amp." One amp is equal to 6.24×10^{18} electrons flowing past a point in 1 second (s). Figure 7-6 shows the number of flowing electrons that equals 1 *ampere* (A). In this diagram each electron that is flowing is counted; 6,250,000,000,000,000,000 electrons passing a point in 1 s equal 1 A. The ampere is used in all electrical equipment, such as motors, pumps, and switches, as the unit for electrical current.

7.1.3 Example of Resistance in a Circuit

Resistance, the opposition to current flow, is present in the wire that is used for conductors and in the insulation

that covers the wires. The unit of resistance is the *ohm*, or Ω (the capital Greek letter omega, which represents the letter "O"). When the resistance of a material is very high, it is considered to be an insulator; if its resistance is very low, it is considered to be a conductor.

It is important for conductors to have low resistance so that electrons do not require a lot of force to move through them. It is also important for insulation that is used to cover wire to have high resistance so that the electrons cannot move out of the wire into another nearby wire. The insulation also ensures that a wire on a hand tool such as a drill or saw can be touched without electrons traveling through it to shock the person using the tool. Examples of materials that have high resistance and can be used as insulators are rubber, plastic, and air.

The amount of resistance in an electrical circuit can be adjusted to provide a useful function. For example the heating element of a furnace is manufactured to have a specific amount of resistance so that the electrons will heat it up when they try to flow through it. The same concept is used in determining the amount of resistance in the filament of a light bulb. The light bulb filament of a 100-W light bulb has approximately 100 Ω of resistance; 100 Ω is sufficient resistance to cause the electrons to work harder as they move through the filament, which causes the filament to heat up to the point at which it glows. The amount of resistance in a circuit designed to convert energy, such as that in a heating element or motor, is referred to as a *load*.

7.1.4 Identifying the Basic Parts of a Circuit

It is important to understand that each electrical circuit must have (1) a supply of voltage, (2) conductors to provide a path for the electrons to flow, and (3) at least one load. The circuit may also have one or more controls. Figure 7-7 shows a typical electrical circuit with the voltage supply, conductors, control, and load identified. In this circuit the voltage source is supplied through terminals L1 and N, the load is a motor, and the two wires connecting the motor to the voltage source are the conductors. The manual switch is a control, because it controls the current flow to the motor. In a large system in a factory it is possible to have more than one load. For

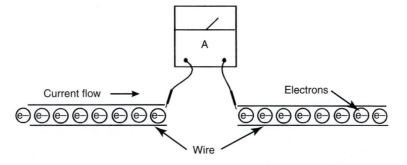

FIGURE 7-6 Electrons passing a point where they are measured. When the number reaches 6,250,000,000,000,000,000 in 1 s, 1 A of current has been measured.

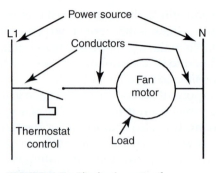

FIGURE 7-7 The basic parts of a circuit: power source, conductors, control, and load.

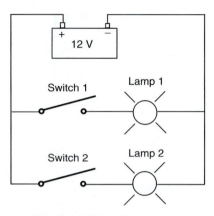

FIGURE 7-9 Ladder diagram of two circuits in parallel, each with one switch and one lamp.

example the system may have multiple conveyors or multiple hydraulic pumps. You will learn more about these components later.

7.1.5 Wiring Diagram and Ladder Diagram

There are two basic ways to diagram electrical circuits. The components can be shown in a wiring diagram or a ladder diagram. The wiring diagram shows the components of a circuit as you would see them if you took a picture of the circuit. The ladder diagram shows the sequence of operation of all the components. Figure 7-8 illustrates a wiring diagram with two switches and a light. The wiring diagram helps you know where the components are located in an electrical cabinet and where the terminals are located. This is important when you are installing components in a circuit. Figure 7-9 shows the same circuit, but this time drawn as a ladder diagram. The ladder diagram clearly shows that the two switches are wired in parallel so either can turn on the light. This is called "showing the sequence of operation." In the wiring diagram it may be difficult to determine if the switches are connected in series or parallel, whereas in the ladder diagram it is very easy to see how the switches operate.

You will learn that one of these diagrams is not necessarily better than the other; rather there is an appropriate time to use each. You will also see that it is easy to convert from either type of diagram to the other.

7.1.6 Equating Electricity to a Water System

It may be easier to understand electricity if you think of it as a water system, with which you may be more familiar. If you examine water flowing through a hose, for example, the flow of water would be equivalent to the flow of current (amps), and the water pressure would be equal to voltage. If you stepped on the hose or bent the end over, it would create resistance that would slow the flow, just as resistance in an electrical circuit slows the flow of current. Figure 7-10 shows

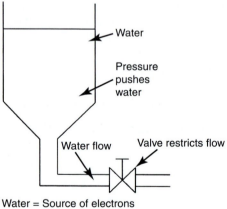

Water = Source of electrons
Water pressure = Voltage that pushes electrons
Water flow = Current flow
Valve restricts flow = Electrical resistance

FIGURE 7-10 A diagram showing similarities between a water system and an electrical system. The flow of water is similar to the flow of current (A), the pressure on the water is similar to voltage, and the resistance from the valve causes the flow of water to slow just as resistance in an electrical system causes current flow to slow.

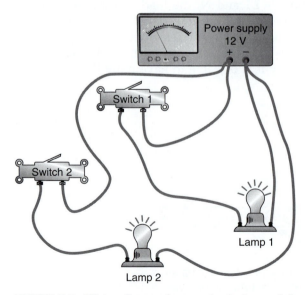

FIGURE 7-8 Wiring diagram for two circuits in parallel.

the similarities between the water system and an electrical circuit.

By definition, 1 V is the amount of force required to push the number of electrons equal to 1 A through a circuit that has 1 Ω of resistance. Also, 1 A is the number of electrons that flow through a circuit that has 1 Ω of resistance when a force of 1 V is applied, and 1 Ω is the amount of resistance that causes 1 A of current to flow when a force of 1 V is applied. You can see that voltage, current, and resistance can all be defined in terms of one another.

7.2 MEASURING VOLTS, AMPS, AND OHMS

It is important to be able to measure the amount of voltage (volts), current (amps), and resistance (ohms) in an circuit in a solar panel to be able to determine when it is working correctly or if something is faulty. If you are checking a tire on your automobile, you can see if it is flat by simply looking at it. Because the flow of electrons through a wire is invisible, it is not possible to simply look at the wire and determine if it has voltage applied to it or if it has current flowing through it, so you must use a meter to make voltage, current, and resistance measurements.

7.2.1 Measuring Voltage

The meter used to measure voltage is called a voltmeter. Figures 7-11 and 7-12 show two types of voltmeters. The

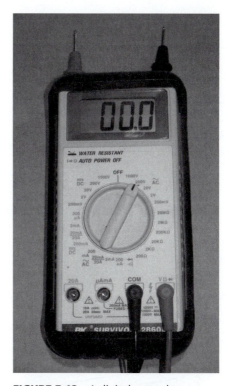

FIGURE 7-12 A digital meter has a digital display to show the amount of voltage, current, or resistance being measured. Some digital meters have a knob to select the range of voltage current and resistance whereas other are self-ranging.

FIGURE 7-11 An analog meter has a needle that is moved by a magnetic field. The face of the meter indicates the amount being measured and the knob on the front is used to determine whether volts, amps, or ohms are being measured and the range of values the meter can measure. (Andrew Scheck/Fotolia, LLC)

meter in Figure 7-11 is an analog-type voltmeter because it uses a needle and a scale to indicate the amount of voltage being measured. The meter in Figure 7-12 is a digital-type voltmeter because it displays numbers to indicate the amount of voltage being measured.

The voltmeter has high internal resistance, approximately 20,000 ohms per volt, so its probes can be safely placed directly across the terminals of the power source without damaging the meter. The range-selector switch, which is the rotary switch in the middle of the meter, adjusts the internal resistance of the meter to set the maximum amount of voltage the meter can measure. The voltmeter can be used to make many different voltage measurements. For example it can measure the amount of supply voltage in a system as well as the amount of voltage provided to each load. Figure 7-13 shows where the probes of a voltmeter should be placed to safely make voltage measurements in a circuit. In Figure 7-13a the voltmeter probes are shown across the terminals of a battery to measure the amount of battery voltage supplied. Figure 7-13b shows the proper location for placing the voltmeter probes to measure the voltage available to lamp 1, and Figure 7-13c shows the proper location for placing the voltmeter probes to measure the voltage available to lamp 2. In each case the voltmeter probes are placed across the terminals where the voltage is being measured.

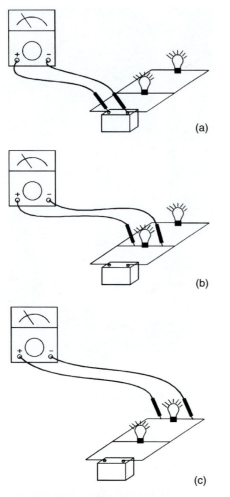

FIGURE 7-13 The proper location to place the probes of a voltmeter to measure the amount of voltage available at the battery. (a) Voltmeter probes placed at the terminals of a battery to measure the amount of battery voltage supplied. (b) The proper location to place the voltmeter probes to measure the voltage available at lamp 1. (c) The proper location to place the voltmeter probes to measure the voltage available at lamp 2.

SAFETY NOTICE

It is important to remember that the voltage to a circuit must be turned on to make a voltage measurement. This presents an electrical shock hazard. You must take extreme caution when you are working around a circuit that has voltage applied so that you do not come into contact with exposed terminals or wires when you are making voltage measurements.

7.2.2 Measuring Large and Small Currents

The meter shown in Figure 7-14 is a clamp-on ammeter, and it is used to measure current in larger AC circuits, up to 100 A (some clamp-on ammeters can measure DC

FIGURE 7-14 A digital clamp-on ammeter that is used to read high currents.

current, too). This type of meter has two claws, or jaws, at the top that form a circle when they are closed. A button on the side of the meter is depressed to cause the jaws to open so that they can fit around a wire without having to disconnect the wire at one end. The jaws of the clamp-on ammeter are actually a transformer that measures the amount of current flowing through a wire by induction. When current flows through a wire, it creates a magnetic field that consists of flux lines. Since AC voltage reverses polarity once every 1/60 of a second, the flux lines collapse and cause a small current to flow through the transformer in the claw of the ammeter. This small current is measured by the meter movement, and the display indicates the actual current flowing in the conductor clamped by the meter. For this reason the clamp-on ammeter can measure current in only one conducting wire at a time. If the clamp-on ammeter were connected around more than one wire at a time, the current in one wire would cancel the current measured in the other wire, which would cause the ammeter to measure zero amps. The clamp-on ammeter for measuring AC current has several ranges so it can measure lower currents (less than 10 A) and currents that are larger than 100 A. If you are trying to measure a very small amount of AC current, you can wrap the wire two or three times around a claw and it will cause the amount of current to multiply by 2 if you use two coils, and multiply by 3 if you use three coils around a claw. The multiplier allows the ammeter to measure smaller currents more accurately.

There is also a version of the clamp-on ammeter that can be specifically designed to measure DC current.

Another type of ammeter is built inside a VOM (Volt Ohm Milliampere meter), which allows you to measure voltage, ohms, or current in a range from milliamps up to 10 A. This ammeter is actually designed to read very small amounts of current, such as 1/1000 A, which is a milliamp and is also written as 0.001 A. The prefix *milli-* means

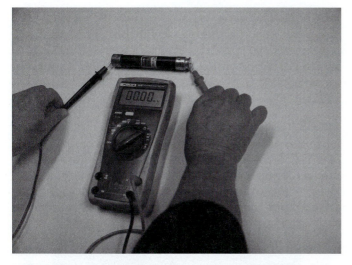

FIGURE 7-15 A true RMS (root mean square) digital VOM (volt ohm milliammeter) that can read voltage, resistance, and current from milliamps through 10 A and measures temperature and capacitance.

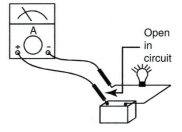

FIGURE 7-16 The proper way to place the meter probes to make a current measurement.

one-thousandth. The analog VOM meter shown in Figure 7-11 has a meter movement that causes a needle to deflect when it senses current flow. This means that when a voltage is applied to a VOM meter, it must be routed through resistors that convert it to current that actually flows through the meter movement. When the VOM meter is set to measure current or resistance, the same meter movement is used. The only part of the meter that changes as it is switched from a voltmeter to an ammeter or ohmmeter is the arrangement of resistors to limit the amount of current flow through the meter movement.

7.2.3 Measuring Milliamps in Solar Energy Installations

At times when you are working with solar energy equipment, you will need to measure the current in a circuit in milliamps, since the amount of current is less than 1 A. For example you may need to measure the amount of current used by a gas valve in a furnace that is used as a backup heat source for a solar heating system, so that you can set the heat anticipator on a heating thermostat. You will also need to make milliamp measurements on the solid-state control boards found in many newer furnaces and air conditioners that are used with solar energy systems. The most important part about reading milliamps is that the ammeter must be placed in series with the current it is reading. For the meter to be put in series with the current it is reading, an open must be created in the circuit. Figure 7-16 shows that an open is made in the circuit when a milliamp measurement is made, and the terminals of the ammeter should be placed on the wires where the hole has been made in the circuit. The open is created by removing one of the wires from the battery terminal. One of the meter probes is placed on the battery terminal, and the other is placed on the end of the wire that was

removed from the battery. This places the meter in series with the circuit, and all the electrons flowing in the circuit must go through the meter because the ammeter has become part of the circuit.

Since the ammeter must become part of the circuit to measure the current, it must have very low internal resistance so that it does not change the total resistance of the circuit. When the VOM meter is set to read milliamps, it is vulnerable to damage because of this small internal resistance.

SAFETY NOTICE!

If you mistakenly place the meter leads across a power source as you would to take a voltage reading when the meter is set to read current, the meter will be severely damaged. You can also be severely burned because the meter can explode, since it has a low resistance component when it is set to read current. This would be similar to taking a piece of bare wire and bending it so it could be inserted into the two holes of an electrical outlet. You must always be sure to check how you have the meter set when you are making voltage, current, or resistance measurements.

As a technician you will be expected to use both the milliamp meter and the clamp-on ammeter. Milliamp meters are used to measure small amounts of current and clamp-on meters are used to measure larger currents such as the amount of current a motor is using. If a motor is using (pulling) too much current, you will be able to determine that it is *overloaded* and will soon fail. Later chapters in this book will explain how to use the clamp-on ammeter to measure motor current.

Digital VOM meters and some analog VOM meters have circuits to measure both DC and AC current. These types of meters have several ranges for measuring small current in the range of milliamps up to 10 A.

7.2.4 Making Measurements with Digital VOM Meters

When you make a measurement with a digital VOM meter, you should follow the same rules as when you use an analog VOM meter. The major difference is that the measurement is presented as a digital number on the meter's display. All you need to do is observe the range-selector switch to determine the units that the meter is measuring and use the value that is displayed as the measurement. Be sure to observe polarity ($+/-$) if the voltage or current is DC. Some digital meters also provide an audible signal for resistance measurements, which is useful when you are locating wires in multi-conductor cables

7.2.5 Measuring Resistance and Continuity with a Digital VOM

You can use a digital VOM to measure resistance for several reasons. For example you can use the digital meter to measure continuity, which is a test for zero resistance or infinite resistance in a component such as a fuse or switch. A fuse with zero resistance is considered good; one with infinite resistance is considered open. When you are using the digital voltmeter to measure continuity, you should begin the test by first checking to see if the meter leads and terminals are good. You can start by touching the two meter leads together; this represents zero resistance to the meter and the meter display should show zero or near zero ohms. When you hold the two leads apart, the meter is reading infinite resistance, and you should see the indication for infinite resistance. In some digital meters the display will show OL and for other digital meters it will display a large number such as 9999. You can test your digital meter for reading zero and infinite by setting it to one of the ohms positions and touching the leads together and observe the reading for zero ohms (continuity) and then allow the leads to not touch and observe that reading of the display for infinite. Figure 7-17 shows a digital meter measuring infinite

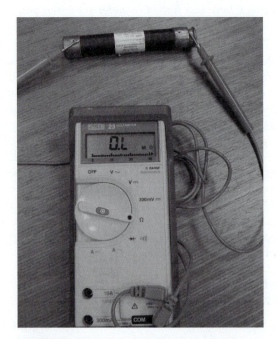

FIGURE 7-17 A digital meter showing high resistance (infinite resistance) in an open fuse.

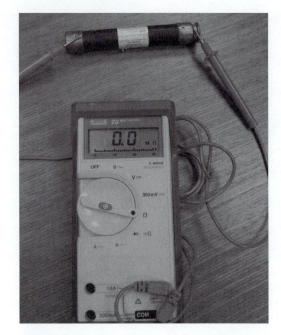

FIGURE 7-18 Meter showing low resistance when fuse is good.

resistance, which indicates the fuse is open. Notice the display is showing OL. Figure 7-18 shows a digital meter measuring zero resistance, which indicates the fuse is good. Notice the display is showing 0.0.

Measuring continuity means that you are looking for zero resistance or infinite resistance. The continuity test is good for measuring the resistance through a switch, fuse, or piece of wire to determine if it has low or infinite resistance. A component is said to have continuity if its resistance is low. If the resistance is high (infinite) the wire, switch or fuse is said to be *open*. Some digital meters have an audible beep that can be selected with the continuity test to provide an audible signal when the resistance is low and there is continuity. This is quite useful when testing fuses, switches, or wires, and it allows you to make the test quickly because you are not looking at the display for a number; rather you are listening for the audible beep.

The third type of measurement that you will make with the ohm meter is to measure actual resistance in a motor winding, or the resistance of a variable resistor. The motors you will be working with in applications where pump motors are used to circulate cooling fluid through a solar heating system have a number of windings, which are basically large coils of wire. If the motor in this type of application is a 110-V motor, it will have a start and a run winding and the amount of resistance in the start winding will always be larger than the amount of resistance in the run winding. In this case you are actually trying to identify the specific amount of resistance in the run and start windings so you can tell which winding is which. In this test you will need to accurately read the amount of resistance the meter is measuring . When you use the meter for this test, its display will show the number of ohms the meter is measuring. Another test in which you will need to measure the exact number of ohms in a coil is used to compare the windings in a *transformer,* so you will need to know the exact number of ohms of resistance in the winding. When you make this type of resistance measurement, you must be sure that no power is applied to the circuit, and that the component you are measuring has its leads isolated so that it does not have interference from nearby resistance. When you touch the meter leads to the two ends of the coil, the amount of resistance shown in the display of the meter is the amount of resistance the meter is measuring. The reason the power needs to be off to the components you are measuring the resistance is that the meter uses a small internal battery as a power supply for measuring resistance.

7.3 USING OHM'S LAW TO CALCULATE VOLTS, AMPS, AND OHMS

When you are working in a circuit, you may need to calculate or estimate the volts, amps, or ohms that the circuit should have. When you are troubleshooting an electrical problem, you will also need to make measurements of the volts, amps, and ohms to determine if the circuit is operating correctly. The problem with taking a measurement is that you will not be sure if the values are too high or too low or if they are just about right. A calculation can tell you if the measurements are within the estimated values the circuit should have.

A relationship exists between the volts, amps, and ohms in every DC circuit that can be identified by Ohm's law, which states simply that the amount of voltage in a DC circuit is always equal to the amps multiplied by the resistance in that circuit. If a circuit has 2 A of current and 4 Ω of resistance, the total voltage is 8 V.

The formula can be changed to determine the amount of amps or ohms. Because we know that $2 \times 4 = 8$, it stands to reason that 8 divided by 2 equals 4, and 8 divided by 4 equals 2. Thus, if you measure the volts in a circuit and find that you have 8 V and if you measure the amps in a circuit and find that you have 2 A, you can determine that you have 4 Ω by dividing 8 by 2. You can also determine the number of amps you have in the circuit by measuring the volts and ohms and calculating: 8 V divided by 4 Ω is 2 A.

7.3.1 Ohm's Law Formulas

When scientists, engineers, and technicians do calculations, they use letters of the alphabet to represent units such as volts, amps, and ohms. The letters are accepted and used by everyone.

In Ohm's law formulas, the letter E is used to represent voltage. The E is derived from first letter of the term electromotive force. The letter R is used to represent resistance. The letter I is used to represent current (amps); it has been derived from the first letter of the word *intensity.*

Ohm's law is represented by the formula $E = IR$. When you use Ohm's law, the basic rule of thumb is that you should use E, I, and R to represent the unknown values in a formula when you begin calculations and you should use V for volts, A for amps, and Ω for ohms as units of measure when you have determined an answer or anytime the values are known or measured.

7.3.2 Using Ohm's Law to Calculate Voltage

When you are solving for voltage, you need to know the amount of current (amps) and the amount of resistance (ohms). If you do not know these two values, you cannot calculate voltage. If you are asked to calculate the amount of voltage when the current is 20 A and the resistance is 40 Ω, you can multiply 20 by 40 to get 800 V. To solve a problem, start with the formula and substitute the values of I and R. Then you can continue the calculation. Notice that the units A (for amps) and Ω (for ohms) are added when the values are known.

$$E = IR$$
$$E = 20 \text{ A} \times 40 \, \Omega$$
$$E = 800 \text{ V}$$

7.3.3 Using Ohm's Law to Calculate Current

The formula for calculating an unknown amount of current in a circuit is

$$I = \frac{E}{R}.$$

To calculate the current, you divide the voltage by the resistance. For example, if you have measured the circuit voltage and found that it is 50 V and if you have measured the resistance and found that it is 10 Ω, the current is found by dividing 50 V by 10 Ω, which equals 5 A.

$$I = \frac{E}{R}$$
$$I = \frac{50 \text{ V}}{10 \, \Omega}$$
$$I = 5 \text{ A}$$

7.3.4 Using Ohm's Law to Calculate Resistance

The formula for calculating an unknown amount of resistance in a circuit is

$$R = \frac{E}{I}.$$

To calculate the resistance of a circuit, you divide the voltage by the current. For example if you have 60 V and 12 A, the resistance is found by dividing the voltage by the current. Again, start with the formula and substitute values to get an answer:

$$R = \frac{E}{I}$$

$$R = \frac{60 \text{ V}}{12 \text{ A}}$$

$$R = 5 \, \Omega$$

7.3.5 Using the Ohm's Law Wheel to Remember the Ohm's Law Formulas

A learning aid has been developed to help you remember all the Ohm's law formulas. Figure 7-19 shows this aid, which is in the shape of a wheel or pie. The top of the pie has the letter E, representing voltage; the bottom left has the letter I, representing current (amps); and the bottom right has the letter R, representing resistance (ohms).

Figure 7-20 shows how to apply the formulas. If you want to remember the formula for voltage, put your finger over the E and the letters I and R show up side by side. The vertical line separating the I and R is the "multiply line" because that is the math function you use with I and R. This represents the formula $E = IR$.

If you want to know the formula for current (I), put your finger over the I; the remaining letters are E over R. The horizontal line that separates the E and R is called the "divide-by line," because to determine I you divide E by R. This represents the formula $I = \frac{E}{R}$.

If you want to know the formula for resistance (R), put your finger over the R; the remaining letters are E over I. This represents the formula $R = \frac{E}{I}$.

7.3.6 Calculating Electrical Power

Electrical power (P) is work that electricity can do. The units for electrical power are watts (W). Watts are determined by multiplying amps times volts ($P = IE$). Most electrical loads that are resistive in nature, such as heating elements, are rated in watts. For example if a heating element is rated for 5 A and 110 V, it uses 550 W of power. You can calculate the amount of power

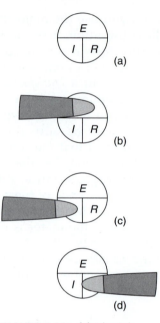

FIGURE 7-20   (a) Ohm's law pie is a memory tool for remembering all of the Ohm's law formulas. (b) Put your finger over the E when you want the formula to find voltage. (c) Put your finger over the I when you want the formula to find current. (d) Put your finger over the R when you want the formula to find resistance.

by using Watt's law, $P = IE$: $5 \text{ A} \times 110 \text{ V} = 550 \text{ W}$. This basic formula can be rearranged in the same way as the Ohm's law formula. You can calculate current by dividing the power (550 W) by the voltage (110 V), and you can calculate voltage by dividing the power (550 W) by the current (5 A). Figure 7-21 shows a pie that is a memory aid for remembering all of the formulas for Watt's law. From this pie you can see that the original formula, $P = IE$, can be found by placing your finger over the P. The formula for solving for current $\left(I = \frac{P}{E} \right)$ can be found by placing your finger over the I, and the formula to solve for voltage $\left(E = \frac{P}{I} \right)$ can be found by placing your finger over the E.

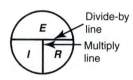

FIGURE 7-19   Ohm's law pie, a memory tool for remembering all the Ohm's law formulas.

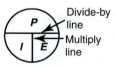

FIGURE 7-21   Formulas for calculating power, current, and voltage.

7.3.7 Presenting All the Formulas

When you are troubleshooting, you may be able to get a voltage reading easily. If you can find a data plate that provides the wattage for a heating element or other load, you will be able to use one of the formulas that was presented to calculate (estimate) the current or the resistance. This value is an estimate because it was calculated from design data and it will give you an idea of what the values should be. When you actually measure the values, you can compare the measurement to the calculation so that you can determine if the system is working correctly. When you are in the field making these measurements, it may be difficult to remember all the formulas, so Figure 7-22 is a chart with all the formulas. The chart is in the shape of a wheel with spokes emanating from an inner ring. A cross is shown in the middle of the inner wheel, separating it into four sections. These sections are represented by the letters P (power, in watts), I (amps), E (volts), and R (ohms).

The formulas in each section of the outer ring correspond to either P, I, E, or R. The formulas for wattage are shown emanating from the inner section marked with P. All the formulas for current are shown emanating from the inner section marked with I. All the formulas for voltage are shown emanating from the inner section marked with E, and all the formulas for resistance are shown emanating from the inner section marked with R.

If you are trying to calculate the power in a circuit or the power used by any individual component, you could use any of the three formulas shown on the wheel emanating from the section identified by the letter P. Thus, you could use $P = EI$, $P = I^2R$, or $P = E^2/R$. Your choice of formula would depend on the two values that are given.

7.3.8 Using Prefixes and Exponents with Numbers

At times you have seen numbers that take up a lot of room to print out—such as 5,000,000 or 0.0000003. When

Prefix	Symbol	Number	Exponent
Mega	M	1,000,000	10^6
Kilo	k	1,000	10^3
Milli	m	0.001	10^{-3}
Micro	μ	0.000001	10^{-6}

FIGURE 7-23 Prefix symbols and exponents commonly used in electricity and found on electrical equipment.

you need to write a long number on the face of a meter or in a calculation, the number may need to be written in a form that does not take up so much room. Several standards have been adopted throughout the fields of mathematics and electricity for this purpose. The chart in Figure 7-23 shows several ways to present the number using less space. For example if you wanted to shorten the number that represents the value for 5,000,000 (5 million) ohms, you could use the prefix *mega* and the value would be expressed as 5 mega ohms or you could use the symbol 5 M Ω. If you wanted to work with this number in a calculator, you could use its exponent form, which would be 5^{e6}. The exponent may be expressed verbally as "5 times 10 to the 6th power."

If you need to shorten the value 7000 (7 thousand) ohms, you would use the prefix *kilo*, and the value would be expressed as 7 kilo ohms, or you could use the symbol 7 kΩ. If you used the exponent form in a calculator, you would use 7^{e3} Ω. When you express the exponent verbally, you would say "7 times 10 to the 3rd power."

If you wanted to shorten the value 0.005 amp, you would use the prefix *milli*, and the value would be expressed as 5 milliamps, or you could use the symbol 5 mA. If you used the exponent in a calculation, you would use 5^{e-3} A. To express the exponent verbally, you would say "5 times 10 to the negative 3rd power." If you wanted to shorten the value 0.000002 A, you would use the prefix *micro*. This number would be expressed as 2 micro amps, or you could use the Greek symbol for micro (μ) and the answer would be 2μA. If you used the exponent in a calculator, you would use 5^{e-6} A. To express the exponent verbally, you would say "5 times 10 to the negative 6th power."

7.4 FUNDAMENTALS OF ELECTRICAL CIRCUITS

A basic simple circuit consists of a power source, conductors (wires), one or more switches (controls) that can close to allow current to flow through them or open to stop current flow, and a load such as a motor. The basic diagram can be drawn in two ways. Figure 7-24 shows these components drawn as a wiring diagram, and Figure 7-25 shows the components in a ladder diagram. The wiring diagram is used to show the location of components in a circuit, whereas the

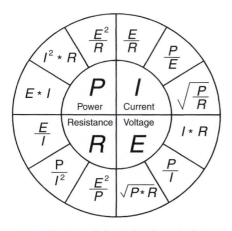

FIGURE 7-22 All formulas for calculating power, current, voltage, and resistance.

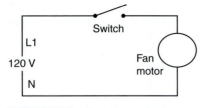

FIGURE 7-24 A power source, switch, and fan motor shown in a wiring diagram. The function of the wiring diagram is to show the location of components in a circuit.

schematic (ladder) diagram is used to show the sequence of operation. The reason the diagram in Figure 7-25 is called a ladder diagram is that as additional loads are added they are shown on additional "rungs," which makes the overall appearance of this diagram look like a ladder. Basically the load in the first lines of the diagram will energize before the loads in the second or third lines (rungs). You will find that wiring diagrams and schematic (ladder) diagrams are used throughout this book. Equipment manufacturers will provide either the schematic (ladder) diagram or wiring diagram or both to help you install and troubleshoot the system.

Some basic terms are provided here to help you understand how electricity works. For example, a *switch* can have two basic conditions, open or closed. When a switch is open in a circuit, it is called an *open switch* and electrons cannot flow past the open contacts and all current in the circuit stops and the load is turned off. For example, if the switch is controlling a motor, when the contacts are opened, the current cannot flow through the switch and the motor is turned off. When the switch is closed, it is called a *closed switch* and the current is allowed to flow and the motor will run.

7.4.1 Relationships Among Voltage, Current, and Resistance in a Series Circuit Using Resistors

Simple circuits consisting of resistors will be used to explain the relationships among voltage, current, and resistance in a series circuit. Resistors are small components

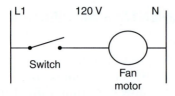

FIGURE 7-25 A power source, switch, and fan motor shown in a schematic (ladder) diagram. Notice the overall appearance of the diagram looks like a ladder and shows sequence of operation.

that are common in modern electronic circuit boards that are used to control systems. The motors and other loads in systems will react in some part as resistors, so it is important for you to understand the relationships among multiple resistors in series circuits. Understanding resistors in series circuits and in parallel circuits is also essential for understanding the solid-state components that are commonly used in solid-state controls found in control systems such as the battery charge controller or the inverter.

As a technician you must understand the effects of switches in series circuits and electrical loads in series circuits. You will learn about the relationships between switches and loads that are connected in series. The second part of this section will explain the effects of electrical switches and loads that are connected in parallel.

The relationships among voltage, current, and resistance in series parallel circuits will also be introduced. It will be easier to understand these relationships when a known value of resistance is used. A resistor is used to provide the resistance for these simple series circuits. Each resistor has a color code consisting of colored bands that are painted on the resistor to identify its resistance value. Ohm's law will also be used to help you understand how voltage and current should react in all types of series circuits. The basic rules discussed in this chapter will help you troubleshoot larger circuits that are commonly found in control systems.

7.5 EXAMPLES OF SERIES CIRCUITS

All electrical circuits have at least a power source, one load, and conductors to provide a path for current. As circuits become more complex, switches are added for control. These switches can provide a variety of functions depending on the way they are connected in the circuit. Figure 7-26 shows a simple circuit that includes a power source, three switches for control, and a motor as the load. This circuit is called a *series circuit* because the current has only one path to travel to the load and back to the power source.

When additional switches are added to the circuit to control the load, they can be connected so that all current continues to have only one path by which to travel around the circuit. This circuit is called a series circuit because the current has only one path by which to travel from L1 through the switches to the load and back to N. If any of the three switches is opened, current is interrupted and the motor will be turned off. When all the switches are closed, the motor will run.

Since all three of these switches are connected in series, each is capable of opening and causing current to stop flowing to the motor. If additional operational or safety switches need to be added to this circuit, they will be connected in series with the original switches. It is

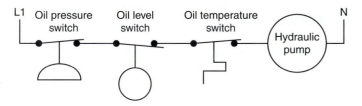

L1   Oil pressure   Oil level   Oil temperature   N
         switch   switch   switch

Hydraulic
pump

FIGURE 7-26 Series circuit with oil pressure switch, oil level switch, and oil temperature switch controlling a hydraulic pump that controls a tracking system on a very large solar panel.

important to remember that when switches are connected in series, any one of them can open and stop current flow in the entire circuit.

7.5.1 Other Components That Are Connected in Series

A *fuse* is another electrical component that is connected in series with all wiring, motors, and controls. The fuses are designed to protect the circuit for the maximum amount of current the wire can handle. If for some reason the amount of current exceeds the normal amounts, the fuse will overheat and melt its fusible link and create an open. Since the fuse is in series with all loads and wires in the control system, it will cause the power to the system to be turned off. Having the fuses wired in series makes them able to open and protect the entire circuit because an open anywhere in the series circuit will cause current to stop in all parts of the circuit.

A *fuse disconnect* is a switch that is combined with the fuses. The disconnect is connected in series with the fuses so it allows you to turn off all power to the unit when you need to change out an electrical component in they system. In this way the disconnect switch is a safety device that is connected in series with every part of the electrical system, and since it is connected in series it will allow you to open it and stop current flow and voltage potential to every part of the system.

7.5.2 Adding Loads in Series

In electrical systems such as those found in solar energy system equipment loads such as motors are not connected in series with other loads; instead, they are connected in parallel with one another so that they all receive the same voltage. Some heating elements for electric heating are connected in series. Resistors are also connected in series to create useful voltage drop circuits for electronic circuit boards and components used in solar energy electronic controls.

Loads are not generally connected in series because they would split the amount of applied voltage. For example if you connected two light bulbs that have the same amount of resistance in their filaments in series and supplied the circuit with 120 V, each light bulb would receive 60 V. Each bulb would glow half as brightly as it would if it were to receive the entire 120 V.

Some loads are connected in series, such as in two electric heating elements that may be used as backup heat for a solar heating system. If the amount of resistance for each heating element is equal, the supply voltage will be split equally between them. For example if the supply voltage to the heaters is 240 V, each heating element will receive 120 V. This means that each heating element must be rated for 120 V. This type of circuit allows smaller heaters to be used on larger voltage sources.

Another problem with connecting loads in series is that if one of the loads has a defect and develops an *open circuit,* the current to all the loads in the circuit will be interrupted. An example of this type of circuit is a string of lights that is used to decorate a Christmas tree. If all the lights are connected in series, they will all go out if one of the lights burns out. The advantage of connecting 50 lights in series, however, is that each of the light bulbs will receive approximately 2.4 V when the circuit is plugged into a 120-V power source.

7.5.3 Using Ohm's Law to Calculate Ohms, Volts, and Amps for Resistors in Series

Some of the loads in electronic circuits are resistors. These resistors are sized so that the supply of voltage will drop into smaller increments. These types of circuits are widely used in electronic circuits for solar energy quipment. Several such circuits will be used to explain how voltage and current are affected by changes in resistance. Figure 7-27 shows an example of three resistors connected in series with a voltage source (battery). Since the resistors are connected in series, there is only one path for current. The resistors are numbered R_1, R_2, and R_3 and the supply voltage is identified as E_T. The formula for calculating total resistance in a series circuit is $R_T = R_1 + R_2 + R_3$. If more than three resistors are used in the circuit, the additional resistors are added to the first three to obtain the total resistance. The arrows show "conventional" current flow. ("Electron" flow would show the flow of electrons [current] moving from the negative battery terminal to the positive terminal. This text will use conventional current flow in its explanations and diagrams. If you would

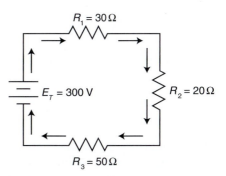

$R_1 = 30\,\Omega$

$E_T = 300\ V$

$R_2 = 20\,\Omega$

$R_3 = 50\,\Omega$

FIGURE 7-27 A series circuit with three resistors connected in series to create one path for current.

rather use electron-flow theory, the current would flow in the direction opposite to the arrows.)

The series circuit in Figure 7-27 has three resistors: R_1 is 30 Ω, R_2 is 20 Ω, and R_3 is 50 Ω. The total resistance for this circuit can be calculated by the following formula:

$$R_T = R_1 + R_2 + R_3$$
$$R_T = 30\,\Omega + 20\,\Omega + 50\,\Omega$$
$$R_T = 100\,\Omega$$

7.5.4 Solving for Current in a Series Circuit

The total current can be calculated by the Ohm's law formula $I = \dfrac{E_T}{R_T} - I = \dfrac{300\,V}{100\,\Omega}$ — or $I = 3$ A. The total current in a series circuit is the same everywhere in the circuit because there is only one path for the current. This means that once you find the total resistance of a series circuit, you can divide total voltage by total resistance and find the total current, and you also have determined the current that flows through each resistor. The current (3 A) has been listed with each resistor in the diagram. Since the current is the same everywhere in the series circuit, it may be identified as I_T or I_1 where it is shown at resistor R_1.

7.5.5 Calculating the Voltage Drop Across Each Resistor

The voltage drop across each resistor can be calculated by the Ohm's law formula $E = IR$. A voltage drop is the actual amount of voltage you would measure if you placed the leads of a voltmeter on the two leads of the resistor. It is important to remember that the current in the series circuit is the same in all places, so if I_T is equal to 3 A, then I_1, I_2, and I_3 are also equal to 3 A. In this case the voltage drop across resistor R_1 is calculated by the formula.

If you placed the probes from a voltmeter on each side of resistor R_1, you would measure 90 V. This is the actual voltage that the resistor is causing to drop when current is flowing through it. The voltage drop across resistor R_2 is calculated by the formula

$$E_2 = I_2 R_2 \quad E_2 = 3\,A \times 20\,\Omega \quad E_2 = 60\,V$$

The voltage drop across resistor R_3 is calculated by the formula

$$E_3 = I_3 R_3 \quad E_3 = 3\,A \times 50\,\Omega \quad E_3 = 150\,V$$

From these calculations you can see that the voltage dropped across R_1 is 90 V, across R_2 is 60 V, and across R_3 is 150 V. If you add these voltage drops, they equal to the supply voltage. This means that $E_1 + E_2 + E_3 = E_T$ (90 V + 60 V + 150 V = 300 V).

EXAMPLE 7-1

This circuit has three resistors connected in series and has a power source of 450 volts. Resistor R_1 is 40 Ω, R_2 is 30 Ω, and R_3 is 20 Ω. Calculate the total resistance for this circuit. After you have calculated the total resistance, divide the total voltage by the total resistance to calculate the total current. When you have determined the total current for this circuit, calculate the voltage drop that would be measured across each resistor.

SOLUTION:

R_T is calculated by the formula $R_T = R_1 + R_2 + R_3$.

$$R_T = 40\,\Omega + 30\,\Omega + 20\,\Omega\ R_T = 90\,\Omega$$

I_T is calculated by the formula $I_T = \dfrac{E_T}{R_T}$.

$$I_T = \frac{450\,V}{90\,\Omega}\ I_T = 5\,A$$

Since current is the same in all parts of the circuit:

$$I_T = 5\ A \quad I_1 = 5\ A \quad I_2 = 5\ A \quad I_3 = 5\ A$$

The voltage that is dropped across each resistor from the current flowing through it is calculated by the formulas

$$E_1 = R_1 \times I_1 \quad E_1 = 40\,\Omega \times 5\,A \quad E_1 = 200\,V$$
$$E_2 = R_2 \times I_2 \quad E_2 = 30\,\Omega \times 5\,A \quad E_2 = 150\,V$$
$$R_3 = R_3 \times I_3 \quad E_3 = 20\,\Omega \times 5\,A \quad E_3 = 100\,V$$

You can check your answers by adding all the drops to see that they equal the total supply voltage.

$$E_T = E_1 + E_2 + E_3 \quad E_T = 200\ V + 150\ V + 100\ V$$
$$E_T = 450\ V$$

7.5.6 Calculating the Power Consumption of Each Resistor

The power consumed by each resistor or the power consumed by the total circuit can be calculated by any of the three formulas for power (wattage): $P = EI$, $P = I^2 R$, and $P = E^2/R$. We previously calculated the voltage, current, and resistance for each resistor in the circuit. This means that we can use any of the three formulas to calculate the power for any resistor. For this example we will use all three formulas to show that they all give the same result. We will use the values for R_1: $E_{R1} = 200\,V$, $I_{R1} = 5\,A$, and $R_1 = 40\,\Omega$.

$$P = EI \qquad P = 200\ V * 5\ A \qquad P = 1000\ W$$
$$P = I^2 R \qquad P = (5\ A)^2 * 40\,\Omega \qquad P = 1000\ W$$
$$P = E^2/R \qquad P = (200\ V)^2/40\,\Omega \qquad P = 1000\ W$$

Since all three formulas are equivalent, you can use any of the three that you choose. The major factor in deciding which formula to use will be the values that you have been given or that you can determine by measuring.

7.5.7 Calculating the Power Consumption of an Electrical Heating Element

The power formula can also be used to calculate the power consumption of any other type of resistance used in a circuit. The electric heating elements used in an electric heating system are large resistance loads. If you know the amount of resistance in the element and the amount of voltage applied to the circuit, you can calculate the current the element uses and the amount of power it consumes. For example if the heating element has 10 ohms of resistance and it is connected to 240 V, it will draw 24 A, and it will consume 5760 W.

EXAMPLE 7-2

A heating element in an electric heating system has a resistance of 2.5 Ω. Calculate the current and the power consumption of this heating element if the heating system is connected to 240 V AC.

SOLUTION:

The amount of current is calculated as $240\,\text{V}/2.5\,\Omega = 96\,\text{A}$. The amount of power can be calculated by the formula $P = I^2R$ or $P = IE$. Using $P = I^2R$ we have $(96)^2 \times 2.5\,\Omega = 23{,}040\,\text{W}$.

Using $P = IE$, we obtain $96 \times 240 = 23{,}040\,\text{W}$.

(Note! This answer can be expressed as 23.040 kilowatts, or 23.040 kW.)

7.6 PARALLEL CIRCUITS

Parallel circuits are used frequently in electrical systems. The difference between the series circuit and the parallel circuit is that all the loads in a parallel circuit have the same voltage supplied, whereas in a series circuit each load has a different amount of voltage depending on its resistance. All the load components in electrical systems, such as pump motors and fan motors in a solar energy heating system, must be provided with the same voltage, so they must be connected in parallel when they are in the same circuit. Each point where a resistor is connected in parallel in this circuit is called a *branch circuit*. The parallel circuit can have any number of branch circuits. A parallel circuit allows current to return to the power supply by more than one path. These paths are identified by the arrows in a diagram.

The current in a parallel circuit is additive, and the formula to calculate total resistance is $I_T = I_1 + I_2 + I_3 + \ldots$ (remember the $\ldots$ means that additional currents could be

added to the total). Another point to understand is that the current in a parallel circuit gets larger as more loads are added to it. The parallel circuit also provides a means for disconnecting any load from the power source while still supplying voltage to the remaining loads. This is accomplished by placing a switch in each branch circuit just ahead of each resistor.

When a switch is placed in the branch circuit, it becomes a *series-parallel circuit* in which some parts of the circuit are series in nature and other parts are parallel. For example the fuse and disconnect for the circuit must be able to interrupt all power that is supplied to the loads in a circuit, so the fuse and disconnect must be connected in series with the power supply. The important point to remember when working with series-parallel circuits is that you use the formulas for series circuits for the part of the circuit that is in series, and the formulas for parallel circuits for the part of the circuit that is in parallel. It is also possible to combine parts of the circuit to make it a simplified series or simplified parallel circuit.

7.6.1 Calculating Voltage, Current, and Resistance in a Parallel Circuit

Voltage, current, and resistance can be calculated in a parallel circuit just as in a series circuit by using Ohm's law. Figure 7-28 shows a circuit with the voltage, current, and resistance calculated at each point in the circuit. The supply voltage is 300 V. With this information, you can determine that the voltage of each branch circuit across each resistor is also 300 V because the voltage at each branch circuit in a parallel circuit is the same as the supply voltage.

The current that is flowing through each resistor can be calculated by using Ohm's law formula for current: $I = E/R$. The current in each branch circuit is calculated from this formula and is placed in the diagram beside each resistor.

$$I_1 = E/R_1 \qquad I_1 = 300\,\text{V}/30\,\Omega \qquad I_1 = 10\,\text{A}$$
$$I_2 = E/R_2 \qquad I_2 = 300\,\text{V}/20\,\Omega \qquad I_2 = 15\,\text{A}$$
$$I_3 = E/R_3 \qquad I_3 = 300\,\text{V}/60\,\Omega \qquad I_3 = 5\,\text{A}$$
$$I_T = E_T/R_T \qquad I_T = 300\,\text{V}/10\,\Omega \qquad I_T = 30\,\text{A}$$
$$I_T = I_1 + I_2 + I_3 \quad I_T = 10\,\text{A} + 15\,\text{A} + 5\text{A} \quad I_T = 30\,\text{A}$$

If the voltage and current are given and the resistance needs to be calculated, the Ohm's law formula for resistance can be used: $R = E/I$.

$$R_1 = E/I_1 \qquad R_1 = 300\,\text{V}/10\,\text{A} \qquad R_1 = 30\,\Omega$$
$$R_2 = E/I_2 \qquad R_2 = 300\,\text{V}/15\,\text{A} \qquad R_1 = 60\,\Omega$$
$$R_3 = E/I_3 \qquad R_3 = 300\,\text{V}/5\,\text{A} \qquad R_3 = 60\,\Omega$$
$$R_T = E/I_T \qquad R_T = 300\,\text{V}/30\,\text{A} \qquad R_T = 10\,\Omega$$

If the total resistance and the total current in a circuit are known, you can calculate the voltage for the

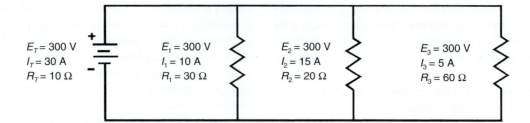

FIGURE 7-28 A parallel circuit with the voltage resistance and current shown at each load.

circuit. If you know the current and resistance at any branch circuit, you can also calculate the voltage using the Ohm's law formula: $E = IR$. The good part about calculating voltage is that once you determine the voltage at any branch circuit, the same amount of voltage is at every other branch circuit and at the supply. The same is true if you calculate the voltage at the supply; you do not need to calculate the voltage at any other branch because it will be the same.

$$E_T = I_T R_T \qquad E_T = 30 \text{ A} \times 10 \text{ } \Omega \qquad E_T = 300 \text{ V}$$
$$E_1 = I_1 R_1 \qquad E_1 = 10 \text{ A} \times 30 \text{ } \Omega \qquad E_1 = 300 \text{ V}$$
$$E_2 = I_2 R_2 \qquad E_2 = 15 \text{ A} \times 20 \text{ } \Omega \qquad E_2 = 300 \text{ V}$$
$$E_3 = I_3 R_3 \qquad E_3 = 5 \text{ A} \times 60 \text{ } \Omega \qquad E_3 = 300 \text{ V}$$

EXAMPLE 7-3

Use the circuit in Figure 7-29 to calculate the individual branch currents, total current, and total resistance.

SOLUTION:

Use the following formulas to calculate the current for each branch:

$$I_1 = E/R_1 \qquad I_1 = 240 \text{ V}/60 \text{ } \Omega \qquad I_1 = 4 \text{ A}$$
$$I_2 = E/R_2 \qquad I_2 = 240 \text{ V}/30 \text{ } \Omega \qquad I_2 = 8 \text{ A}$$
$$I_3 = E/R_3 \qquad I_3 = 240 \text{ V}/15 \text{ } \Omega \qquad I_3 = 16 \text{ A}$$

Use the following formula to calculate the total current I_T:

$$I_T = I_1 + I_2 + I_3 \quad I_T = 4 \text{ A} + 8\text{A} + 16 \text{ A} \quad I_T = 28 \text{ A}$$

Use the following formula to calculate total resistance R_T:

$$R_T = E_T/I_T \quad R_T = 240 \text{ V}/28 \text{ A} \quad R_T = 8.57 \text{ } \Omega$$

At times you will need to calculate the total resistance of a parallel circuit when only the branch resistance and supply voltage are provided. You can calculate the individual currents at each branch circuit and then calculate the total with the formula $I_T = I_1 + I_2 + I_3$. After you have determined the total current, you can use the formula $R_T = E_T/I_T$.

Another method called the "product over the sum" method can be used to calculate total resistance in a parallel circuit. This method is called the product over the sum method because you multiply the two resistors to get the product and then add the two resistors to get the sum. The third step in the calculation includes dividing the product by the sum (product over sum). The formula is written as follows:

$$R_T = \frac{R_1 \times R_2}{R_1 + R_2}$$

You should notice that with this formula you can calculate the total resistance of only two resistors at a time. Since this circuit has three resistors, you would need to find the total of the first two resistors in the branch and then use their total with the third resistor in the formula again to find the "grand" total resistance. We will use this method to calculate the total resistance for these resistors that are connected in parallel: $R_1 = 30 \text{ } \Omega$, $R_2 = 20 \text{ } \Omega$, $R_3 = 60 \text{ } \Omega$.

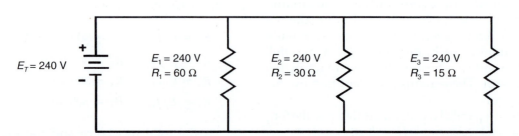

FIGURE 7-29 Parallel circuit showing voltage and resistance for circuit in Example 7-3.

Since we can use only two resistances at a time in this method, in the first step we will find the resistance of R_1 and R_2. The amount of the resistance of R_1 and R_2 in parallel will be called "equivalent resistance." Since this is the first equivalent resistance, we will call this resistance R_{eq1}. Next we will find the parallel resistance of R_{eq1} and R_3. Since we have only three resistors, the amount of the resistance of R_{eq1} and R_3 will be the total resistance (R_T) for the circuit.

$$R_{eq_1} = \frac{R_1 \times R_2}{R_1 + R_2} \quad R_{eq_1} = \frac{30\,\Omega \times 20\,\Omega}{30\,\Omega + 20\,\Omega} \quad R_{eq_1} = 12\,\Omega$$

$$R_T = \frac{R_{eq_1} \times R_3}{R_{eq_1} + R_3} \quad R_T = \frac{12\,\Omega \times 60\,\Omega}{12\,\Omega + 60\,\Omega} \quad R_T = 10\,\Omega$$

From these calculations you can see that the total parallel resistance is 10 Ω. It is important to understand that in all parallel circuits, the total resistance will always be smaller than the smallest branch circuit resistance. You can see that in this circuit, the smallest resistance in the branch circuits is 20 Ω, and the total resistance is 10 Ω.

The next calculation shows the total resistance calculated from the formula:

$$R_T = \frac{1}{\dfrac{1}{R_1} + \dfrac{1}{R_2} + \dfrac{1}{R_3}} \quad \text{or} \quad \frac{1}{R_T} = \frac{1}{R_1} + \frac{1}{R_2} + \frac{1}{R_3}$$

This formula is designed to be used with a calculator. If you do not have a calculator, it is recommended that you use the previous method. If you use a calculator, you should use the following keystrokes to get an answer using this formula:

$$R_T = \frac{1}{\dfrac{1}{R_1} + \dfrac{1}{R_2} + \dfrac{1}{R_3}} \quad R_T = \frac{1}{\dfrac{1}{30\,\Omega} + \dfrac{1}{20\,\Omega} + \dfrac{1}{60\,\Omega}}$$

Notice you might have a key identified as INVERSE, 1/X or a key might be identified as RECIPROCAL.

EXAMPLE 7-4

Use the parallel circuit in Figure 7-30 to calculate the current through each resistor, the total current, and the total resistance for this circuit. The supply voltage is 100 V.

SOLUTION:

$$\begin{array}{lll}
I_1 = E/R_1 & I_1 = 100\,\text{V}/20\,\Omega & I_1 = 5\,\text{A} \\
I_2 = E/R_2 & I_2 = 100\,\text{V}/10\,\Omega & I_2 = 10\,\text{A} \\
I_3 = E/R_3 & I_3 = 100\,\text{V}/5\,\Omega & I_3 = 20\,\text{A} \\
I_T = I_1 + I_2 + I_3 & I_T = 5\,\text{A} + 10\,\text{A} + 20\,\text{A} & I_T = 35\,\text{A} \\
R_T = E_T/I_T & R_T = 100\,\text{V}/35\,\text{A} & R_T = 2.86\,\Omega
\end{array}$$

$$R_T = \frac{1}{\dfrac{1}{R_1} + \dfrac{1}{R_2} + \dfrac{1}{R_3}} \quad R_T = \frac{1}{\dfrac{1}{20\,\Omega} + \dfrac{1}{10\,\Omega} + \dfrac{1}{5\,\Omega}}$$

$$R_T = 2.86\,\Omega$$

7.6.2 Calculating Power in a Parallel Circuit

The formula for calculating power in a parallel circuit is the same as the formula for a series circuit: $P_T = P_1 + P_2 + P_3$. The formula for power at each individual resistor is found from the original Watt's law formula $P = IE$. This means that you must calculate the power consumed by each branch resistor and then add them all together to get the total power consumed. For example, in the parallel circuit shown in Figure 7-28, the voltage at R_1 is 300 V and the current is 10 A, the voltage at R_2 is 300 V and the current is 15 A, and the voltage at R_3 is 300 V and the current is 5 A. The following calculations are used to determine the power consumed by each individual resistor and the total power used by the whole circuit:

$$\begin{array}{lll}
P_1 = I_1 E_1 & P_1 = 10\,\text{A} \times 300\,\text{V} & P_1 = 3000\,\text{W} \\
P_2 = I_2 E_2 & P_2 = 15\,\text{A} \times 300\,\text{V} & P_2 = 4500\,\text{W} \\
P_3 = I_3 E_3 & P_3 = 5\,\text{A} \times 300\,\text{V} & P_3 = 1500\,\text{W} \\
P_T = P_1 + P_2 + P_3 & P_T = 3000\,\text{W} \times 4500\,\text{W} + 1500\,\text{W} \\
& \quad\quad P_T = 9000\,\text{W}
\end{array}$$

You could also calculate the total power consumed by this circuit by using the formula

$$P_T = I_T E_T \quad P_T = 30\,\text{A} \times 300\,\text{V} \quad P_T = 9000\,\text{W}$$

7.7 INTRODUCTION TO MAGNETIC THEORY

The operation of all types of DC and AC generators and motors can be explained with several simple magnetic theories. As a technician or maintenance person that

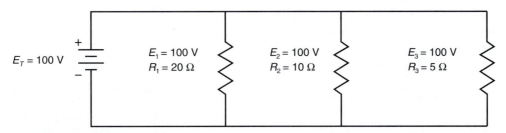

FIGURE 7-30   Parallel circuit for Example 7-4.

works on wind energy systems, you will need to fully comprehend all magnetic theories so that you will understand how these components operate. You must understand how a generator is supposed to operate before you can troubleshoot it and perform tests to determine if it has failed. Understanding magnetic theory will make this job easier. It is also important to understand that some of the magnetic theories rely on AC voltage. These theories are introduced in first part of this chapter, and more detail about AC voltage and magnetic theories that use AC voltage follows in last part of this chapter.

"Magnet" is the name given to material that has an attraction to iron or steel. This material was first found naturally about 4000 years ago as a rock in a city called Magnesia. The rock was called magnetite and was not usable at the time it was discovered. Later it was found that pieces of this material could be suspended from a wire and would always orient themselves so that the same ends always pointed the same direction, toward the earth's North Pole. Scientists soon learned from this phenomenon that the earth itself is magnetic. At first the only use for magnetic material was in compasses. It was many years before the forces caused by two magnets attracting or repelling could be utilized as part of a control device or motor.

As scientists gained more knowledge and as equipment became available to study magnets more closely, a set of principles and laws evolved. The first of these showed that every magnet has two poles, a north pole and a south pole. When two magnets are placed end to end so that similar poles are near each other, the magnets repel each other. It does not matter if the poles are both north or both south, the result is the same. When the two magnets are placed end to end so that the south pole of one magnet is near the north pole of the other magnet, the two magnets attract each other. These concepts are called the "first and second laws of magnets."

When sophisticated laboratory equipment became available, it was found that this phenomenon is due to the basic atomic structure and electron alignments. By studying the atomic structure of a magnet, scientists determined that the atoms in the magnet were grouped in regions called "domains," or "dipoles." In material that is not magnetic or that cannot be magnetized, the alignment of the electrons in the dipoles is random and usually follows the crystalline structure of the material. In material that is magnetic, the alignments in each dipole are along the lines of the magnetic field. Because each dipole is aligned exactly like the ones next to it, the magnetic forces are additive and are much stronger. In material where the magnetic forces are weak, it was found that the alignment of the dipole was random and not along the magnetic field lines. The more closely this alignment is to the magnetic field lines, the stronger the magnet is. Today we refer to a piece of soft iron in which all dipoles are aligned as a "permanent magnet." This term is used because the dipoles remain aligned for and

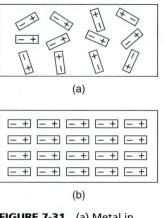

(a)

(b)

FIGURE 7-31 (a) Metal in which dipoles are randomly placed, which makes a weak magnet. (b) Metal in which the dipoles are aligned to make a strong magnet.

the magnet will retain its magnetic properties long periods of time. Figure 7-31a shows nonmagnetic metal, in which the dipoles are randomly placed, and Figure 7-31b shows a piece of metal that is magnetic, in which the dipoles align to make a strong magnet.

7.7.1 A Typical Bar Magnet and Flux Lines

Figure 7-32 shows a bar magnet that is made of soft iron that has been magnetized. The magnet is in the shape of a bar, and its north and south poles are identified. Because the bar remains magnetized for a long period of time, it is a permanent magnet. The magnet produces a strong magnetic field because all its dipoles are aligned. The magnetic field produces invisible *flux lines* that move from the north pole to the south pole along the outside of the bar magnet. The figure shows these flux lines as lines of force that form a slight arc as they move from pole to pole.

Because the flux lines are invisible, you will need to perform a simple experiment to allow you to see that the flux lines do exist and what they look like as they

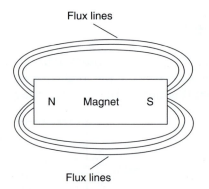

FIGURE 7-32 Example of a bar magnet. Notice the poles are identified as north (N) and south (S). Flux lines emanate from the north pole to the south pole.

surround the bar magnet. For this experiment you will need a piece of clear plastic film, such as the plastic sheets used for overhead transparencies, and some iron filings. Place the plastic sheet over a bar magnet, making the plastic as flat as possible, and sprinkle iron filings over it. The filings will be attracted by the invisible flux lines as they extend in an arc from the north pole to the south pole along the outside of the magnet. Because the flux lines begin at one pole and stretch to the other, the highest concentration of flux lines will be near the poles. The iron filings will also concentrate around the poles, but a definite pattern of flux lines can be seen along each side of the bar magnet. If an overhead projector is available, the image of the flux lines can be projected onto a projector screen or blackboard so that they can be seen more easily. The pattern of these filings will look similar to the diagram in Figure 7-32. The number of flux lines around a magnet is directly related to the strength of the magnet. A stronger magnet will have more flux lines than a weaker magnet. The strength of a magnet's field can be measured by the number of flux lines per unit area. Because the strength of a magnet's field is based on the alignment of the magnetic dipoles, the number of flux lines will increase as the alignment of the magnetic dipoles increases.

Some materials, such as alnico and Permalloy, make better permanent magnets than iron, because the alignment of their magnetic domains (dipoles) remains consistent even after repeated use. You may find these materials used in some expensive controls and motors, but normally the permanent magnet will be made of soft iron. The reason permanent magnets are useful in many types of controls, especially in motors and generators, is that the soft iron produces residual magnetism for long periods of time over many years. Permanent magnets have several drawbacks, however. One is that the magnetic force of a permanent magnet is constant and cannot be turned off if it is not needed. This means that if something is attracted to a magnet, it will remain attracted until it is physically removed from the force of the flux lines. Another problem with a permanent magnet's flux field being constant is that it cannot easily be made stronger or weaker if circumstances so require.

7.7.2 Electromagnets

An *electromagnet* is produced when current flows through a *coil* of wire. One type of electromagnet is made by connecting a coil of wire to an electric cell (battery). The electromagnet has properties that are similar to those of a permanent magnet. When a wire conductor is connected to the terminals of the battery, current will begin to flow, and magnetic flux lines will form around the wire like concentric circles. If the wire is placed near a pile of iron filings while current is flowing through it, the filings will be attracted to the wire just as if the coil were a permanent magnet. Figure 7-35 shows two diagrams indicating the

location of magnetic flux lines around conductors. Figure 7-33a shows flux lines will occur around any wire when current is flowing through it. You can set up several simple experiments to demonstrate these principles. In one experiment you can insert a current-carrying conductor through a piece of cardboard and place iron filings around the conductor on the cardboard. When current is flowing in the wire, the filings will settle around the conductor in concentric circles, showing where the flux lines are located. As the amount of current is increased, the number of flux lines will also increase. The flux lines will also concentrate closer and closer to the wire until the current reaches saturation. When the flux lines reach the saturation point, additional increase of current in the wire will not produce any more flux lines.

When a straight wire is coiled up, the flux lines will concentrate and become stronger. Figure 7-33b shows flux lines around a coil of wire that has current flowing through it. Because the flux lines are much stronger in a coil of wire, most of the electromagnets that you will encounter will be in the form of coils. For example, coils are used in transformers, relays, solenoids, and motors.

One advantage an electromagnet has over a permanent magnet is that the magnetic field can be energized and deenergized by interrupting the current flow through the wire. The strength of the magnetic field can also be varied by varying the strength of the current flow through the conductor that is used for the electromagnet. This theory is perhaps the most important theory of magnetism, because it is used to change the strength of magnetic fields in motors, which causes the motor shaft to

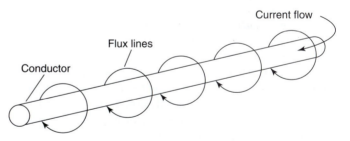

(a) Few flux lines around conductor that is not coiled

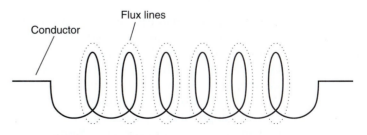

(b) Flux lines become more concentrated when wire is coiled

FIGURE 7-33 (a) Flux lines around a wire carrying current. (b) Flux lines around a coil of wire that is carrying current. Notice that the number of flux lines increase when the wire is coiled.

produce more torque so it can turn larger loads or for a generator to produce more voltage. Another important theory of the electromagnetic coil is that the magnetic field can be turned off by interrupting the current flow through the coil. When a switch is added to the coil circuit, the magnetic field can be turned on and off by turning the switch on and off to interrupt the flow of current in the coil. When the switch is opened, current is interrupted and no flux lines are produced, so the magnetic field will not exist.

Components such as generators, relays, and solenoids use the principle of switching the magnetic field on and off. When current flows in the coil, the magnetic field is energized. When the current to the coil is interrupted, the magnetic field is turned off. This principle is also used to turn generators on and off. When current flows through the coils of a generator, it will generate voltage when its shaft is turned. When current is interrupted, the magnetic field will diminish and the generator will stop producing current.

7.7.3 Adding Coils of Wire to Increase the Strength of an Electromagnet

Another advantage of the electromagnet is that its magnetic strength can be increased by adding coils of wire to the original single coil of wire. The increase of the magnetic field occurs because the additional coils of wire require a longer length of wire, which provides additional flux lines. The magnetic field will be stronger when the coil is more tightly wound because the flux lines are more concentrated. Thus very fine wire is used in some electromagnets to maximize the number of coils. As smaller wire is used, however, the amount of current flowing through it must be reduced so that the wire is not burned open.

You will learn that some motors use this principle to increase their horsepower and torque ratings. These motors have more than one coil that can be connected in various ways to affect the torque and speed of the motor's shaft. *Torque* is defined as the amount of rotating force available at the shaft of a motor. You will also learn that coils can be connected in series or in parallel to affect the torque and speed of a motor.

7.7.4 Using a Core to Increase the Strength of the Magnetic Field of a Coil

The strength of a magnetic field can also be increased by placing material inside the coil to act as a core. The farther the core is inserted into the coil, the stronger the flux field becomes. When the core is removed completely from the coil, it is considered to be an air coil magnet, and the magnetic field is at its weakest point. If a soft iron is used as the core, it will strengthen the magnetic field, but it also creates a problem because it has excessive residual magnetism, which is unwanted. Residual magnetism means the core will retain magnetic properties when current is interrupted in the coil, which will make it like a permanent magnet. This problem can be corrected by using laminated

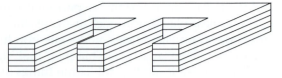

FIGURE 7-34 Examples of thin strips of laminated steel pressed together for use in a coil.

steel for the core. The laminated-steel core is made by pressing sheets of steel together to form a solid core. Figure 7-34 shows layers of laminated steel pressed together to form a core. When current flows through the conductors in the coil, the laminated-steel core enhances the magnetic field in much the same way as the soft iron, and when current flow is interrupted, the magnetic field collapses rapidly because each piece of the laminated steel does not retain sufficient magnetic field.

7.7.5 Reversing the Polarity of a Magnetic Field in an Electromagnet

When current flows through a coil of wire, the direction of the current flow through the coil will determine the polarity of the magnetic field around the wire. The polarity of the magnetic field around the coil of wire is important because it determines the direction a motor shaft turns in an AC or DC motor. If the direction of current flow is reversed, the polarity of the magnetic field is reversed, and the direction a motor shaft is turning is reversed. In some motors used in factory systems, such as fan motors and pump motors, the direction of rotation is very important. In these applications you will be requested to change the connections for the windings in the motor or the supply voltage for a three-phase motor to make the motor rotate in the opposite direction. The changes you are making take advantage of changing the direction of current flow through a coil or changing the polarity of the supply voltage with respect to the other phases so that the motor will reverse its rotation.

Another way to think of reversing the polarity is when a wire is moved through a magnetic field to generate a voltage as in a generator or alternator. When a wire is moved through a magnetic field in one direction, it will cause current to flow in one direction. If the wire moves back through the magnetic field in the opposite direction, the current flow in the wire will be reversed. This is the operational theory that is used in generating current with a generator or alternator.

7.8 ALTERNATING CURRENT

In an alternating current (AC) circuit the electrons travel in one direction and then change direction and move in the other direction. This movement by the electrons can best be shown in its characteristic waveform. The AC waveform is shown in Figure 7-35, and it is called a sine wave. The sine wave has a positive half-cycle in which

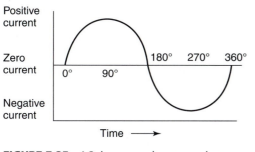

FIGURE 7-35   AC sine wave shown moving through 360°.

the electrons flow in one direction, and a negative half-cycle in which the electrons flow in the reverse direction.

7.8.1 Frequency of AC Voltage

AC voltage has a sinusoidal waveform that is positive 1/60th of a second and then is negative for 1/60th of a second. This oscillation from positive to negative is called *frequency*, and it is the most important feature of AC voltage that is different from DC voltage. Frequency is the rate the electrical current changes from flowing in the positive direction to reversing and flowing in the negative direction. The reason the electrical current flows in the positive direction and then in the negative direction in AC voltage is that it is produced by moving a wire through a magnetic field in one direction and then in the opposite direction. This is accomplished by forming the wire in the shape of a coil and pressing the coil onto a rotating shaft. When the shaft is rotated, the wire in the coil automatically moves in one direction cutting the magnetic lines of force when the shaft rotates through 180°, and the wire moves down through the magnetic field. As the shaft continues to rotate through the remainder of one revolution, from 180° to 360°, the wire moves back upward through the magnetic lines of force and electrons move in the opposite direction. This cycle continues as the shaft rotates and the electron flow is created in the positive direction and then in the negative direction. The typical frequency for AC voltage in the United States and much of the rest of North America is 60 Hz. The frequency of 60 Hz is determined by the rotating speed of the alternator when the voltage is generated. Frequency is more technically defined as the number of cycles (sine waves) that occur in 1 second. Figure 7-36 shows a number of sine waves occurring in 1 second(s).

The "period" of a sine wave is the time it takes one sine wave to start from the zero point and pass through 360°, as seen in Figure 7-37. A period represents one complete cycle, which is one complete revolution of the alternator. Because the shaft of the alternator rotates one full circle to produce the sine wave, we equate one complete cycle to 360°. Thus the sine wave can be described in terms of 360°. In Figure 7-37 the sine wave starts at 0°, reaches a positive peak at 90°, returns to 0° at 180°, reaches a negative peak at 270°, and finally returns to 0° at

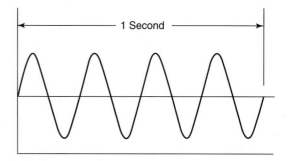

FIGURE 7-36   The frequency of AC voltage is calculated from the number of cycles that occur in 1 second.

360°. The number of degrees will be used to identify points of the sine wave in future discussions. The period of a sine wave can be described as $P = \dfrac{1}{\text{frequency}}$, and frequency can be defined as $F = \dfrac{1}{\text{period}}$.

7.8.2 Capacitors and Capacitive Reactance

A *capacitor* is an electrical device that has two terminals and the capability of storing a charge. It is made of two conducting plates that are separated by an insulator called a dielectric. If the capacitor is an electrolytic capacitor, one of the terminals is negative and the other is positive. In all other capacitors the two terminals do not have polarity.

The operation of the capacitor can best be described by applying DC voltage to it and allowing it to charge and then to discharge through a switch. The negative terminal of the battery (DC power source) is connected to the negative terminal of the capacitor and when the switch is closed electrons (current) begin to flow from the negative terminal of the battery to the negative terminal of the capacitor when the circuit is complete. It is important to understand that current does not flow through the capacitor, since its two plates are separated by a dielectric, which is a good insulator. Electrons continue to flow until the voltage (charge) on the capacitor plate is equal to the

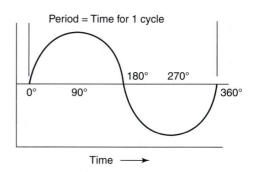

FIGURE 7-37   The period of AC voltage is calculated as the time it takes for one cycle to occur.

battery voltage. At this point there is no longer a difference of potential between the capacitor and the battery, so current flow is stopped. Since the dielectric between the positive and negative plates of the capacitor is a good insulator, theoretically it will remain charged indefinitely when the switch is opened and the capacitor is disconnected from the battery.

Since there is a large potential difference between the positive and negative plates of the capacitor, the electrons will flow from the negative plate to the positive plate until the charge on each plate is equal, and the capacitor is considered discharged. At this point if the switch is moved to position A again, the capacitor will charge again. When the capacitor is placed in a circuit with DC voltage, the capacitor will become charged and remain charged until the circuit is changed to allow the capacitor to discharge. If the capacitor is placed in a circuit with AC voltage, it will charge and discharge at the frequency of the voltage.

The opposition caused by a capacitor is called "capacitive reactance," and the opposition caused by an inductor is called "inductive reactance." The combined opposition caused by capacitive reactance, inductive reactance, and resistance in an AC circuit is called "impedance." The main difference between capacitive reactance, inductive reactance, and resistance is that a phase shift occurs between the voltage waveform and current waveforms. Figure 7-38a shows a capacitor in an AC circuit with a resistor, Figure 7-38b shows the voltage and current waveforms for this circuit, and Figure 7-38c shows the vector diagram that is used to calculate the amount of phase shift. A vector diagram is a graph derived from a trigonometry calculation used to determine phase angle.

When voltage encounters a capacitor in a circuit, it will take time to charge up the capacitor and then discharge it. This causes the reactance, which becomes an opposition to the voltage waveform. The capacitor does not affect the current waveform.

7.8.3 Calculating Capacitive Reactance

The total amount of opposition caused by the capacitor is called capacitive reactance, X_C. Even though you may never need to calculate capacitive reactance, you should understand the effects of changes in capacitance and frequency on the amount of capacitive reactance. Capacitive reactance can be calculated by the following formula:

$$X_C = \frac{1}{2\pi FC}$$

where $\pi = 3.14$

F = frequency

C = capacitance in microfarads

You must know the value of the capacitor and the frequency of the AC voltage. For example if a 40-μF capacitor is used in a 60-Hz circuit, the amount of opposition (capacitive reactance) is 66.35 Ω. Notice that

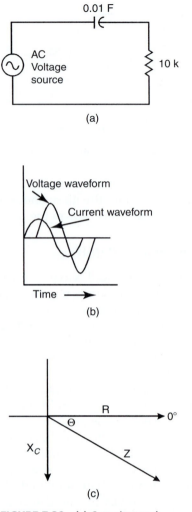

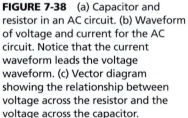

FIGURE 7-38 (a) Capacitor and resistor in an AC circuit. (b) Waveform of voltage and current for the AC circuit. Notice that the current waveform leads the voltage waveform. (c) Vector diagram showing the relationship between voltage across the resistor and the voltage across the capacitor.

because the capacitive reactance is an opposition, its units are ohms. This calculation is as follows:

$$X_C = \frac{1}{2\pi FC} = \frac{1}{2 \times 3.14 \times 60 \times 0.000040} = 66.35\,\Omega$$

EXAMPLE 7-5

Calculate the capacitive reactance of a 60-Hz AC circuit that has a 90-μF capacitor.

SOLUTION:

$$X_C = \frac{1}{2\pi FC} - \frac{1}{2 \times 3.14 \times 60 \times 0.000090} = 29.49\,\Omega$$

7.8.4 Calculating the Total Opposition for a Capacitive and Resistive Circuit

The amount of total opposition (impedance) caused by a capacitor and resistor in an AC circuit can be calculated. Figure 7-39a shows the diagram that is used to determine the impedance for this type of circuit. For example if a circuit has 70 Ω of resistance and 40 Ω due to capacitive reactance, the total impedance must be calculated using the distance formula, because the voltage and current in the circuit are out of phase. The formula for calculating impedance of this circuit is

$$Z = \sqrt{R^2 + X_C^2}$$
$$Z = \sqrt{70^2 + 40^2}$$
$$Z = 80.62\,\Omega$$

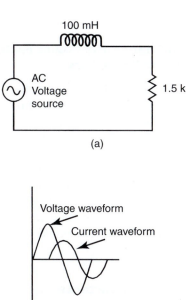

(a)

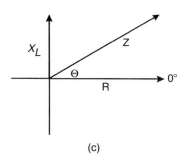

(b)

(c)

FIGURE 7-39 (a) An inductor and a resistor in an AC circuit. (b) Waveforms for the voltage and current in the inductive and resistive circuit. (c) Vector diagram showing voltage across the inductor leading the voltage across the resistor.

7.9 RESISTANCE AND INDUCTANCE IN AN AC CIRCUIT

When inductors (coils of wire) are used in an AC circuit, they will create an opposition that is similar to resistance. The opposition caused by an inductor is called inductive reactance. The main difference between capacitive reactance, inductive reactance, and resistance is that a phase shift between the voltage and current waveforms occurs. Figure 7-39a is a diagram of an inductor in an AC circuit with a resistor, Figure 7-39b shows the voltage and current waveforms for this circuit, and Figure 7-39c shows the vector diagram that is used to calculate the amount of phase shift for the circuit.

When current encounters an inductor in a circuit, it will take time to charge up the inductor and then discharge it. This causes the reactance, which becomes an opposition to the current waveform. The inductor does not bother the voltage waveform. In Fig. 7-41b you can see that the current waveform *lags,* or starts later than, the voltage waveform.

7.9.1 Calculating Inductive Reactance

The total amount of opposition caused by the inductor is called inductive reactance, X_L. As in the case of capacitive reactance, inductive reactance can be calculated, and you should remember that even though you do not calculate reactance in the field when you are troubleshooting, it is important that you understand the effect that changing the size of the inductor or changing the frequency has on the total amount of inductive reactance. Inductive reactance can be calculated by the following formula:

$$X_L = 2\pi FL$$

where $\pi = 3.14$

      F = frequency
      L = inductance in Henries, or H

You must know the value of the inductor and the frequency of the AC voltage. For example if an 80-H inductor is used in a 60-Hz circuit, the amount of opposition (inductive reactance) is 30,144 Ω (30.144 kΩ). Notice that because the inductive reactance is an opposition, its units are ohms. This calculation is as follows:

$$X_L = 2\pi FL = 2 \times 3.14 \times 60\,\text{Hz} \times 80\,\text{H} = 30,144\,\Omega$$

EXAMPLE 7-6

Calculate the inductive reactance of a 60-Hz AC circuit that has a 20-H inductor.

SOLUTION:

$$X_L = 2\pi FL = 2 \times 3.14 \times 60 \times 20 = 7,536\,\Omega$$

7.9.2 Calculating the Total Opposition for an Inductive and Resistive Circuit

The amount of total opposition (impedance) caused by an inductor and resistor in an AC circuit can be calculated. For example if a circuit has 70 Ω of resistance and 50 Ω due to inductive reactance, the total impedance must be calculated with a vector diagram similar to the ones shown in Figures 7-38c and 7-39c, because the voltage and current in this circuit are out of phase. The formula for calculating impedance of this circuit is

$$Z = \sqrt{R^2 + X_L^2}$$
$$Z = \sqrt{70^2 + 50^2}$$
$$Z = 86.02\,\Omega$$

7.10 TRUE POWER AND APPARENT POWER IN AN AC CIRCUIT

In a DC circuit you could simply multiply voltage times the current and determine the total power of the circuit. In an AC circuit you must account for the current caused by any resistors and calculate it separately from the power caused by resistors and capacitors or resistors and inductors. The reason for this is that when current is caused by a resistor, it is called true power (TP), and current caused by capacitive reactance and inductive reactance is called apparent power (AP). Apparent power does not take into account the phase shift caused by the capacitor or inductor. The power that occurs when current flows through a resistor is called true power because the resistor does not cause a phase shift between the voltage waveform and the current waveform.

The main point to remember is that true power can be used to determine the heating potential in an AC circuit. Thus if you have 1000 W due to true power, you can determine that you can get 1000 W of heating power from this circuit.

7.11 CALCULATING THE POWER FACTOR

The amount of true power and the amount of apparent power in an AC circuit combine to become a ratio called the power factor (PF). The formula for power factor is

$$PF = \frac{TP}{AP}$$

EXAMPLE 7-7

Calculate the power factor for a circuit that has 600 VA of apparent power and 500 W of true power. Note that when volts are multiplied by amps in a reactive circuit, the units for apparent power are VA, which stands for volt-amps.

SOLUTION:

$$PF = \frac{TP}{AP} = \frac{500\,W}{600\,VA} = 0.83, \quad or \quad 83\%$$

The true power in a circuit is always smaller than the apparent power, so the power factor is always less than 1. In a pure resistance circuit, the true power and the apparent power are the same, so the power factor is 1. When the power factor becomes too low, the power company adds a penalty to the electric bill that increases the bill substantially. When a factory has an electric bill of more than $20,000 per month, this penalty becomes important. In these types of applications, the power factor can be corrected by adding extra capacitance to an inductive circuit (having large or multiple motors) or by adding extra inductance to a capacitive circuit. Commercial power factor-correction systems are available for these applications.

7.11.1 How to Change the Power Factor with Inductors or Capacitors

When you are working with a power factor that is caused by induction, capacitors can be added to bring the power factor back to 1, which is also called "unity." If the power factor is caused by excessive capacitance, you can add inductance coils to correct the power factor. When solar PV systems are connected to the grid, the power factor must be corrected prior to when the connection is made. The power factor for most AC voltage generated by solar PV systems has a power factor that needs corrected by adding capacitance. Capacitor banks are used for the correction.

7.11.2 Volt Ampere Reactance (VAR)

The units of apparent power are VAR, which stands for volt ampere reactance. When the value of VAR is very large, it is measured in thousands of VAR and the designation kVAR will be used. The units of VAR are the same whether the apparent power is caused by capacitance or by inductance.

7.12 THEORY AND OPERATION OF A RELAY AND CONTACTOR

A *relay* is a magnetically controlled switch that is the main control component in an electrical system. Figure 7-40 is a picture of a typical relay, and Figure 7-41 shows a cutaway picture of a smaller plug-in type relay with its contacts and coil identified. The relay can consist of a single coil and a number of sets of contacts. The coil becomes an electromagnet when it is energized, and its magnetic field causes each set of normally open contacts to close and each set of normally closed contacts to open.

FIGURE 7-40   A typical larger relay.

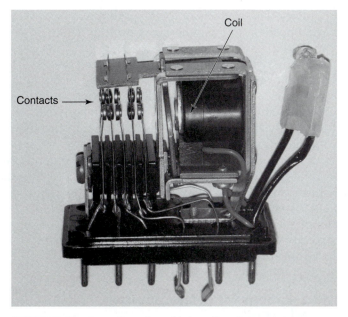

FIGURE 7-41   A smaller relay with its coil and contacts shown.

The contacts are basically a switch that is operated by magnetic force. The part of the relay that moves and causes the contacts to move is called the armature. Power is applied to the coil of the relay first, and the magnetic flux causes the armature to move and causes the contacts to change position. The coil is part of the control circuit, and the contacts are part of the load circuit.

You need to envision its operation as two separate pieces, the coil and contacts, even though they are mounted near each other and operate almost simultaneously. It is important to understand that the coil must be energized first, and a split second later the magnetic field built up in the coil will cause the contacts to move.

7.13.1 Pull-In and Hold-In Current

When voltage is first applied to a coil of a relay, it draws excessive current because the coil of wire presents resistance only to the circuit when current first starts to flow. As the flow of current increases in the coil, inductive reactance begins to build, which causes the current to become lower. The current creates a strong magnetic field around the coil, which will cause the armature to move. When the armature has moved, it causes the induction in the magnetic coil to change, so that less current is required to maintain the position of the armature.

Figure 7-42 shows a diagram of the pull-in current and the hold-in current. The pull-in current is also called the inrush current, and the hold-in current is also called the seal-in current. The pull-in current is typically three to five times larger than the hold-in current.

7.13.2 Normally Open and Normally Closed Contacts

A relay can have *normally open* contacts or *normally closed* contacts. It is important to understand that the use of the word "normal" for contacts indicates the position of the contacts when no voltage is applied to the coil. The contacts can be held in their normal position by a spring or

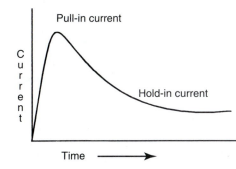

FIGURE 7-42   Diagram of pull-in and hold-in currents for relay coils. Notice that the pull-in current is approximately three to five times larger than the hold-in current.

by gravity. The contacts of a relay will move from their normal position to their energized position when power is applied to its coil.

Some types of contacts can be changed or converted from normally open to normally closed in the field. Other types are manufactured in such a way that they cannot be changed. The contacts can be converted in the field by the technician by simply removing them from the relay and turning them upside down. When a set of normally open contacts is inverted, the contacts become normally closed, and when normally closed contacts are inverted, they become normally open contacts. This means that a technician in the field can change the contacts in a relay to get the exact number of normally open or normally closed contacts needed for the application.

7.13.3 Ratings for Relay Contacts and Relay Coils

When you change a relay that is worn or broken, you must ensure that the voltage rating of the coil of the new relay matches the voltage of the control circuit exactly. This means that if the voltage for the control circuit is 24 V AC, the coil must be rated for 24 V. If the control voltage is 70 V AC, the coil must be rated for 70 V. The voltage rating for a relay coil is stamped directly on the coil. The coil may also have a color code. If the coil is rated for 24 V AC, it will be color coded black. If the coil is rated for 70 V AC, it should be color coded red or have a red-colored stamp on the coil. If the relay coil is rated for 208 or 240 V AC, it will be color coded green or identified with a green stamp or green printing on the coil. DC coils are color coded blue. It is important to understand that the current rating for a relay coil is seldom listed on the component. If it is important to know the current rating for the coil, you can look for it in the catalog or on the specification sheet that is shipped with the new relay. If you change a relay, you must also make sure that the rating for the contacts meets or exceeds the current rating and the voltage rating of the load to which it will be connected.

Contact ratings are grouped by voltage and by current. The voltage ratings are generally broken into two groups of 300 V and 600 V. This means that if you are using the contacts to control 240 or 208 V AC, you would use contacts that have a 300-V rating. If you are using the contacts to control a 480 V AC motor, you would need to use 600-V-rated contacts.

The current rating of contacts is listed in amps or horsepower. The current rating or horsepower rating must exceed the amount of current the relay is controlling. This means that if the relay is controlling 7 A, the contacts will need to be rated for more than 7 A of current. The current rating of the contacts and the voltage for the c_____ will be printed directly on the con-on the _____

7.13.4 Identifying Relays by the Arrangement of Their Contacts

Some types of contact arrangements for relays have become standardized so that they are easier to recognize when they are ordered for replacement or when you are trying to troubleshoot them. The diagrams in Figure 7-43

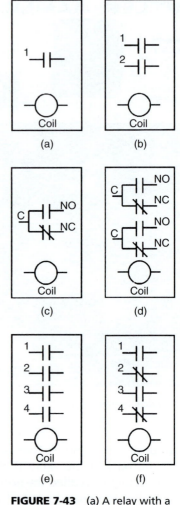

(a)   (b)

(c)   (d)

(e)   (f)

FIGURE 7-43 (a) A relay with a single set of normally open contacts. This type of relay is called a single-pole, single-throw (SPST) relay. (b) A relay with two individual sets of contacts. This relay is called a double-pole, single-throw (DPST) relay. (c) A relay with two sets of contacts that are connected at one side at a point called the common (C). The output terminals are identified as normally open (NO) and normally closed (NC). This type of relay is called single-pole, double-throw (SPDT). (d) A relay with two sets of SPDT contacts. This relay is called a double-pole, double-throw (DPDT) relay. (e) A relay with multiple sets of normally open contacts. (f) A relay with a combination of normally open and normally closed contacts.

are examples of some of the standard types of relay arrangements. Figure 7-43a shows a relay with a set of normally open contacts. This type of relay can also have a single set of normally closed contacts instead of normally open contacts. Since this relay has only one contact and it can close or open only this type of relay, it is called a single-pole, single-throw (SPST) relay. The word "pole" in this term refers to the number of contacts, and the word "throw" refers to the number of terminals to which the input contacts can be switched. Since the contact in this relay has one input and it can be switched to only a single output terminal, it is said to have a single throw.

Figure 7-43b shows a relay with two sets of normally open contacts. Since this relay has two sets of single contacts, it is called a double-pole, single-throw (DPST) relay. Double-pole refers to the two individual sets of normally open contacts, and single-throw refers to the one output terminal for each contact. When the coil is energized, both sets of contacts will move from their normally open position to the normally closed position.

Figure 7-43c shows a relay with a set of normally open and a set of normally closed contacts that are connected on the left side. The point where this connection is made is called the common point and is identified with the letter "C." When the relay coil is energized, the normally open part of the contacts will close, and the normally closed part of the contacts will open. Since these contacts have a common point at the input terminal, and two output terminals (one normally open [NO] and one normally closed [NC], the relay is called a single-pole, double-throw (SPDT) relay. The most important feature of this relay is that the contacts have two terminals on the output side, so it is called a double-throw relay. The single-pole, double-throw relay is used when two exclusive conditions exist and you do not want them to occur at the same time.

Figure 7-43d has two sets of single-pole, double-throw contacts, so it is called a double-pole, double-throw (DPDT) relay. In this case double-throw is used because two sets of normally open/normally closed contacts are provided. Each set has a common point on its left side (input side) and a terminal that is connected to the normally open (NO) set and a terminal that is connected to the normally closed (NC) set on its right side. This type of relay is used when exclusion is needed and the loads are 208 V AC or 240 V AC and they need power from both L1 and L2. In this type of application L1 will be connected to the common terminal (C) of one set of contacts, and L2 will be connected to the common terminal (C) of the other set. This will cause L1 and L2 to be switched the same way in both conditions.

Figure 7-43e shows multiple sets of normally open contacts. This type of relay can have any number of sets of normally open contacts. The additional sets of contacts can be added to the original contacts in some types of relays. If the original relay is manufactured with this provision, you can purchase the additional contacts and add them to the original relay by placing them on top of the original relay and tightening the mounting screws to make the additional contacts operate with the relay armature. The contacts for this type of relay can all be normally closed if the application requires it. The main feature of this type of relay is that it can have any number of contacts.

Figure 7-45f shows a relay with multiple sets of individual normally open and normally closed contacts. This type of relay is similar to the one shown in Figure 7-45e, except in this type of relay the contacts can be any combination of normally open or normally closed sets. In most cases the contacts in this type of relay are convertible in the field. As a technician, you can add sets of contacts and change them from normally open to normally closed or vice versa as needed. In most installations, the relays are provided in the original equipment and you will need to identify them only for installation and troubleshooting purposes.

7.13.5 The Difference Between a Relay and a Contactor

A contactor is similar to a relay in that it has a coil and a number of contacts. The main difference is that the contactor is larger and its contacts can carry more current. A relay is generally defined as magnetically controlled contacts that carry a current of less than 15 A. A *contactor* is defined as having contacts that are rated for 15 A or more. Some manufacturers do not follow the 15-A rating, so you may find a relay that has a current rating for its contacts in excess of 15 A, and you also may find contactors with contact ratings of less than 15 A. In general, the main difference is that a contactor is specifically designed so that its contacts can carry a larger amount of current, up to 2250 A. Contactors are rated by the National Electrical Manufacturers Association (NEMA), and their sizes range from 00 to 9. They may also be rated by using horsepower (hp).

In most systems the motor is controlled by a contactor instead of a relay because the amount of current the motor draws is usually in excess of 15 A.

7.13.6 NEMA Ratings for Contactors

Figure 7-44 presents a table of NEMA ratings for three-phase magnetic contactors and motor starters, showing that a size 00 contactor is rated to safely carry up to 9 A for a continuous load. You should also notice that the contactors are rated for up to 575 V. You should remember that the current rating for contactors depends on the size of the contacts, and the voltage rating of the contactors depends on the way the relay is manufactured so that arcs do not jump between different terminals. This means that contactors that are rated for a higher voltage have more plastic or insulating material between the sets of contacts so that arcs do not jump from the contacts to other parts of the relay or to other sets of contacts.

TABLE 7-1   Horsepower (Hp) and Locked-Rotor Current (LRA) Ratings for Three-Phase Single-Speed Full-Voltage Magnetic Controllers for Limited Plugging and Jogging-Duty

Size of Controller	Continuous Current Rating (Amperes)	At 200V 60Hz		At 230V 60Hz		At 380V 50Hz		At 460V 60Hz		At 575V 60Hz		Service-Limit Current Rating* (Amperes)
		Hp	LRA	Hp	LRA	Hp	LRA	Hp	LRA	Hp	LRA	
00	9	1.5	46	1.5	40	1.5	30	2	25	2	20	11
0	18	3	74	3	70	5	64	5	53	5	42	21
1	27	7.5	151	7.5	140	10	107	10	88	10	70	32
2	45	10	255	15	255	25	255	25	210	25	168	52
3	90	25	500	30	500	50	500	50	418	50	334	104
4	135	40	835	50	835	75	835	100	835	100	668	156
5	270	75	1670	100	1670	150	1670	200	1670	200	1334	311
6	540	150	3340	200	3340	300	3340	400	3340	400	2670	621
7	810	. . .		300	5000			600	5000	600	4000	932
8	1215	. . .		450	7500			900	7500	900	6000	1400
9	2250	. . .		800	13400			1600	13400	1600	10700	2590

*See clause 4.1.2

table also identifies the load as a maximum horsepower rating. This means that you can identify the load from its current rating or its horsepower rating and select the proper contactor size to safely control the load.

You should notice from this table that the next larger size of contactor is a size 0. This contactor is rated for up to 18 A on a continuous basis. The size 9 contactor is the largest, and its current rating is 2250 A.

Questions

1. Explain why the voltage at each branch of a parallel circuit is the same as the supply voltage.
2. Explain the symbols M, k, m, and μ and provide an example of how each is used.
3. Discuss the advantage of connecting switches in series with a load.
4. Explain infinite resistance.
5. List two things you should be aware of when making resistance measurements with an ohm meter.

Multiple Choice

1. In a series circuit the current in each part of the circuit is _____
   a. always zero.
   b. the same.
   c. equal to the voltage.
2. Milliamps are _____
   a. 1/1,000 of an ampere.
   b. 1/1,000,000 of an ampere.
   c. 1 million amperes.
3. If a temperature control switch, high-pressure switch, and oil level switch are connected in series with a hydraulic motor and the oil level switch is opened because of low oil level, the motor will _____
   a. still run because two of the other switches are still closed.
   b. not be affected because no other loads are connected to it in series.
   c. stop running because current flow will be zero.
4. Voltage in a parallel circuit _____
   a. increases as additional resistors are added in parallel.
   b. decreases as additional resistors are added in parallel.
   c. stays the same across parallel branches as additional resistors are added in parallel.
5. Resistance in a parallel circuit _____
   a. increases as additional resistors are added in parallel.
   b. decreases as additional resistors are added in parallel.
   c. may increase or decrease when resistors are added in parallel, depending on their size.

Photovoltaic (PV) Controllers and Inverters

OBJECTIVES FOR THIS CHAPTER

When you have completed this chapter, you will be able to:

- Explain the function of the combiner box.
- Draw a diagram of an inverter and explain its operation.
- Identify the major parts of an inverter and explain their functions.
- Identify three types of inverter circuits (variable-voltage input, pulse-width modulation, and current-source input) and explain the operation of each.
- Explain the function of a battery charge controller.
- Identify the major components in a solar energy control system.

TERMS FOR THIS CHAPTER

Absorption Stage Charging
Anti-Islanding Circuit
Boost Regulator
Buck Converter
Buck-Boost Regulator
Bulk Stage Charging
Charge Controller
Chopper
Circuit Breaker
Combiner Box
Contactor
Converter
Current-Source Input (CSI)
DC Interface Enclosure
Disconnect Enclosure
EMI (Electromagnetic Interference)
Float Stage Charging

Forward Converter
Full-Bridge Converter
Half-Bridge Converter
Inverter
Linear Power supply
Maximum Power Point Tracking (MPPT)
Pulse-Width Modulation (PWM)
Push-Pull Converter
SCADA (Supervisory Control and Data Acquisition)
Six-Step Inverter
Solar Charge Controller
Switch-Mode Power Supplies (SMPS)
Transformer
Two-Stage Battery Charging
Variable-Voltage Input (VVI)

8.0 OVERVIEW OF THIS CHAPTER

Photovoltaic systems produce DC voltage at a variety of voltages and currents. The voltage from these systems can be used in stand-alone residential systems in which the voltage can remain as DC voltage. When these systems use batteries, a charge controller must be used to ensure that the batteries are charged to their peak potential and that they are not allowed to overcharge. The charge controller also protects the batteries against being completely discharged.

In other residential and commercial applications, the DC voltage must be converted to AC voltage through an inverter so it can be used by existing AC equipment such as lighting, heating and air conditioning equipment, pumps, and fans. In industrial applications, the AC three-phase voltage must be the same quality (voltage, phase, and frequency) that is provided by the grid so that sensitive equipment will run the same whether it is using voltage from the grid or from the photovoltaic panels. Other solar applications are designed to tie the system to the grid, which means that the DC voltage that the photovoltaic panels produce must be converted through an inverter into single-phase or three-phase AC voltage at 60 Hz, and this voltage must be consistent and pure so that it can be used with existing voltage on the grid. If the voltage is three phase, its phasing must match the three-phase voltage on the grid.

The photovoltaic panels will produce different amounts of DC voltage depending on the intensity of the sunlight that strikes them. The voltage difference must be controlled to ensure the system is working to its optimum capacity. A tracking system is used on some photovoltaic systems to move the surface of the panels so that they harvest the maximum amount of solar energy.

In some applications, the voltage from the photovoltaic panels is stored directly into a bank of batteries so the voltage can be used at a time when solar energy is not charging the system such as nighttime or on cloudy days when the amount of voltage produced cannot provide all the voltage that is needed. When the photovoltaic panels are not charging the battery, the controller must manage the discharge rate and ensure that the battery is not completely discharged or that the battery bank is not damaged.

Some larger systems that provide solar power for industrial applications are basically grid-tied systems and do not use any batteries. When the photovoltaic panels are producing energy, the electrical power is used directly in the industry electrical bus that provides electrical power to the entire plant, and if more electrical power is produced than is being consumed in the factory, it is sent back into the grid. If the photovoltaic panels are not producing sufficient electricity, additional power is

used to measure the electrical power that goes into or comes from the grid.

A typical system may include the photovoltaic panels, combiner boxes, the charge controller, the batteries, and the inverter. This chapter will explain the basic theory of operation for all the components in the system and provide in-depth information about inverters, controllers, and the components and circuits that make them operate. It will also explain typical voltage controllers for solar applications. The chapter will provide typical applications and diagrams that show how the voltage control is used to efficiently operate photovoltaic systems. Other parts of the chapter will show more complex control systems and how they operate. The emphasis in this chapter will be the theory of operation of these systems and controls and how to troubleshoot and repair them.

8.1 TYPES OF APPLICATIONS THAT NEED CHARGE CONTROLLERS

The *charge controller* is required on any photovoltaic system that uses batteries. The charge controller protects the life of the batteries and prevents the solar panels from overcharging the batteries. The charge controller also stops the batteries from being drained by the solar panels at night. Some charge controllers also record information about the photovoltaic system's performance.

The charge controller may have a circuit called a low-voltage disconnect, which prevents the batteries from being discharged too deeply. The charge controller has many special features that must be matched to each system to ensure that the system is operating efficiently. For example, the charge controller needs to be matched to the type of batteries that are used, such as gel or lead-acid. The voltage level of the system (12, 24, 48, or 60 V DC) must match the maximum amps the system can generate at that voltage. In most cases, the charge controllers can be oversized to allow the solar array to be expanded in the future.

Some systems that require a controller include smaller residential systems that store the electrical energy in a battery or bank of batteries. The charge controller is used in commercial systems that have battery backup and use batteries to store excess electricity rather than sell it back to the grid. Some other types of stand-alone systems use solar panels to charge batteries and use the electricity to power lights in parking lots or power remote systems for wind instruments on highways or other remote weather instruments.

Some charge controllers are also designed to change the position of the photovoltaic panels so that they harvest the maximum amount of electrical power. This type of charge controller tracks the electrical maximum power point of a PV array to deliver the maximum available current for charging batteries.

8.1.1 Solar Photovoltaic Panels Combiner Box

The *combiner box* allows multiple solar PV panels to be connected and fused as a single voltage source before the power is sent to the charge controller. The enclosure can be rated NEMA 3R, which means that it is rainproof and sleet resistant for outdoor use or it can be rated NEMA 4, which means that it is watertight for indoor uses. With these ratings, the NEMA 3R combiner box can be installed directly on the rooftop where it is exposed to all types of outdoor weather, or the NEMA 4 box can be installed inside a partial building enclosure on the roof or immediately inside the building where the power is brought in. The terminal connections in the combiner box are rated for 600 volts, which makes the panel usable for residential or industrial usage. The box has 10 circuits that are rated for a total of 120 A. Each of the circuits is fused for 15 A and provides a connection for one solar panel. Larger combiner boxes can be rated for up to 200 A. Figure 8-1 shows a typical combiner box with its door open, and Figure 8-2 shows that the connections for each PV array are on the panel board and they are set for making connections at fuse blocks. The next part of the photovoltaic control system is the charge controller.

8.2 BASIC OPERATION OF A SOLAR CHARGER CONTROLLER

The *solar charge controller* can be designed to charge small batteries on boats, RVs, or other small residential solar systems that produce electrical power for directly charging batteries. The solar charger can also be designed to control the larger industrial systems in which batteries are

FIGURE 8-2 The fuse block inside a junction box. Fuses could also be inside a combiner box. (Courtesy Ron Swenson, www.SolarSchools.com)

charged as backup energy. Figure 8-3 shows the Schneider Electric Xantrex XW60 MPPT battery charge controller. This is a larger solar charge controller. The next sections will explain the operation of the charge controller and how it provides protection for the batteries and other components in the system.

FIGURE 8-1 Combiner box that allows multiple PV panels to be connected. (Courtesy Ron Swenson, www.SolarSchools.com)

FIGURE 8-3 Schneider Electric Xantrex XW60 MPPT battery charge controller. (Courtesy of Schneider Electric Renewable Energies Business Unit.)

8.2.1 Maximum Power Point Tracking System

Some large industrial charge controllers have a circuit that helps in tracking the sun and continually moves the solar array so that it is always receiving the maximum amount of solar energy. If the PV solar array is mounted in a fixed position, it will not be able to harvest the maximum amount of solar energy all day long as the sun moves across the sky. The most efficient PV arrays are movable and have a tracking system that allows the panels to change position as the sun moves to ensure that the panels are receiving the maximum amount of solar energy. This tracking system is called the *maximum power point tracking* (MPPT) system. When charging, the controller regulates battery voltage and output current based on the amount of energy available from the PV array and level of charge of the battery. During the day the sun will track across the sky, and when the controller senses a change in the amount of maximum solar energy that is coming from the sun because it has moved, the MPPT system will adjust the position of the panels. This control system uses an algorithm (mathematical formula) designed to maximize energy harvest from the PV array. The MPPT constantly adjusts the operating points of the array to ensure that they stay on the maximum power point. On some brands of controllers, the tracking can be accomplished without stopping the energy harvest to sweep the array. This feature is beneficial in all sunlight conditions, especially in areas with fast-moving cloud cover and quickly changing solar conditions. The next section will explain how the charge controller manages the level of charge for the batteries to ensure that they receive the maximum charge and do not overcharge the battery.

8.2.2 Two-Stage and Three-Stage Charging States for the Charge Controller

The solar controller is basically a battery charger and is designed to control how the batteries are charged by the DC source (the PV array). Battery charging can be accomplished in two or three stages. The first stage of charging is called *bulk stage charging* and occurs when the maximum amount of electrical power is sent to the batteries for charging. If the batteries are discharged, the controller operates in constant current mode, delivering its maximum current to the batteries until they are nearly fully charged.

The second charging state is called the *absorption stage*, the point at which the battery voltage is nearly at the full charge point. When the battery reaches the absorption stage, the controller then operates in constant voltage mode, holding the battery voltage at the absorption voltage setting for a preset time limit (the default time such as 3 hours). During this time, current falls gradually as the battery capacity is reached.

The third stage of charging is called the *float stage charging.* During this stage, the battery is nearly fully charged, the voltage is controlled, and the current is limited. During the float stage, the battery controller lowers the amount of current flowing to the battery but keeps the voltage just below maximum level so it does not overcharge the battery. When the charge controller goes into the float stage charging, it begins to limit the amount of current that is sent to the battery, and charging tapers off.

The charge controller transitions to float stage charging when any one of three criteria are met:

1. The charge current allowed by the batteries falls below the exit current threshold, which is equal to 2% of battery capacity (for a 500 amp-hour battery bank, this would be 10 A), for 1 minute.
2. The battery voltage has been at or above the float voltage (which it reached during the bulk stage) for 8 hours.
3. The battery voltage has been at the bulk/absorption voltage setting for a preset time limit (the maximum absorption time).

8.2.3 Two-Stage Battery Charging

Some solar battery controllers provide only the *two-stage battery charging* process. The two stages include the bulk stage charge and the absorption stage charge. The float stage is not used; instead when the battery is at the full charge state, the controller does not allow any current to flow to the battery. As current is used by the load, the controller begins to lower the battery voltage, resorts to the absorption stage, and begins to charge the battery again. When the battery is at near charge and very little load current is being used, the two-stage charge controller changes between the absorption stage and the full charge mode. A voltage-monitoring circuit is used to continually monitor the charge state of the battery and determine the amount of charge that is needed. The different types of batteries and the types of charges that they need are discussed in Chapter 7, and you can refer back to that chapter to review this information.

8.2.4 Automatic Night Disconnect for the PV Array

At times, especially during the night or on cloudy days, the output voltage level of the PV array will drop below the voltage level of the batteries. When this occurs, the voltage from the batteries can begin to flow back into the PV array. The charge controller has a circuit specifically designed to senses the low voltage output from the PV array, and this circuit opens an internal relay to prevent battery current from flowing back to the PV array. When the controller is in this mode, it draws minimal power from the batteries. Since the relay contact opens the circuit completely, it eliminates the need for a blocking diode between the battery and the PV array. The controller may need a diode if the PV array consists of amorphous solar modules or thin-film to prevent damage when cloud cover is intermittent and occasionally the output voltage of the solar PV array drops below the battery-charging level.

8.3 BASIC CONTROL DIAGRAMS FOR PHOTOVOLTAIC SYSTEMS

Figure 8-4 shows a block diagram of all the components needed to control a photovoltaic array that is connected to the controller. The solar panel array is in the top left. The diagram shows that the voltage from the solar panels can be used directly through an inverter to provide power to a residence or small commercial application, along with power from the grid. If the solar panels provide more power than the residence or commercial application needs at any time, the extra power flows through a net meter back into the grid. If batteries are used for storage, the DC voltage provided from the solar array can be sent through a charge controller into the batteries where it can be stored. The controller ensures the batteries are not overcharged and are not allowed to be drawn down through discharge to an unsafe level. Figure 8-5 shows an electrical block diagram of a typical solar energy controller. The entire solar energy system consists of all the parts shown in the diagram from the photovoltaic panels on the far right side to the output section on the left.

8.3.1 Solar PV Panels

The electrical block diagram shows that the solar panels are the energy converter for the system. The photovoltaic panels convert sunlight to DC electricity. The solar panels can be a single panel or an array of multiple panels. The panels are connected to positive and negative terminals on the controller. It is important to remember that the polarity of the connections must be followed closely. The solar array is grounded at a single-point earth ground. The solar panels are grounded as are other parts of the system. The electrical circuit can have a positive ground or a negative ground to meet local or other codes.

8.3.2 DC Interface Enclosure

The next section of the system is the *DC interface enclosure:* a terminal board (TB3), a DC disconnect switch, and a DC contactor. The disconnect switch is a manual switch, and the contactor is controlled by a magnetic coil. The DC interface provides a means of disconnecting all DC power from the solar array from the main controller. Any time major repair work must be accomplished, the disconnect will allow that part of the system to be disconnected and locked out.

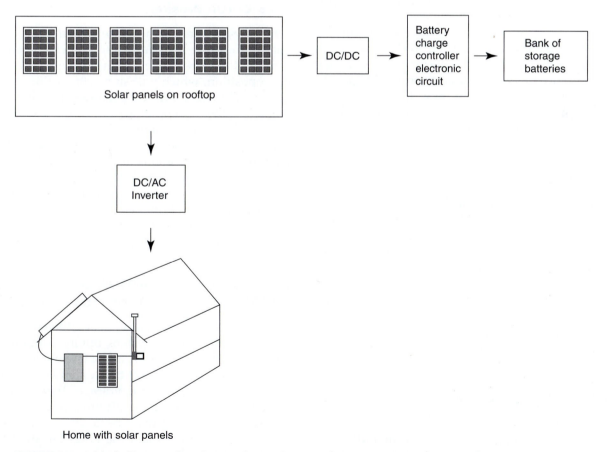

FIGURE 8-4 A block diagram of a solar panel array that provides power directly to a residence or commercial establishment through an inverter, or the power can be stored in batteries for later use.

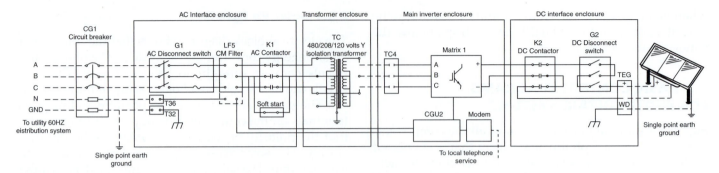

FIGURE 8-5 A block electrical diagram of six parts of a simple photovoltaic system and controls. (Courtesy Schneider Electric Renewable Energies Business Unit.)

8.3.3 Main Inverter Enclosure

The main *inverter* section includes the inverter electronic circuit that changes DC voltage to AC voltage. The DC voltage enters the inverter section as a pure DC voltage without any frequency. The amount of voltage can be any value from the minimum to the maximum that the PV array can produce. The amount of voltage will vary with the amount of sunlight that is harvested. For example, in the early morning and late afternoon, the voltage will be less than the amount the array produces at midday near noon. The amount of voltage will also be less on cloudy days. The amount of AC voltage that comes out of the inverter is proportional to the DC voltage that goes into the inverter.

A modem is provided to send data about the amount of energy that is converted to data collection center. The data that are sent track the time of day and the amount of energy that is produced from the PV panels. These data also can indicate any faults or other symptoms that can be used to determine the health of the system. Inverters are explained in more detail in a later section.

8.3.4 Transformer Enclosure

The transformer section takes the AC voltage and steps it up to the usable voltage for 3 phase industrial voltage at 480 or 208 V which can also provide 120 AC line to neutral for a residences. The neutral can be grounded or ungrounded for the lower voltage. The *transformer* shown in the diagram is a three-phase wye connected transformer, which provides isolation from the primary to the secondary side. The secondary voltage in this controller can be 208 or 480 voltage. The VA rating is 100kVA, which means that the transformer is large enough to handle over 200 A at 480 V.

8.3.5 AC Interface Enclosure

The AC interface enclosure houses the AC disconnect switch, which is a manual switch and an EMI filter. The *EMI (electromagnetic interference)* filter helps limit interference that may be injected into the power system from

the electronics in the circuit or from other components. The AC contactor is a large relay that is controlled by a relay coil. The relay coil can receive a signal from any computer-controlled device or programmable logic controller (PLC), which means that the system can be controlled from a distance from a *SCADA* (supervisory control and data acquisition) system. This allows the disconnect to be opened manually or from an electrical control signal. Any time the system requires maintenance, the AC manual disconnect can be move to the open position and locked out for safety while the work is completed.

8.3.6 Circuit Breakers

Circuit breakers protect the entire system from an overload condition. If the load causes too much current to flow through the system, the circuit breakers will trip to the open position, which will stop current flow in all three circuits. The circuit breaker is a three-phase breaker, but it will trip if the current in any one of the legs exceeds the maximum current amount. When the circuit breaker trips, all three of the circuits open at the same time. The load in these applications may consist of the entire plant that has the potential to use much more electrical power than the PV system can supply, so it is important that the PV equipment and circuits are protected by fuses or circuit breakers to prevent an overload condition to occur that draws too much amperage from the system.

8.4 POWER DISTRIBUTION DIAGRAM FOR A LARGER INDUSTRIAL PHOTOVOLTAIC SYSTEM

The larger photovoltaic systems will have similar components to the smaller system, but in addition they will have capacitors, inductors, and a few other components that are required to protect the system and make it fully compatible with the grid voltage. It is important to remember that any time a photovoltaic system is connected to the grid, it must be certified that its voltage meets all the standards of the grid, that the PV equipment is protected from damage from the grid system, and that the PV system does not have a failure that can

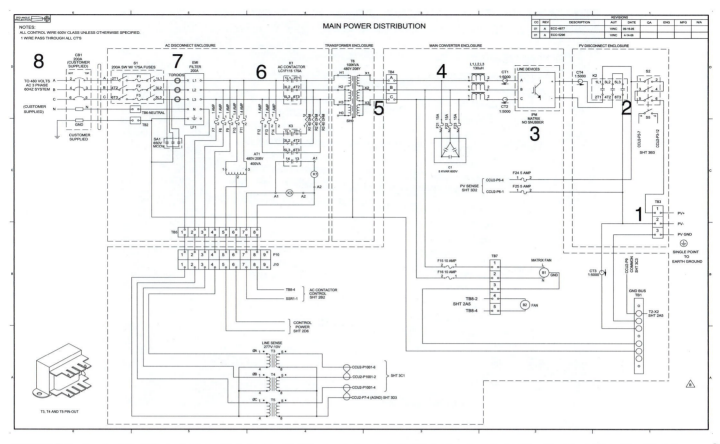

FIGURE 8-6 An electrical block diagram that shows all of the parts for an industrial solar energy circuit. (Courtesy Schneider Electric Renewable Energies Business Unit.)

affect the grid. Figure 8-6 shows an electrical block diagram of all the sections for a commercial-grade solar energy system. The sections are numbered so it is easier to see what part of the circuit is being discussed. This section will explain all the parts and in the system and show some of the components. In some systems, all the equipment in the diagram is mounted in one cabinet. In other systems, the equipment is mounted in multiple cabinets around the rooftop. Figure 8-7 shows a typical solar energy system installed on the rooftop of a large industrial site.

8.4.1 Connection for PV Panels

Number 1 on the diagram shows the PV panels and where they are connected. Since industrial photovoltaic systems are typically larger than residential systems, the panels are connected in arrays. Notice the connections are made at terminal board, TB3. Notice that terminal 1 on TB3 is for the positive (+) lead and terminal 2 on TB3 is for the negative (−) lead. If multiple arrays of photovoltaic panels are used, they must be combined with their polarity connected correctly.

If multiple photovoltaic panels are connected into an array, the connections are made in a combiner box. Figures 8-8 and 8-9 show a combiner boxes and

junction boxes mounted on the rooftop directly under multiple solar panels and near solar arrays on the roof. Weather-tight electrical conduits provide protection for all the interconnected wires as they are routed between the solar panels and the combiner box.

8.4.2 PV Disconnect Enclosure and Solar System Meter

The next section of the system is identified with the number 2 in Figure 8-6. This part of the system is the photovoltaic (PV) *disconnect enclosure*. This part of the system has a manual switch (S2) that disconnects the voltage from the PV panels a contactor (K2) that is controlled by a relay coil. There is also a CT4, which is a current transformer that indicates the amount of DC current the system is producing. Figure 8-10 shows the solar system meter and the disconnects mounted on the ground where the systems ties into the incoming power that originally supplied the building prior to the installation of the solar energy system. The solar electrical system disconnect is identified in the picture by the letter *a*; the solar electrical meter is letter *b*; the disconnect switches for two small inverters are letter *c*; the solar electrical system splice box is letter *d*; two isolation transformers are shown at letter *e*; and two smaller

FIGURE 8-7   Large array of solar panels on the rooftop of a large industrial building. (Courtesy Ron Swenson, www.SolarSchools.com)

FIGURE 8-8   A combiner box mounted directly under a set of solar panels. Notice the weather-tight electrical conduit feeding the combiner box at the bottom.   (Courtesy Ron Swenson, www. SolarSchools.com)

inverters are shown at letter *f*. The meter is a net meter that can measure the power coming into the system from the grid if the solar energy system cannot provide all the power needed during the day and at night. The disconnect switches allow the system to be isolated whenever work needs to be completed on the system, or if the power from the solar energy system needs to be disconnected from the remainder of the electrical power system. If the system is using data acquisition, the communications box can be connected through the same equipment boxes.

8.4.3 Converter Section

The next section in this system is shown at number 3 in Figure 8-6. The converter has an inverter circuit that converts DC voltage to AC voltage. The actual circuit inside the inverter will be explained in detail in the next sections of this chapter. The inverter takes DC voltage that comes from the photovoltaic array and converts it to AC voltage at exactly 60 Hz. The voltage from the solar array will change as the amount of sunlight changes. Figure 8-11 shows a rooftop mounted inverter in the cabinet and a disconnect switch mounted near it. Some systems use a number of smaller inverters near

FIGURE 8-9 Additional junction boxes mounted around the rooftop near solar arrays. (Courtesy Ron Swenson, www.SolarSchools.com)

the PV arrays, whereas other systems may use one or two larger inverters that are mounted in a central location.

8.4.4 Capacitor and Inductors

The next components in the system shown in the electrical block diagram are capacitors and inductors, which are shown at number 4 in the diagram. The capacitors and inductors filter the AC power. The capacitors filter the AC voltage, and the inductors filter the AC current. The capacitors can also be used to help correct the power factor for the system. You will learn more about power factor and the problems that are created when it is not corrected in Chapter 12.

8.4.5 Transformer Section

The transformer is shown as number 5 in the diagram and it is a 100kVA transformer that provides 480 V.

225 KW Inverter   DC Disconnect

FIGURE 8-11 Inverter and disconnect for solar photovoltaic system.   (Courtesy Ron Swenson, www.swenson.com)

FIGURE 8-10 (a) Solar electrical system disconnect, (b) solar electrical meter, (c) the disconnect switches for two small inverters, (d) solar electrical system splice box, (e) two isolation transformers, (f) two smaller inverters.   (Courtesy Ron Swenson, www. SolarSchools.com)

The transformer steps up the voltage from the inverter to 480 V. The transformer is called an isolation transformer since there is no connection between the primary and secondary windings. Figure 8-12 shows the isolation transformer mounted on the left side of a 225 KW inverter where the DC voltage from the PV array is changed to AC voltage. The isolation transformer is used to step up the AC voltage.

8.4.6 AC Contactor

The AC *contactor* is shown as number 6 in the diagram. The contacts of the contactor are controlled by a coil, and the voltage for the coil can be provided by a computer or PLC controller that is connected to a SCADA system. The contactor provides complete isolation between the output circuit where the load is connected and the inverter.

8.4.7 EMI (Electromagnetic Interference) Filter

The EMI filter is number 7, and it provides filtering of any unwanted signals or interference that may get into the AC voltage or current. The filters help keep the signal clean so that it can be used for computer loads or other sensitive loads.

8.4.8 Circuit Breaker and Output Section

Number 8 shows the circuit breaker and output section for the system. The circuit breaker is a three-phase circuit breaker that will protect any load using the voltage from over current. The circuit breaker can be reset and used again if it trips. If one of the legs of voltage senses an overcurrent, all three circuits will be opened at the same time.

8.5 ANTI-ISLANDING CIRCUITS AND OTHER PROTECTION CIRCUITS

Any time a PV system is connected to the grid, a number of problems may occur that put the PV equipment at risk if there is a problem on the grid, such as a lightening strike or other major failure. There can also be problems for the grid if the electrical equipment such as transformers or inverters in the solar energy equipment fails while it is connected to the grid. The term "islanding" describes a condition in which parts of the grid disconnect and leave other parts still connected. Sometimes this occurs by accident, and other times the sections of the grid are isolated so that one or more of them can operate and supply power to a section of the grid while other parts are disconnected. For example, when there is a hurricane, tornado, or ice storm, a part of the grid that is damaged may need to be disconnected from the main grid and power may need to be turned off to this part so major repairs can be made. In these cases, islands are created intentionally. There are other cases when a failure causes parts of the grid to go down and small energy suppliers such as solar energy or wind energy equipment ends up being the sole provider of electrical power, which may allow it to overload and become damaged. Another problem that can occur when parts of the grid are de-energized and other power generation equipment such as solar energy systems are allowed to continue to produce electrical power is that the electrical power can back-feed or back energize parts of the grid that workers are trying to isolate so they can work on them safely with power removed. In these applications, some islands are allowed, if the appropriate steps are taken to be sure the grid itself is not being back-fed. But unintentional islands that provide uncontrolled voltage to be fed back into the grid are absolutely forbidden. Most switch gear and electronic controls used with solar farms and solar panels have a circuit called an *anti-islanding circuit* that prevents them from continually adding electrical power to the grid if the section of the grid they are connected to is taken down by problems or on purpose.

FIGURE 8-12 Inside view of inverter during installation. (Courtesy Ron Swenson, www.SolarSchools.com)

Islanding conditions are monitored by electronic circuits in most inverters by some combination of the following:

1. A change in the level of voltage magnitude that is sudden and unexpected.
2. A change in system frequency caused by one or more generation sources.
3. A sudden increase in active output power (kW) well beyond the typical level that would indicate a short circuit somewhere in the system.
4. A change in power factor or reactive output power (kVAR) that is different from existing conditions.

Any of these conditions may indicate a failure or trip condition by a small generator disconnecting or some section of the load becoming disconnected from the grid. These conditions can occur with smaller residential systems or larger commercial or industrial systems. If any of these conditions occur, your solar energy system may need to disconnect quickly from the grid to protect it and to prevent its generated power from back-feeding the grid.

The IEEE Std. 1547 covers and explains the conditions in which anti-islanding protection should operate, and what levels of parameters define abnormal conditions. In individual inverters, the safety circuits usually include *phase locked loop* (PLL) circuits that ensure that the system voltage is constantly monitored for problems. The IEEE Std. 1547 specifies the types of protection for anti-islanding that must be included in any solar energy equipment that is connected to the grid.

8.6 BASIC OPERATION OF AN INVERTER

The inverter circuit has been discussed in general in several of the previous sections. This section will provide a more detailed discussion about the different types of inverters and the components in them. The input and output circuits for the different types of inverter circuits will also be shown and discussed.

In solar energy systems, the inverter is considered a complete electrical controller and does much more than just convert DC voltage to AC voltage. Inverters generate AC electricity with one of three wave characteristics, which include a sine wave, a quasi-sine wave, or a square wave. The sine wave inverter is usually more expensive than the others, but it may be required to provide AC voltage to some sensitive electronics loads and it is also typically recommended for large systems. The quasi-sine wave and the square-wave inverters are less expensive, so they are usually used when the load is typically a lighting load or another load that does not require a pure sine wave.

Another way to discuss the three types of inverters is to include the circuits they use, such as the variable-voltage

input (VVI), pulse-width modulation (PWM), and current-source input (CSI). These will be discussed later in this section. Today, factories and other industrial applications are using solar energy systems and the power they use from the solar energy system must have voltage that has specific voltages and frequencies for these industrial applications.

8.6.1 Single-Phase Inverters

The simplest inverter to understand is the single-phase inverter, which takes a DC input voltage and converts it to single-phase AC voltage. The main components of this inverter can be either four silicon-controlled rectifiers (SCRs) or four transistors. This circuit was originally called a DC-link converter, now it is simply called an inverter.

The diagram in Figure 8-13 shows four SCRs used in the inverter circuit. SCR_1 and SCR_4 are fired into conduction at the same time to provide the positive part of the AC waveform and SCR_2 and SCR_3 are fired into conduction at the same time to provide the negative part of the AC waveform. The waveform for the AC output voltage is shown in this figure, and you can see that it is an AC square wave. A phase-angle control circuit is used to determine the firing angle, which provides the timing for turning each SCR on so that it provides the AC square wave. The load is attached to the two terminals (A and B) where the AC square-wave voltage is supplied.

FIGURE 8-13 An isolation transformer on the left and a 225-KW inverter on the right. (Courtesy Ron Swenson, www.SolarSchools.com)

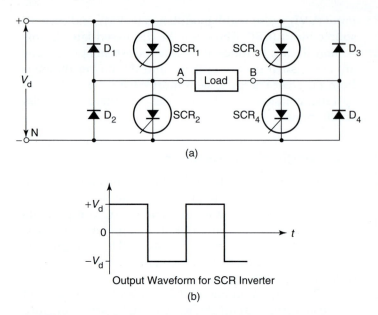

(a)

Output Waveform for SCR Inverter

(b)

FIGURE 8-14 (a) Electrical diagram of a typical inverter circuit that uses four silicon-controlled rectifiers (SCRs). (b) output waveform for SCR inverter.

8.6.2 Using Transistors for a Six-Step Inverter

Figure 8-15 shows the electrical diagram of an inverter that uses four transistors instead of four SCRs. Since the transistors can be biased to any voltage between saturation and zero, the waveform of this type of inverter can be more complex and can look more like the traditional AC sine wave. The waveform shown in this figure is a six-step AC sine wave. Two of the transistors will be used to produce the top (positive) part of the sine wave, and the remaining two transistors will be used to produce the bottom (negative) part of the sine wave.

When the positive part of the sine wave is being produced, the transistors connected to the positive DC voltage are biased in three distinct steps. During the first step, the transistors are biased to approximately half-voltage for one-third of the period of the positive half-cycle. Then these transistors are biased to full voltage for the second-third of the period of the positive half-cycle. The transistors are again biased at the half-voltage for the remaining third of the period. This sequence is repeated for the negative half-cycle. This means that the transistors that

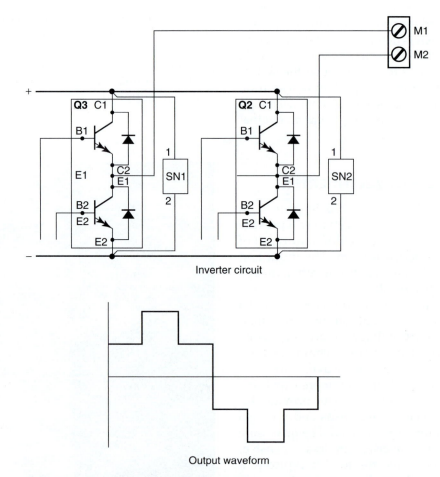

Inverter circuit

Output waveform

FIGURE 8-15 Electronic diagram of a transistor inverter with the output waveforms for the AC voltage.

are connected to the negative DC bus are energized in three steps that are identical to the steps used to make the positive half-cycle.

Since six steps are required to make the positive and negative half-cycles of the AC sine wave, this type of inverter is called a *six-step inverter*. The AC voltage for this inverter will be available at the terminals marked M1 and M2. Even though the AC sine wave from this inverter is developed from six steps, the motor or other loads see this voltage and react to it as though it were a traditional smooth AC sine wave. The timing for each sine wave is set so that the period of each is 16 milliseconds (ms), which means it will have a frequency of 60 Hz. The frequency can be adjusted by changing the period for each group of six steps if the inverter were being used on a 50-Hz system.

The amount of AC voltage available at the M1 and M2 terminals can be varied to any value from zero to the maximum provided by the DC voltage since the transistors can be turned on or biased to allow or control the amount of voltage through them.

8.6.3 Three-Phase Inverters

Three-phase inverters are much more efficient for industrial applications for which large amounts of voltage and current are required. The basic circuits and theory of operation are similar to the single-phase transistor inverter. Figure 8-16 is a diagram of a three-phase inverter with three pairs of transistors. Each pair of transistors operates like the pairs in the single-phase six-step inverter. This means that the transistor of each pair that is connected to the positive DC bus voltage will conduct

to produce the positive half-cycle, and the transistor that is connected to the negative DC bus voltage will conduct to produce the negative half-cycle.

The timing for these transistors is much more critical since they must be biased at just the right time to produce the six steps of each sine wave, and they must be synchronized with the biasing of the pairs for the other two phases so that all three phases will be produced in the correct sequence with the proper number of degrees between each phase.

8.6.4 Variable-Voltage Inverters (VVIs)

A *variable-voltage inverter (VVI)* is basically a six-step, single-phase or three-phase inverter. The need to vary the amount of voltage to the load became necessary when these inverter circuits were used in AC variable-frequency motor drives and welding circuits. Originally these circuits provided a limited voltage and limited variable-frequency adjustments because oscillators were used to control the biasing circuits. Also many of the early VVI inverters used thyristor technology, which meant that groups of SCRs were used with chopper (switching) circuits to create the six-step waveform. After microprocessors became inexpensive and widely used, they were used to control the biasing circuits for transistor inverters to give these six-step inverter circuits the ability to adjust the amount of voltage and the frequency through a much wider range. Motors needed the adjustable frequency to increase or decrease their speeds from their rating that was determined by the number of poles the motor has when it is manufactured. The voltage of the

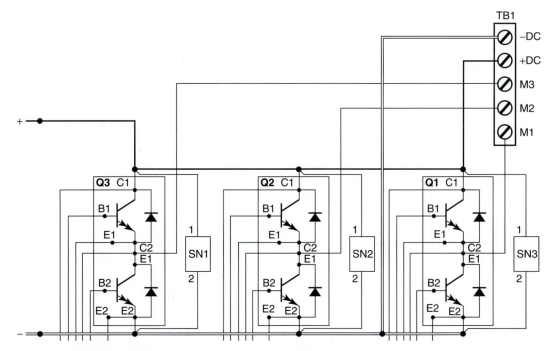

FIGURE 8-16 Electrical diagram of a three-phase inverter that uses six transistors.

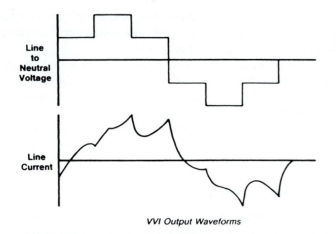

FIGURE 8-17 Voltage and current waveforms for the variable-voltage input (VVI) inverter.

drive needed to be constantly adjusted as the frequency was adjusted so that the motor received a constant ratio of voltage to hertz to keep the torque constant. This became a problem at very low speeds at which motors tended to loose torque.

Figure 8-17 shows the voltage and current waveform for the VVI inverter. The voltage is developed in six steps and the resulting current looks like an AC sine wave. These are the waveforms that you would see if you placed an oscilloscope across any two terminals of this type of inverter.

8.6.5 Pulse-Width Modulation Inverters

Another method of providing variable-voltage and variable-frequency control for inverters is to use *pulse-width*

modulation (PWM) control. This type of control uses transistors that are turned on and off at a variety of frequencies. This provides a unique waveform that makes multiple square-wave cycles that are turned on and off at specific times to give the overall appearance of a sine wave. The outline of the waveform actually looks very similar to the six-step inverter signal. An example of this type of waveform is provided in Figure 8-18. The overall appearance of the waveform is an AC sine wave. Each sine wave is actually made up of multiple square-wave pulses that are caused by transistors being turned on and off very rapidly. Since the bias of these transistors can be controlled, the amount of voltage for each square-wave pulse can be adjusted so that the entire group of square waves has the overall appearance of the sine wave. If you look at the voltage waveform for the PWM inverter, you will notice that the outline of the AC sine wave still looks like the six-step sine wave originally used in the VVI inverters. The height of the steps of the AC sine wave is also increased when the voltage of the individual pulses is increased. This increases the total voltage of the sine wave that the PWM inverter supplies.

The width (timing) of each square-wave pulse can also be adjusted to change the period of the group of pulses that makes up each individual AC sine wave. When the width of the sine wave changes, it also changes the period for the sine wave. This means that the frequency is also changed and is controlled for the PWM inverter by adjusting the timing of each individual pulse. Since adjusting the voltage and frequency is fairly complex, the PWM inverter uses a microprocessor to control the biasing of each transistor. If thyristors are

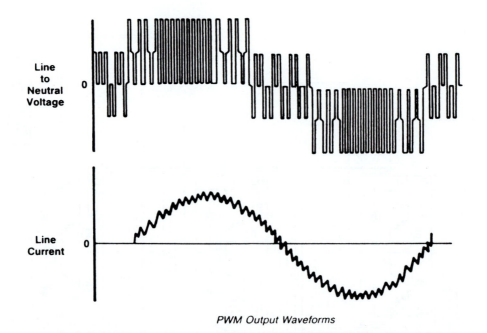

FIGURE 8-18 Voltage and current waveforms for the pulse-width modulation (PWM) inverter. Notice that the overall appearance of each waveform is an AC six-step sine wave and that it is actually made of a number of square-wave pulses.

used as in SCRs, the microprocessor will control the phase angle for the firing circuit.

Early PWM circuits used thyristors such as SCRs to produce the square-wave pulses. The control circuit included triangular carrier waves to keep the circuit synchronized. This sawtooth waveform was sent to the oscillator circuit that controlled the firing angle for each thyristor. Today, the PWM inverters use transistors mainly because of their ability to be biased from zero to saturation and back to zero at much higher frequencies. Modern circuits will more than likely use transistors for these circuits because they are now manufactured to handle larger currents that are well in excess of 1500 A.

8.6.6 Current-Source Input (CSI) Inverters

The current-source input (CSI) inverter produces a voltage waveform that looks more like an AC sine wave and a current waveform that looks similar to the original on/off square wave of the earliest inverters that cycled SCRs on and off in sequence. This type of inverter uses transistors to control the output voltage and current. The on time and off time of the transistor are adjusted to create a change in frequency for the inverter. The amplitude of each wave can also be adjusted to change the amount of voltage at the output. This means that the CSI inverter like the previous inverters can adjust voltage and frequency usable in variable-frequency motor drive applications or other applications that require variable voltage and frequency. Figure 8-19 shows the voltage and current waveform for the CSI inverter.

8.6.7 DC-To-DC Control (Converters and Choppers)

The information provided in this section may be found in only a limited number of more complicated control systems for solar energy equipment. The conversion of DC voltage to a different value of DC voltage may be required in some systems and one way to accomplish this is with a circuit called a *chopper*. The chopper was originally specifically designed to convert a fixed DC voltage

into variable DC voltages primarily used to control the speed of DC motors. Since DC voltage is not readily available as a supply source in industry, these circuits must rely on a rectifier circuit to change AC supply voltage to DC. In this section, we will treat these circuits from the point at which DC voltage is supplied to them from the rectifier circuits. In this sense, they will be classified as a DC-to-DC converter. When troubleshooting these circuits, you must test the rectifier circuits to ensure that a sufficient supply of DC voltage is available, even though the circuit diagram does not include the AC-to-DC rectification. Today, conditions may exist that require the change of DC voltage from one level to another. There will also be times when a DC and AC uninterruptible power supplies are needed.

8.7 OVERVIEW OF DC-TO-DC VOLTAGE CONVERSION

DC-to-DC voltage conversion is widely used in power supply circuits because every piece of equipment that has an electronic board in it requires a wide variety of DC voltage supplies. In these circuits, the DC voltage is changed to AC voltage through an inverter, and then the AC voltage is manipulated back to DC. This may seem a strange way to accomplish the change, but in larger power supplies, the voltage and current may be easier to manipulate to provide the change in voltage and current levels if the voltage goes through a circuit in which it is in the form of AC. When ancillary equipment is used with the solar energy equipment, each circuit may need its own voltage supplied from a power supply. This means that computers, PLCs, and all other electronic equipment may require separate DC power supplies to operate correctly. Today, the older chopper circuits have been modified into the new power supply technology with newer types of circuits and they are all more commonly called *converters*. Most often you will find converter circuits today in *switch-mode power supplies* (SMPS).

8.7.1 Linear Power Supplies

The operation of a *linear power supply* is simple, but its efficiency is quite poor, in the range of 30% to 40%. Figure 8-20 is the electronic diagram for a typical linear power supply. The first part of the power supply is exactly like the rectifier sections presented earlier in this chapter. The power supply uses a transformer and a four-diode full-wave bridge rectifier to produce the pulsing DC output waveforms. A capacitor is used as a filter to smooth out the DC. The remainder of the circuit contains a regulator and the load. The regulator is the part of the circuit that makes a linear power supply different from the newer switch-mode power supplies.

The regulator in this circuit acts as a voltage divider between the regulator and the load. To understand this

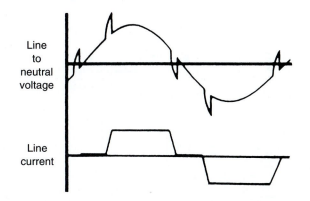

FIGURE 8-19 Voltage and current waveform for the current-source input (CSI) inverter.

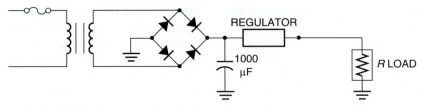

FIGURE 8-20 Example of a simple linear power supply. (Courtesy NXP Semiconductors.)

operation you need to think of the load as a fixed resistance and the regulator as a variable resistance. As you know, when two resistances are connected in series, such as the regulator and the load resistance, the amount of voltage supplied to them will be shared. The amount of voltage will be split by the ratio of the resistance. For example if the ratio of the resistance is 2:1, the regulator will have twice as much voltage measured across it than the load. If the power supply delivered 30 V, the regulator would have 20 V dropped across it, and the load would have 10 V measured across it.

If the resistance of the regulator was changed so that the ratio of resistance with the load was 1:1, the voltage across the regulator would drop to 15 V, and the voltage measured across the load terminals would be increased to 15 V. This means that the voltage to the load terminals will change any time the resistance in the regulator changes. This type of circuit is simple to operate, which makes the linear power supply easy to manufacture and troubleshoot. The problem with this type of power supply is that all the voltage that is dropped across the regulator is wasted energy. If the regulator drops 10% of the voltage and the load gets 90%, the power supply is operating somewhat efficiently. If the regulator drops 90% and the load gets 10%, this is a tremendous amount of wasted energy. The other drawback of the linear power supply is that since it must drop a portion of the supply voltage through the regulator, the

components in the regulator must be sized large enough to handle the excess heat that is generated. This tends to make the linear power supply up to two times larger (and heavier) than the new switch-mode power supplies.

8.7.2 Switching Power Supplies

Switch-mode power supplies (SMPS), also called switching power supplies, have become more popular than linear power supplies in the past 10 years because they provide a regulated voltage with more efficiency and they do not require the larger transformers and filtering devices that the linear power supplies require. For example, the linear power supplies generally have average conversion efficiencies of 30%, whereas the SMPS have efficiencies up to 80%. Since the SMPS do not need the larger components, they are more usable in modern circuits in which cabinet space and board space are at a premium. Designers are continually trying to reduce the size and weight of electronic controls, and one easy way has been to change to SMPS.

Figure 8-21 shows an electronic block diagram of a switch-mode power supply. This diagram will help you understand how the SMPS converts a DC input voltage to a new value of DC voltage that is filtered and regulated. The first block of the power supply is the rectifier and filter section; it is shown in the diagram as a diode and capacitor, indicating the AC voltage is rectified to

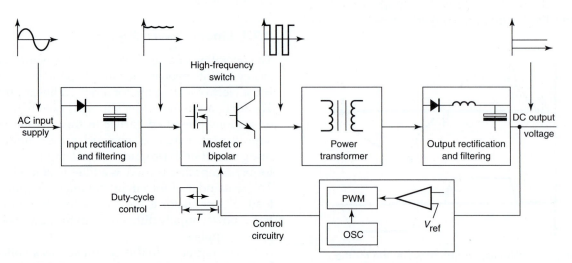

FIGURE 8-21 A switch-mode power supply. (Courtesy NXP Semiconductors.)

pulsing DC and then filtered to reduce the amount of ripple. The second block in the diagram shows the symbols of a MOSFET (metal oxide semiconductor field effect transistor) and bipolar transistor. This section is the high-frequency switching section and it uses either MOSFETs or bipolar transistors to convert the DC voltage to a high-frequency AC square wave. The high-frequency AC square wave can be 20–100 kHz. The incoming AC voltage is rectified to DC and then the high-frequency switching section changes it back to AC for several reasons. First, the incoming voltage is always fluctuating and it is full of transient voltages that could be damaging to solid-state components if they are allowed to reach them. The two-step conversion helps isolate these fluctuations and transients. The second reason is that higher frequencies allow for higher conversion efficiencies.

The next section of the SMPS is the power transformer section. The power transformer will isolate the circuits and step up or step down the voltage to the level required by the DC voltage. The output of the transformer is sent to a second rectifier section. Since the first rectifier section was for the input voltage, it is called the "input rectifier," and since the second rectifier is used for supplying output voltage, it is called the "output rectifier" section. The output rectification section is different from the input rectifier in that the frequency of the voltage in the second section will be very high (20–100 kHz). This means that the output ripple of the high-frequency voltage will be nearly filtered naturally because of the number of overlaps between each individual output pulse. Since the ripple is small, the actual capacitors in the filter section will be rather small.

The final section of the SMPS is the control and feedback block, which contains circuitry that provides pulse-width modulated output signal. The pulse-width modulation provides a duty cycle that can vary pulse by pulse to provide an accurate DC output voltage. This block of the power supply uses an operational amplifier to compare the output voltage to a reference voltage and to continually make adjustments to the output voltage. The oscillator in this circuit provides the frequency for the duty cycle.

The final three blocks (power transformer, output rectification, and control/feedback) can have different circuitry called "topology" that is simpler or more complex than this example. Each of the topologies has advantages and disadvantages. The more common topologies will be presented in the next section.

8.7.3 The Buck Converter

The *buck converter* circuit is the basis for several other similar circuits called forward converters. The buck converter is a step-down DC to DC converter; it is part of a switch-mode power supply that uses a transistor, capacitor, and inductor with a diode to regulate voltage levels below the supply

voltage level. The advantage of the buck converter is that it provides DC power more efficiently than a power supply that uses a resistance as a voltage drop to regulate voltage at a reduced level. When a resistance voltage drop is used to regulate voltage, power is converted to heat when it is bled off by the resistance. In the buck converter, the voltage is regulated by charging and discharging the capacitor and inductor in the circuit, which store the power that is bled off and reuse it rather than converting it to heat, which is a power loss. The buck converter may be as much as 95% efficient.

The buck converter circuit and the input and output voltages for this circuit are shown in Figure 8-22. This circuit would be connected directly after the power transformer. This circuit is fairly simple in that it consists of a transistor, inductor, diode, and capacitor. When the transistor is turned on, power will flow directly to the output terminals. This voltage must also pass through the inductor, which will cause current to build up in it in much the same way that a capacitor charges. When the transistor is switched off, the stored current in the inductor will cause the diode to become forward bias, which will let it freewheel and allow the current to be delivered to the load that is connected to the output terminals.

The waveforms at the bottom of the diagram show the square wave of the input voltage in the top line. The waveform on the bottom line shows the effects of the inductor putting the stored current back into the circuit to the load. This current is shown as the dashed line that occurs during the square wave off cycle. This type of

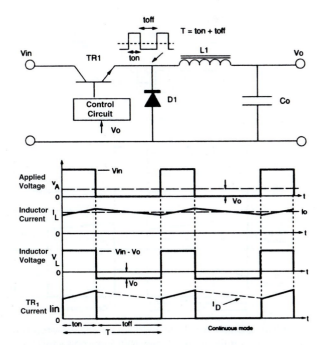

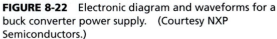

FIGURE 8-22 Electronic diagram and waveforms for a buck converter power supply.   (Courtesy NXP Semiconductors.)

circuit is a step-down buck converter because the output voltage will always be smaller than the input. Voltage regulation for this circuit is controlled by duty cycle. If the off time for the duty cycle is lengthened, the average voltage to the output will be lower, and if the off time for the duty cycle is shortened, the average power will increase.

8.7.4 The Boost Regulator

The *boost regulator* is a second type of fundamental regulator circuit for the switch-mode power supply. It is a step-up regulator. The electronic diagram and waveforms for this type of converter are shown in Figure 8-23. The transistor has been moved to a point after the inductor, and it is now connected directly across the positive and negative lines of the output. The freewheeling diode is connected in series and reverse bias with the inductor. The capacitor remains in parallel with the output voltage terminals to provide filtering. It is important to remember that this circuit is also connected directly to the secondary windings of the power transformer, just like the buck regulator.

When the transistor is switched on, current flows in this circuit and builds up in the inductor. When the transistor is turned off, the voltage that was built up across the inductor due to the stored current is returned to the circuit because it is reverse bias to the applied voltage. Since this voltage is reverse bias, it will be allowed to pass through the diode to the load when it builds to a level that is larger than the applied voltage. When the diode goes into conduction, it will pass the power stored in the inductor along with the supply voltage. This means that

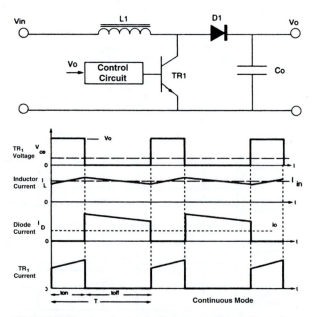

FIGURE 8-23   An electronic diagram and waveforms for a boost regulator.   (Courtesy NXP Semiconductors.)

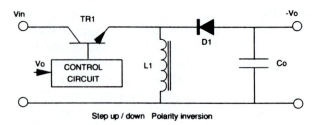

FIGURE 8-24   The electronic circuit for the buck-boost regulator.   (Courtesy of NXP Semiconductors.)

the output voltage of the boost converter will always be larger than the input voltage, hence, the name "boost." The amount of voltage at the output will be regulated by adjusting the duty cycle of the circuit.

8.7.5 The Buck-Boost Regulator

A combination of the buck regulator and boost regulator is shown in Figure 8-24. This combination circuit is called the *buck-boost regulator* and it utilizes the strong points of both of the previous regulators. The transistor is connected in series like the buck converter, and the inductor has been moved to a position where it is connected in parallel with the output terminals. The freewheeling diode is connected as it was in the boost regulator.

The transistor controls the voltage to the output in this circuit. When it is turned on, the inductor will store energy. When the transistor is turned off, the stored energy will be large enough to forward bias the diode and pass voltage to the output terminals. Since this circuit has the basic operation of both the buck and boost regulators, the output voltage can be regulated both above and below the input voltage level. For this reason, the buck-boost regulator is more popular. The waveforms for this type of circuit are similar to those of the boost regulator.

8.7.6 The Forward Converter

The *forward converter* is basically a buck converter with a transformer and a second diode added to allow energy to be delivered directly to the output through the inductor during the transistor on time. Figure 8-25 shows the electronic diagram and waveforms for the forward converter. The transistor is connected in series with the primary of the additional transformer. The second transformer provides a phase shift that causes the polarity of its voltage to flow to the output while the transistor is in conduction. This allows a better flow of energy to the output through the transistor. This means that the output current is continuous with the result that the ripple will be minimal. This also means that the filtering

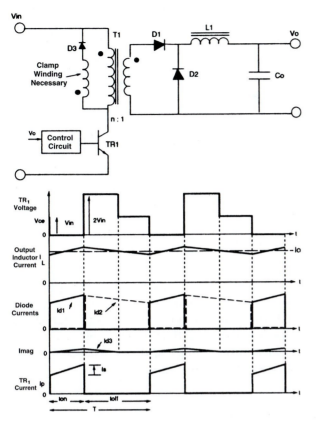

FIGURE 8-25 Electronic diagram and waveforms for the forward converter. (Courtesy NXP Semiconductors.)

capacitors can be smaller, which makes the entire power supply smaller.

8.7.7 The Push-Pull Converter

As the switch-mode power supply has evolved, additional adjustments to the original circuits have been made to get more power from smaller components. This means that the efficiency for the system must be increased. One simple way to do this is to use a center-tapped transformer that utilizes both the top and bottom half-cycles. Figure 8-26 is a diagram of the *push-pull converter* in which the converter

utilizes a center-tapped transformer for both the primary and secondary windings. The primary winding is controlled by two transistors, which allows one of them to conduct during each half-cycle, so the output is receiving voltage directly through one of them at all times. This means that the efficiency of this configuration is approximately 90%. This allows the overall size for the power supply to be smaller for a comparable power supply whose efficiency is 75% to 80%.

8.7.8 The Half-Bridge Converter

One of the problems with the push-pull converter is that the flux in the two sections of the center-tapped transformer primary and secondary windings can become unbalanced and cause heating problems. Another problem is that each transistor must block twice the amount of voltage than other converters block. The *half-bridge converter* provides several advantages over the push-pull converter. Figure 8-27 shows the electronic diagram for the half-bridge converter. The more expensive center-tapped transformer is replaced with a traditional transformer. This circuit still uses two transistors and two sets of diodes, as does the push-pull circuit. The main difference of the half-bridge converter is that it utilizes two large bulk capacitors (C1 and C2). These capacitors are connected so that each one is in series with one of the transistors. This means that power can be transferred to the output during the on time for each transistor, which increases efficiencies to the 90% range. Since center-tapped transformers are not used, the problem with flux unbalance is also eliminated. These advantages also allow this type of converter to be utilized in power supplies up to 1000 watt.

8.7.9 The Full-Bridge Converter

The *full-bridge converter* adds two additional transistors to the half-bridge converter. This means that four transistors are available to provide power to the output section, so this type of converter is used in power supplies in excess of 1000 W. Figure 8-28 shows the electronic diagram of

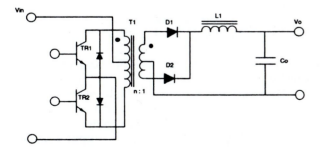

FIGURE 8-26 Electronic diagram for the push-pull converter. (Courtesy NXP Semiconductors.)

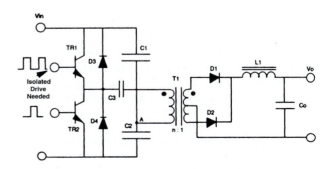

FIGURE 8-27 Electronic diagram of the half-bridge converter. (Courtesy NXP Semiconductors.)

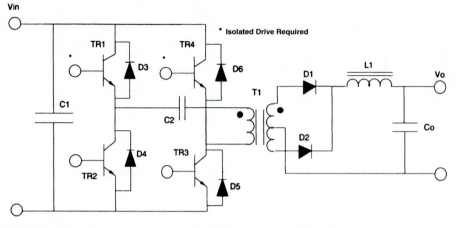

FIGURE 8-28 Electronic diagram of the full-bridge converter. (Courtesy NXP Semiconductors.)

the full-bridge converter. Each transistor has a clamping diode connected across its collector-emitter terminals and they are driven alternately in pairs. Transistors T1 and T3 are energized together for one half-cycle, and transistors T2 and T4 are energized together for the other half-cycle. One advantage of the full-bridge converter is that it requires only one capacitor for smoothing the output voltage, whereas the half-bridge converter requires two. The full-bridge converter will be used in larger power supplies usually more than 1000 W. Since this type of converter is more complex, it is normally used in the largest types of power supplies.

Questions

1. Explain the operation of a simple DC-to-AC inverter and draw its input and output waveforms.
2. Identify the major components in the solar energy control system shown in Figure 8-5 and explain their function.
3. Identify the major components in the industrial solar energy control system shown in the diagram in Figure 8-6 and explain their function.
4. Explain the function of the capacitor and inductor in solar energy controller.
5. Explain the function of a combiner box and why it is used.

6. Explain the problem that occurs when a grid fault occurs and an island is created.
7. Explain the operation of an inverter.
8. Identify the major parts of an inverter and explain their function.
9. List three types of inverter circuits and explain the output waveform of each.
10. Explain the difference between a linear power supply and a switch-mode power supply.

Multiple Choice

1. Power flow from the photovoltaic panel to the point where usable AC voltage is provided in the following order:
   a. DC voltage from the panels to the combiner boxes and on to the inverter that produces DC voltage.
   b. DC voltage from the panels to the combiner boxes and on to the inverter that produces AC voltage.
   c. AC voltage from the panels to the combiner boxes and on to the inverter that produces AC voltage.
2. The combiner box is a box in which _____
   a. the output of several inverters is connected.
   b. the wires from the solar panels are connected.
   c. batteries are connected.
3. A pulse-width modulation (PWM) inverter provides an output waveform that can have _____
   a. its voltage, current, and frequency varied.

   b. only its voltage varied.
   c. only its current and frequency varied.
4. An Inverter _____
   a. changes AC voltage to DC voltage.
   b. changes DC voltage to DC voltage.
   c. changes DC voltage to AC voltage.
5. A switch-mode power supply (SMPS) consists of the following circuits:
   a. An input inverter circuit, a high-frequency switch, a power transformer, and an output inverter circuit.
   b. An input rectification circuit, a high-frequency switch, a power transformer, and an output rectification circuit.
   c. An input rectifier circuit, an input voltage regulator, a high-frequency switch, an output voltage regulator, and an output rectification circuit.

Storing Electrical Energy and Batteries

OBJECTIVES FOR THIS CHAPTER

When you have completed this chapter, you will be able to

- Explain the difference between a primary-type cell and a secondary-type cell.
- Explain what a deep-discharge battery is.
- Explain the advantages and disadvantages of an absorbed glass mat (AGM) battery.
- Explain the advantages and disadvantages of a lead-acid battery.
- Explained what an electrolyte is for a battery.
- Explain how the specific gravity of electrolytes helps determine the condition of a battery's charge.
- Demonstrate how batteries are connected in series and parallel.

TERMS FOR THIS CHAPTER

Absorbed glass mat (AGM) battery
Actual (Available) Capacity
Amp-Hour
Battery Cycle
Cold Cranking Amps
Constant Current Charge
Constant Voltage Charge
Cycle Life
Deep-Cycle Battery
Depth of Discharge
Discharge (of a Battery)
Dry Cell Battery
Electrolyte
Float (Floating) Charge
Flooded Cell Battery
Galvanic Cell
Gel Cell Battery

Hydrometer
Lead-Acid Battery
Lithium-Ion Battery
Nickel Cadmium (NiCad) Battery
Nickel Metal Hydride Battery
Open Circuit Voltage Level
Primary Battery Cells
Reserve Capacity
Sealed Lead-Acid (SLA) Battery
Secondary Battery Cells
Shallow-Discharge/Shallow-Cycling Battery
Specific Gravity
SLI (Starting, Lighting, and Ignition) Battery
State of Charge
Trickle Charge
Valve-Regulated Lead-Acid battery (VRLA)
Wet Cell Battery

9.0 OVERVIEW OF THIS CHAPTER

One of the problems with converting solar energy to electricity is that the power from the sun is limited to the daylight hours, and electrical energy in a residence or commercial establishment may be consumed throughout the 24-hour day. One way to combat this problem is to store electrical energy produced by the solar photovoltaic panels in batteries, so it can be used at a later time. In many solar PV applications, the solar panels produce more electricity than can be used at any given time, and the excess electrical power is stored in one or more batteries for later use.

This chapter will explain how a basic battery uses a chemical reaction to produce DC electricity. It will also explain how batteries can store electrical power until it is needed. There are several types of batteries used in solar applications, including wet cells and gel cells. The advantages and disadvantages of these types of batteries will be explained. The information provided in this chapter will help you make decisions in selecting the type of batteries that you use to store power from solar PV panels.

This chapter will also explain the terms such as voltage, current, power, ampere-hour, kilowatt-hour, and many other terms used with batteries so you can understand the basic terminology used to define battery operation and life. You will also learn about different ways that batteries are charged and discharged. Different types of batteries such as deep-cycle lead-acid, shallow-cycle batteries, lithium-ion batteries, nickel-cadmium batteries, and nickel-metal hydride batteries will be explained in detail and advantages and disadvantages of each will be discussed.

The final section of this chapter will explain how multiple batteries called strings are connected to make larger battery storage systems for residential, commercial, and industrial applications. Diagrams will be provided to show how batteries can be connected to create larger storage cells, and how they can be connected with the original grid-supplied voltage for an application. The information in this chapter will help you feel more comfortable with selecting the correct batteries for an application and installing and maintaining them throughout their life. Since batteries will eventually wear out, they will need to be safely recycled. Sections in this chapter will explain some of the problem areas for recycling that some of the materials used in batteries present.

9.1 WHAT A BATTERY IS AND HOW IT STORES ELECTRICAL ENERGY

A battery is a device that produces DC electricity from a chemical reaction. Batteries are typically called storage devices because the chemical elements in them can be stored for a long period of time, up to several years, and yet produce the chemical reaction that creates the DC electricity. Batteries have been around for well over 100 years. A typical battery consists of one or more galvanic cells that create electricity and then store it in a chemical state. To keep this explanation as simple as possible, we will discuss only a single galvanic cell. In reality the battery is made up of multiple galvanic cells. The *galvanic cell* in its simplest form consists of two types of metals that allow ions from one metal to move to the other inside the battery, which will release electrons that become electrical current. By definition, an "ion" is an atom that changes its electrical charge by adding or losing one or more electrons. You should remember from Chapter 7 that an electron has a negative charge, and when electrons are freed and allowed to flow, they become electrical current.

In its simplest form a battery cell has three main parts: a positive electrode (terminal) and a negative electrode that are not connected to each other and a chemical that is in liquid or gel form and is called the *electrolyte*. One of the simplest batteries to explain uses zinc and copper, as shown in Figure 9-1. This type of battery consists of a copper metal electrode and a zinc metal electrode. The zinc metal electrode is in a container that is made of a porous material and it contains a zinc metal solution. The zinc metal electrode is mounted in the porous container so it is somewhat isolated from the copper solution. The zinc metal electrode in its porous container is mounted in a glass beaker with the copper metal electrode. A solution that contains copper ions is used to fill the glass beaker, which causes the chemical reaction to begin. During the chemical reaction ions from the copper electrode move through the solution over to the zinc electrode and begin to coat it with a thin layer. After a short period of time the zinc electrode will have a thin coating of copper. For the battery to begin producing an electron

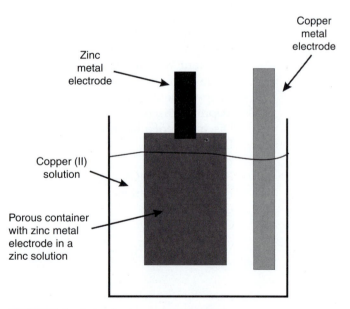

FIGURE 9-1 A simple zinc and copper battery.

flow, it is important that the zinc electrode and the copper electrode are in the same container (the beaker) and submerged in a liquid solution (an acid, the electrolyte) that contains copper ions. Since the solution completely surrounds both the copper electrode and the zinc electrode, the copper ions can move through the electrolyte solution and evenly coat the zinc electrode. The liquid electrolyte (acid) becomes a conductive medium.

At the zinc electrode, the zinc metal combines with oxygen atoms and is oxidized (electrons are given up) as it enters the acidic solution. The hydrogen ions in the acid solution are reduced, which means they take on electrons. The negative terminal of the battery, which is the copper cathode, begins to produce an excess number of electrons. When an electrical load such as a lamp is connected between the negative terminal (copper electrode) and the positive terminal (zinc electrode), electrons will flow from the negative terminal through the load and back to the positive battery terminal. If sufficient numbers of electrons can flow, they will produce a usable current that is large enough to illuminate the lamp. The electromotive force is a potential that is called voltage, and it exists between the negative terminal and the positive terminal and causes the electrons to move from the negative terminal to the positive terminal.

A DC voltmeter can be used to measure the potential difference between the positive and negative terminals of the battery cell. The copper zinc cell produces 1.1 V of potential difference. Different types of metals produce different amounts of voltage and these materials will be discussed later. In order for a battery to be useful, the number of individual cells must be connected in series to produce the desired amount of voltage, such as 6, 12, 24, or 48 V, and the cells must be connected in parallel to produce the amount of current that will be needed.

Some batteries are specifically designed to be recharged, so that they can store energy over and over. Other smaller type batteries are specifically designed to be used one time and then disposed of. When the chemical acid in the battery breaks down to begin to move electrons, the acid becomes weak and at some point it is not strong enough to produce a sufficient electron flow. At this point the battery is considered completely discharged.

9.2 TERMS USED WITH BATTERIES

Before we can begin to understand fully how batteries operate, we need to define some terms that relate to their operation. This section will discuss many of the terms used with batteries and provide explanations of them.

9.2.1 Ampere-Hour and Actual Capacity

The unit of measure for the amount of electrical storage capacity a battery has is called an *ampere-hour* or a *amp-hour*. This unit of measure can also be used to define the total capacity the battery has in a single charge. One ampere-hour is equal to one ampere of electrical current flowing through a load for one hour. On smaller batteries this may be identified as a milliampere, which is 1/1000th of an ampere. Another important term is the *actual capacity*, which is also called the *available capacity*, which is the total battery capacity expressed as ampere-hours.

9.2.2 Charging the Battery

The process of supplying electrical energy to a battery is called "charging the battery." When electricity is put into the battery, it is converted to chemical energy and is stored until it is ready to be used as electrical power. In a solar energy application the battery is continually being charged when sunlight is converted to electricity in the solar photovoltaic panel. Basically this means the battery is continually being charged throughout daylight hours every day.

The charge for the battery can be a *constant current charge,* which ensures that the current rate during the charging period stays constant. Another type of charge the battery can receive is a *constant voltage charge,* which allows the current to fluctuate. Typically the current draw will be high at the beginning of the charge and lower as the charge is nearly completed. When the charge is a constant current charge, the voltage is allowed to increase until the battery reaches full charge. There are times in the solar photovoltaic cell that the voltage and current will both change due to changing sunlight conditions.

When the battery is being charged, electrical energy is converted back into chemical energy inside the battery. This process will generally cause the electrolyte in the plates in the battery to take on heat and become warm while the battery is being charged. It is important to periodically inspect a battery that is being charged to ensure that it does not overcharge, which would cause the battery to overheat significantly to the point where it might become damaged.

9.2.3 Discharging the Battery

Discharge is the process of removing electrical power (voltage and current) from a battery. Discharge can also be described as the process whereby the battery converts chemical energy back to electrical energy. A battery is usually discharged by connecting an electrical load such as a lamp, electrical heater, or other resistance-type load to the positive and negative terminals of the battery. In a solar energy system, the battery will typically be discharged any time the electrical load in the house or the commercial location is energized. When lights are turned on or other electrical components are used, they will draw electrical current from the battery and cause it to discharge.

The battery will continue to discharge until it reaches its lowest voltage level, called the "cutoff voltage." The cutoff voltage is the lowest voltage at which a battery discharge is considered complete. When the voltage drops to this point or below, the battery will be inoperable and considered completely discharged for useful purposes.

The battery may have some minor voltage and minimal current remaining, but it is not sufficient to provide electrical power for the application to which it was applied.

The cutoff voltage is a value that is usually determined in specifications so that those who identify voltage levels for product usage can match a battery to their product.

Since each battery is designed for specific purposes and applications, and they are made from different materials, the cutoff voltage is different from one battery to another. It is also important to remember that some applications for batteries may drain them very quickly at first, and if the wrong type of battery is specified or being used for that service, it may go dead more quickly and be drained below the cutoff voltage than a different type of battery would.

In disposable batteries the chemical process during discharge is a one-way process, which cannot be reversed through recharging. Once the battery is completely discharged, the chemistry has changed to a point that no more electrical energy can be provided by the battery, so it must be discarded.

9.2.4 Depth of Discharge

Another term in this area is *depth of discharge* (DOD), also referred to as *state of charge.* The depth of discharge is basically the amount of energy that has been removed from the battery at any given time. The depth of discharge is usually measured as a percentage of the total capacity of the battery. For example if the depth of discharge is measured at 40%, it means that 40% of the energy in the battery has been used, and 60% is remaining. If the depth of discharge is measured at 75%, it means that 75% of the battery's energy is being used, and 25% is remaining.

Several other terms are used in connection to discharge of the battery: "discharge rate," which can be specified as high or low, and "deep discharge." A high-rate discharge uses the battery's electrical power at a rate that will drain it below the cutoff voltage point within one hour. If the rate of discharge is such that it will allow the battery to remain above the cutoff voltage point for more than an hour, it is considered a low-rate discharge. It is important to understand that the discharge rate will be directly correlated to the amount of load that is attached to the battery at any given point. For example if the battery is used for solar application on a residence, and only one light is used during the daytime, the discharge rate would be rather low. If the same battery is used in the nighttime, and the household is using multiple lights, the discharge rate may be rather high. Another term that is used when describing discharge is "drain," which means that electricity is being removed (discharged) from the

battery and its available power is decreasing. In some cases when the battery is completely discharged, it is said to be completely drained.

9.2.5 Battery Cycle and Deep-Cycle Charge

Another important term to understand is the *battery cycle,* which refers to the point at which the battery was completely charged and then to the point at which it was completely discharged. A variation of this is the *deep-cycle discharge,* which means that the battery is designed to allowed it to be usable even when its discharge down to 80%. These types of batteries are specifically designed so that as they are discharging, their voltage and current levels become lower at specified design rates that allow them to still remain useful for their applications. Some batteries, such as nickel cadmium (NiCad), are designed for this purpose, and if they are allowed to discharge to a deep level each time they are charged up, they will last much longer.

Typically deep-cycle batteries have thicker plates than car batteries and they can survive a larger number of discharge cycles than starting-type automotive batteries. This is why the deep-cycle battery is better suited for solar energy applications. It generally has a greater amount of long-term energy delivery. The reason batteries designed for automotive and other transportation applications have thinner plates is to make the battery as light as possible.

The opposite condition to the deep discharge is called a *shallow discharge, or shallow cycling.* With a shallow cycle, a battery is continually recharged as soon as any voltage or current is drawn down on it. An example of this type of shallow cycle is the battery used in an automobile. The automobile battery is designed to have an alternator or generator connected to it so that it is continually recharged as soon as any power is used from it. This type of charge is also called a *float charge* or a *floating charge,* since electricity that is removed is continually replaced by the alternator or generator. Another type of charge for batteries is the tapered charge, which occurs when a battery is nearly discharged before it is recharged; when it is recharged, the amount of voltage and current is larger at the beginning of the recharge cycle and slowly reduces (tapers off) during the recharge cycle until it is just a trickle at the end of the charging cycle. When only a small amount of current is allowed to flow during the recharge process, it may be referred to as a *trickle charge.* The shallow discharge battery is not as well suited for solar energy applications as is the deep-cycle battery.

The recharging process for all rechargeable batteries is not the same. For example the lead-acid battery in an automobile or in a solar application is designed to be continually recharged as it is using electrical power. The nickel-cadmium (NiCad) battery is designed to be completely discharged before it is recharged. It will produce useful voltage all the way down to the point of discharge, which is one of its design strengths. If the NiCad battery is constantly recharged as a shallow charge as soon as a small amount of electrical power has been used from it, it will very quickly adopt what is called a "memory," and it will seldom be able to deep discharge after it is shallow charged several times. This creates a problem for the application of the NiCad battery in that once it achieves this shallow discharge memory, it no longer is useful for its original application, where it is expected to have a deep discharge. This means that it is important to allow a NiCad rechargeable battery to go completely dead so that its voltage drops near or below the cutoff voltage before it is charged again. If a NiCad battery adopts a memory that allows discharge to only 10% or 20%, the memory can be changed in some cases by connecting the NiCad battery to a load that ensures that it discharges completely to below the 1-volt level. By continually recharging and then totally discharging the NiCad, you may be able to reverse the memory process and take the battery back to its original state.

To be usable for solar applications, deep-cycle batteries should be able to maintain a cycle life of several thousand cycles under high depth of discharge of 80% or more. Wide differences in cycle performance and cycle life may be experienced when several types of deep-discharge batteries are compared.

9.2.6 Cycle Life

The term *cycle life* applies to rechargeable batteries and refers to the total number of times a battery can be charged and discharged over its lifetime. When a rechargeable battery is continually charged and discharged, its cells slowly deteriorate. When the battery reaches a point at which it can be recharged only at proximately 80%, it is considered to be at the end of its cycle life. The cycle life specification is provided for rechargeable batteries so that you can determine how long the battery will sustain charging and discharging over its lifetime. In the case of batteries for solar energy applications, this is an important specification, since once the battery has existed beyond its cycle life, it will have to be replaced for the system to maintain its efficiency. In many cases the replacement of the complete set of batteries will be a large expense that must be budgeted several times over the life of the project. New research is continually trying to extend the cycle life of batteries to minimize the number of times they need to be replaced during a solar project lifespan.

9.2.7 Open Circuit Voltage Level

The *open circuit voltage level* refers to the amount of voltage that is measured between the positive terminal and the negative terminal of a battery when no load is connected to it. In some batteries the open circuit voltage will be much higher than the voltage level when the load is applied to the battery. If the difference between the open circuit voltage and the voltage when the load is applied is large, it is indicating that the battery is nearly discharged. Typically with new fully charged batteries, there will be little difference between open circuit voltage and the voltage under load. As batteries age, and their cells begin to wear out, the difference between the open circuit voltage and the voltage under load may be larger.

9.2.8 Cold Cranking Amperage (CCA)

The next three terms are primarily used to describe batteries used in automotive and truck applications, but the ratings these terms provide will help you compare one battery to the next to determine its size and capability when you are trying to select a battery for a solar energy application. *Cold cranking amps* (CCA) specify the battery's ability to start an automobile or truck engine at a very cold temperature. This rating is based on a temperature of 0°F or –18°C, and it refers to the number of amperes a new fully charged battery is able to deliver constantly for 30 seconds. A 12-V battery must be able to provide a minimum of 7.2 V continually during the 30-second period for the cold cranking amperage rating. The higher the number of amperes listed in the cold cranking amp specification, the larger the battery is. Even though a battery may be designed to be used in a solar energy storage application, it will still have a cold cranking amp rating so the battery can be compared to other batteries to determine its size and ability to charge and store energy and then deliver it as electrical voltage and current.

9.2.9 Cranking Amps (CA)

Cranking amps describes the number of amperes that are available in a battery at 32°F. This rating is also called marine cranking amps (MCA), which identify how much current is available for larger engines. Hot cranking amps (HCA) is a term that is seldom used any longer, but it measures available amperes when the battery is at 80°F. The cranking amp specification helps you identify the capacity of the battery and allows you to compare it to other similar batteries you are considering for solar energy application.

9.2.10 Specific Energy or Energy Density of a Battery

Specific energy is also called energy density; it refers to the amount of electrical power a battery can produce as compared to its weight and size. Some very small batteries have a high specific energy or energy density that allows them to produce a large quantity of electrical energy in their small size. This capacity is important when size is a consideration in the installation or application. In some cases size limitation means the battery must be small but able to produce a large amount of electrical energy.

9.2.11 Reserve Capacity (RC)

The last term relating to batteries is a specification *reserve capacity* (RC). This is an important rating to help you identify how much amperage the battery can deliver on a continual basis. The reserve capacity is the number of minutes a fully charged battery will discharge 25 A on continual basis at 80°F until its charge drops below a specific level. For a 12-V battery the voltage level is below 10.5 V.

9.3 PRIMARY AND SECONDARY BATTERY CELLS

All batteries (wet cells and dry cells) can be further broken down into two main categories: *primary battery cells* that are non-rechargeable, and *secondary battery cells* that are rechargeable. A wet cell battery is one that has a liquid electrolyte and a dry cell batter is one in which the electrolyte is a paste or chemical mixture that is not a liquid. The chemical reaction that produces electricity in the primary cells is not reversible, so the battery cannot be recharged. These types of batteries are used one time and then they must be disposed of safely because many of them have a variety of chemicals in them. The primary cells are also called "disposable batteries" since they are used one time.

The chemical reaction in a secondary cell can be reversed many times, so the battery is rechargeable. At some point the chemicals in the secondary cell will begin to break down, or the plates in the cell will begin to deteriorate until the battery will no longer be able to accept a charge

9.4 WET CELL AND DRY CELL BATTERIES

A *wet cell battery* uses a liquid electrolyte, whereas a *dry cell battery* uses a paste or dry electrolyte. The electrolyte in a wet cell battery is typically an acid such as sulfuric acid that is mixed with a percentage of water, and the electrolyte completely submerges all the cells in the battery so it can easily transfer electrons to the negative terminal of the battery. The wet cell battery is also called a *flooded cell battery*. When the wet cell battery is charged, it will convert some of the electrolyte to a vapor or gas that is mostly hydrogen, which is very explosive. Care must be taken to allow the hydrogen gas to be safely released from the cell, or the cell will become pressurized and possible explode. Typically a special cap called a hydrocap is used because it is specially designed to allow the hydrogen gas to combine with oxygen during the charging process, which neutralizes the explosive nature of the hydrogen gas.

Another problem that arises with the wet cell battery is that when it is repeatedly charged and discharged, the recharge process causes some of the liquid electrolyte to evaporate into a gas if the battery is being overcharged. The evaporation of the electrolyte will eventually lower the liquid level in the wet cell battery. When the level of the electrolyte starts to lower, the top part of the plates will begin to be exposed to air instead of being fully submerged and covered with electrolyte. When this occurs, the area of the plate that can produce electrical power is reduced, and the battery slowly begins to produce less and less electrical power until it will no longer accept a charge. This condition typically shows up as the battery producing a voltage that is lower than its rating; when this occurs, the low voltage will generally become evident and someone will eventually check the battery and determine that its electrolyte level is low. For this reason, wet cell batteries should be periodically checked. If it is determined that the liquid level is low, distilled water can be added to bring the liquid back to its proper level. Distilled water should be used since it does not have any impurities that would contaminate the battery plates. It is important to ensure that the liquid from the battery does not come into contact with your eyes or skin, as it will cause a severe burn. If it comes into contact with your clothing, it will cause the fabric to deteriorate.

The dry cell battery typically has a paste or similar substance for an electrolyte. These types of battery are generally totally enclosed. They are typically used for smaller applications and come in 1.5-V and 9-V sizes. The battery sizes are identified as AAA, AA, B, C, and D, which are all 1.5-V batteries, and 9 V. Figure 9-2 shows examples of some of these types of batteries.

The more common types of dry cell batteries include alkaline batteries, which are made of zinc manganese dioxide and are basically general-purpose batteries; mercury oxide batteries, which are made from zinc mercury oxide and are used in smaller devices such as hearing aids and watches; nickel-cadmium (NiCad) batteries, which are used in all types of rechargeable applications; and lithium-ion

FIGURE 9-2 Typical 1.5- and 9-V dry cell batteries. Examples of AAA, AA, 9 V, B, and D size batteries.

batteries, which are currently used in computers and other electronic devices. Other less common types of batteries include zinc silver oxide batteries that are also used in small devices such as hearing aids and watches, lithium sulfur dioxide batteries that are used in space probes and security systems, and lithium iodine batteries that are used in cardiac pacemakers.

9.5 SEALED BATTERIES

Each cell in sealed batteries is sealed so that the electrolyte cannot leak out or spill. This ensures that the electrolyte in the battery will not evaporate and leave the plates exposed, which will cause them to become damaged. The sealed battery also ensures that the electrolyte, which is typically an acid, will not spill and cause damage to the surrounding area. Sealed batteries can be broken down into batteries that have a special cap to prevent the electrolyte from leaking or have an electrolyte in the form of a gel. Some sealed batteries may leak slight amounts of gel if they are left inverted for long periods of time. The leakage will occur through the vent in the cap. Generally these types of batteries are maintenance-free because they do not require frequent checking of the electrolyte level.

9.5.1 Valve Regulated Lead-Acid Batteries (VRLA)

The battery that has a special cap to prevent the electrolyte from leaking is called a *valve regulated lead-acid* (VRLA) battery. This type of battery is considered a low-maintenance battery because you do not have to check the level of the electrolyte, since the battery cap for the cell is specifically designed to prevent the electrolyte from evaporating. This type of battery is also called a sealed lead-acid (SLA) battery, since the cap is designed to seal each cell and allow only a small amount of gas to vent.

The way this battery operates is that some hydrogen gas and some oxygen gas are created during the charging process inside the battery, and the battery design ensures that the two gases mix and recombine. When hydrogen and oxygen recombine, they form water droplets, which will mix back with the electrolyte solution. Any remaining gas that does not recombine is allowed to vent through the cap.

9.5.2 Gel Type Batteries (AG) and Sealed Lead-Acid (SLA) Batteries

A *gel cell battery* is a valve regulated lead-acid (VRLA) battery with its electrolyte in the form of a gel. The vents on these batteries are not typically designed to be removed. The gel consists of sulfuric acid mixed with silica gel. When these two components are mixed, the resulting gel is similar to the consistency of jello and it becomes somewhat immobile when it is placed over the cells in the battery. The gel cell battery can be moved to virtually any position because the gel will not run out of it.

The gel cell battery limits the electrolyte evaporation and spillage that are common to the traditional wet cell battery. Since the gel cell battery is sealed, it is typically called a *sealed lead-acid* (SLA) battery. In theory the cell is completely sealed, but in actuality a small amount of gas is allowed to be vented through a valve regulation system. Another difference between the lead-acid battery and the gel cell is that the plates in the gel cell include lead-calcium plates, so sometimes this battery is referred to as a lead-calcium battery.

The gel cell battery must be charged at a slower rate than a typical lead-acid battery or AGM batteries, to keep it from overheating. This is typically not a problem when the gel battery is used with solar applications, since the solar charging can be controlled by regulating the amount of current that is allowed to flow to each cell.

If the gel cell is overcharged, it can develop voids in the gel that will remain forever and will eventually cause a loss in the battery's capacity. In warmer climates the gel will eventually give up water through the gassing off of hydrogen and/or oxygen, which can be enough over several years to permanently damage the gel and the battery's ability to maintain a charge. It is for this and other reasons that gelled cell batteries are no longer sold, and newer AGM (absorbed glass mat) batteries have virtually taken their place since they have many of the advantages but none of the disadvantages of the gel cell.

9.5.3 Absorbed Glass Mat (AGM) Batteries

The *absorbed glass mat* (AGM) battery was designed in the late 1970s, and it was an advance over the gel cell. The main differences between the AGM battery and the gel cell battery are that the AGM battery uses a separator made from spun glass, which is a small (fine) fiber boron-silicate glass mat and it uses a valve in the cap to regulate the oxygen and hydrogen gas to ensure that it recombines into water. This process is called "recombination." During the recharge process oxygen is produced from the positive plate and is directed to the hydrogen gas to create water. The separator material absorbs the acid inside the battery. The material is actually very porous but it can absorb acid up to seven to eight times its weight. When it is full of acid, the absorbent material helps lower the internal resistance of each cell. Lower internal resistance provides a higher performance rate for the battery, since any energy lost from internal resistance makes the battery less efficient.

Since all the electrolyte (acid) is contained in the glass mats, it will not leak or spill, even if the plastic case is broken. It also means that the battery will not freeze and expand since there is no liquid acid. This is an important feature when these types of batteries are used for solar energy applications in northern areas that typically have freezing temperatures in the winter.

Typically these types of batteries last longer since they are not prone to losing electrolyte and allow the battery plates to operate without being coated with electrolyte, which tends to make them wear out sooner.

The AGM material in the battery allows it to be deep cycled, which makes it very usable for solar applications. This battery is also usable for solar applications because the amount of maintenance it requires is minimal, and it does not require periodic inspection to check the electrolyte level as do the wet cell batteries. One problem that the AGM battery does present is that a special circuit must be provided to the charging circuit to ensure that the battery is charged at a slightly lower voltage than would be used with a wet cell battery. The AGM battery may be a sealed regulated valve (SRV) battery, a valve regulated lead-acid (VRLA) battery, or a sealed battery because of the way it is designed.

Another important point is that the AGM battery does not have gel in it, so it can be charged at a rapid rate just like the lead-acid battery without any damage to the cells. The AGM battery does not heat up during charging like the gel cell, so it can be quick charged or slow charged without any problems. This make the AGM battery well suited for the storage component of the solar PV system. One of the disadvantages of the AGM battery is that it is more expensive than the lead-acid battery.

9.5.4 The Starting, Lights, and Ignition (SLI) Battery

The lead-acid batteries used in automobiles are typically used for starting, lighting, and ignition applications on the automobile. For this reason these types of batteries are called *SLI batteries* and are designed to give a lot of current during starting but then can be recharged immediately by the car's alternator. The SLI battery can withstand several starts per hour if necessary and yet receive the recharge current from the alternator. The SLI battery does not work as well for large solar energy applications because deeply discharging the battery will greatly shorten its life. It does work reasonably well for smaller residential solar energy storage applications.

9.6 HOW A LEAD-ACID BATTERY IS MANUFACTURED

One of the more common types of batteries used with smaller residential solar energy storage applications is the *lead-acid battery*, lead-acid also called an accumulator, flooded cell battery, or wet cell battery. It is made up of individual cells that are electrically connected to provide the terminal voltage for the battery. Each cell in the battery has a number of plates that are referred to as a positive electrode that are made of lead dioxide and a number plates that are made of sponge lead that is a very porous

FIGURE 9-3 An empty battery case with its top before any cells are placed into the chambers.

FIGURE 9-4 A cutaway picture of a lead-acid battery.

plate and connected to the negative terminal. The plates in the lead-acid battery are covered with a liquid solution (electrolyte) that is a mixture of sulfuric acid and water. Figure 9-3 shows an empty battery box before the cells are put into it along with the top that has the holes where you can inspect the electrolyte level or add water. Figure 9-4 shows a cutaway picture of a basic lead-acid battery after the cells have been put into each chamber. You can see the lead-acid battery is contained in a plastic case, called the protective casing. The case is made of an insulating material, usually polypropylene plastic, and it protects the cells inside from any outside damage. Each cell has positive and negative plates that are made of different types of lead. You can see each set of plates is separated in the battery by a plastic plate separator so they can operate individually. The liquid electrolyte solution is poured over each cell. Since each cell is separated from the others, the liquid level in one cell can be lower or higher than in any of the others. It is important that the liquid level covers the top of each cell. You can also see that each set of cells has a connection point to the positive and negative terminals and they terminate at the top of the battery at the positive and negative terminals that protrude from the front of the battery on either side.

The 12-V battery is composed of six independent cells that are connected electrically in series. Each individual cell in a lead-acid battery produces about 2.14 V per cell; when six of the cells are connected in series, they produce a total of 12.6 to 12.8 V for a 12-V battery. Each individual cell is physically isolated and produces voltage independently, but they are electrically connected in series inside so that they produce the full 12 V. All six completed cells are placed inside the plastic case, but they are separated by the cell dividers, which isolates them into individual compartments. The battery case is typically made from polypropylene plastic. The polypropylene plastic is not affected by sulfuric acid. Figure 9-5 shows how the cells are connected internally.

After the plates are assembled into the battery, the electrolyte solution is added to each individual cell in the battery through the access holes in the top. There are six holes in the top of the battery (one hole per cell), so the level of the electrolyte can be varied in each of the six cells. This is why you need to check each cell for its electrolyte level in case one cell gasses off more electrolyte solution than another. The vent caps can be individual caps or they can be grouped three together. Vent caps are

FIGURE 9-5 A lead-acid battery showing the internal positive and negative connections of the cells.

located on the top of the battery and they fit tightly into the access holes for each of the cells. The caps fit so tightly that no gas or liquid can escape except through the vent hole in the cap.

Each cell has a number of lead dioxide plates that are connected to a negative terminal and a number of sponge lead plates that are connected to the positive terminal. The sponge lead plate is made only from lead. Figure 9-5 is a cutaway diagram of a battery and you can see how the plates are placed in each cell so that the lead dioxide and sponge lead plates alternate, and there is just enough space between the plates to allow the electrolyte to cover each plate completely. When the electrolyte solution, a mixture of sulfuric acid and water, makes contact with each of the plates, a chemical reaction begins. The chemical reaction occurs when the sulfuric acid in the electrolyte completely covers the lead dioxide plate (the positive plate). This reaction creates a lead sulphate on the plate that gives up electrons. When the lead dioxide gives up electrons, it becomes more positive.

The electrons from the lead dioxide plate travel to the sponge lead plates that are pure lead, and these electrons become the current the battery produces. Since all of the pure lead plates in each of the cells are electrically connected to one another and the negative terminal on the battery, the electrons have a path to flow through the battery and to the negative terminal of the battery.

When the battery is recharged, electrical current is put back into the battery, which causes the lead sulphate in each cell to break down, resulting in the lead dioxide being redeposited on the positive electrode, and lead being replaced on the negative electrode.

The process of charging and discharging can continue over and over during the life of the battery. When the battery is used as the storage medium for a solar energy system, the batteries are charged during the day when the sunlight creates electrical power in the solar photovoltaic panels, and they are discharged during the evening when sunlight is not available. The rate of the discharge is based on the amount of electrical load that is placed on the system by turning on lights and using other electrical loads.

The amount of charge that is in the battery can be easily checked by measuring the *specific gravity* of the electrolyte. The electrolyte solution for the lead-acid battery is a mixture of sulfuric acid and distilled water. The exact percentages of sulfuric acid and water can be measured with an instrument called a *hydrometer*. The hydrometer is a rubber squeeze bulb mounted on the end of a glass tube. The glass tip of the tube is placed under the liquid level of the solution in the battery and when the bulb is collapsed and released a sample of the fluid is sucked up into the tube. When the tube is nearly full, a float in the glass tube will begin to float at a level that indicates the specific gravity of the fluid. The specific gravity of the fluid is an indication of the percentages of the sulfuric acid and distilled water in the mixture. The specific gravity of water is rated as 1.000 and the specific gravity of pure sulfuric acid is 1.850. This means that with a mixture of 35% sulfuric acid and 65% water solution and a fully charged battery, the specific gravity should be 1.265; when the battery is discharged, the specific gravity should be close to 1.155. You will learn more about specific gravity and maintenance of batteries at the end of this chapter.

9.7 HOW AN ABSORBED GLASS MAT (AGS) BATTERY IS MANUFACTURED

Figure 9-6 is a cutaway diagram of the absorbed glass mat (AGM) battery. It is composed of six separate cells. Each cell is composed of two long plates that are made from very thin lead sheets formed into a tight spiral. Each plate has an absorbent glass mat that is a sponge-like material on each side of the plate, so when the plate is wrapped in a spiral, the mat totally encases the plate surface. Since the glass mat is nested between the plates, it will hold the electrolyte in place between the plates where it can make continual contact with the entire surface of each plate. The other advantage of using thin sheet lead is that more surface area can be provided. Typically the thin sheet of lead would need a tremendous amount of support if it were left in the battery as a sheet, whereas once the glass mat is applied to each side of the sheet and the sheet is rolled into a tight spiral, it gains

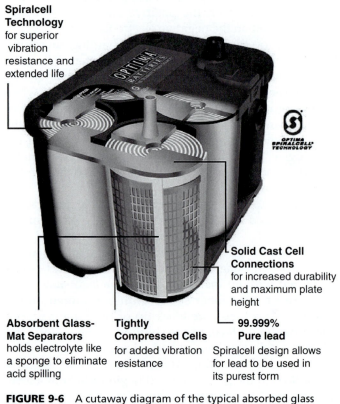

Spiralcell Technology for superior vibration resistance and extended life

Absorbent Glass-Mat Separators holds electrolyte like a sponge to eliminate acid spilling

Tightly Compressed Cells for added vibration resistance

99.999% Pure lead Spiralcell design allows for lead to be used in its purest form

Solid Cast Cell Connections for increased durability and maximum plate height

FIGURE 9-6 A cutaway diagram of the typical absorbed glass mat (AGM) battery. (Courtesy Optima Batteries)

structural stability. This design with the lead sheet placed between glass mat material on each side stops the possibility of plate-to-plate movement.

This type of battery can be designed as a short-term high-current battery like the lead-acid battery for automobiles, and it can be designed as a deep-cycle battery that can be used for solar applications that charge for long periods of time during the day and discharge for long periods of time overnight.

9.8 THEORY OF OPERATION AND ADVANTAGES AND DISADVANTAGES OF DIFFERENT TYPES OF BATTERIES

This section of the chapter will introduce the majority battery types that are available today for use in solar energy storage systems and for test equipment used in solar energy applications. Each battery's theory of operation and advantages and disadvantages will be discussed. Some of the batteries such as the lead-acid and the absorbed glass mat type batteries have been discussed earlier in this chapter, so this section will provide a short review and the advantages and disadvantages of these two types of batteries.

9.8.1 Lead-Acid Battery

The lead-acid battery is composed of two types of plates; one plate is made of pure lead, and other plate is made of lead oxide. A variety of other elements may be added to the plates to change the density, hardness, or porosity of the plates so they can give up electrons or take on electrons more easily and help the battery maintain a longer life. The electrolyte for the lead-acid battery consists of a mixture of 35% sulfuric acid and 65% water. This solution causes a chemical reaction that produce electrons on the positive plate and the electrons move to the negative plate where they flow through the negative battery terminal as battery current. The percentage of water and sulfuric acid can be determined by a test of the electrolyte with a hydrometer, which measures the specific gravity of the fluid. When the battery is discharged, the electrolyte solution is closer to the makeup of water; when the battery is fully charged, the solution is more acidic. When the battery is charging, electricity is put into the battery and the electrons are returned to the plates where they are stored and are ready for the next discharge cycle.

Advantages of lead-acid batteries include that they are one of the least expensive batteries to manufacture and maintain, and they can provide high power and are very reliable. Lead-acid batteries are mass-produced and available in a large variety of voltages, currents, and sizes. The lead-acid battery produces the most amount of electrical energy per weight when compared to other battery technologies. There is continual research on the technology and new lead-acid battery technology is improving the life of the battery. It is also important that recycling facilities are in place and legislation covers the handling of lead and other material in the battery. Lead-acid batteries are among the most recycled of all the battery types.

Disadvantages of lead-acid batteries include problems with low energy output in colder weather. The acid in the lead-acid battery is caustic and can cause burns to the skin of persons handling the battery or it can damage materials it comes into contact with. The lead-acid battery produces sulfur and hydrogen fumes when it is being recharged. These fumes can damage human lung tissue if inhaled. The fumes from overcharging can damage any metal in the battery box structure or near by.

9.8.2 Absorbed Glass Mat (AGM) Battery

The absorbed glass mat battery is made by placing glass mat on both sides of a sheet of thin lead material that makes up the plates. The thin lead material is sandwiched between two layers of the glass mat and then rolled into a tight spiral to give it structural integrity. A 12-V AGM battery will have six of the spirals cells connected electrically in series to produce the total voltage. Each of the cells in the AGM battery produces approximately 2.4 to 2.5 V; when they are connected in series, they produce approximately 12.8 V. The electrolyte solution is poured into each cell, and it is absorbed by the glass mat in much the same way as a baby's disposable diaper absorbs moisture. This means that the electrolyte will seldom if ever leak from the cell because it is held in suspension in the glass mat material. Once each cell is sealed with a special cap that allows the hydrogen and oxygen vapor to recombine into water and return to the glass mat material, the battery becomes spill-proof.

Some of the advantages of the absorbed glass mat battery include that they are virtually maintenance-free since they will not lose electrolyte solution when charged. Since it is a completely sealed battery and does not produce any fumes, it can be shipped by aircraft, which means it can be shipped across the country and throughout the world. The following is a complete list of advantages and disadvantages:

Advantages

- Maintenance-free
- Leak-proof/spill-proof
- Very low gassing unless it is overcharged
- Rechargeable to 90% in approximately 4 hours (full charge in 6 hours), which is within the daylight hours for solar application

Disadvantages

- Total battery life becomes shorter if allowed to completely discharge.
- The charging circuit needs control for temperature compensation and voltage regulation.

- Completely sealed, so electrolyte cannot be replaced if it happens to be overcharged continually.
- Charge must be limited to 14.4 to 14.7 V; otherwise the battery will become overcharged.

9.8.3 Alkaline Batteries

Alkaline batteries are disposable batteries and are used in test equipment for solar energy applications such as meters. The alkaline battery is made from a paste form of manganese dioxide that has a large number of molecules that are converted into manganese oxide and hydroxyl ions. The hydroxyl ions then react with zinc to form zinc oxide and water, releasing electrons. The alkaline battery has a carbon rod that connects the cell material to the external battery terminals. The free electrons inside the battery move toward the carbon rod, through the battery terminal, and out to the circuit that has the load that is connected to the battery. The alkaline battery stops producing electrical power when all the manganese dioxide is used up. Alkaline batteries are available in a variety of sizes. The alkaline battery lasts longer than zinc carbon batteries but costs more.

9.8.4 Lithium-Ion Batteries

Lithium is a lightweight metal that easily forms ions, which makes it a good material to use to make batteries. The newer *lithium-ion batteries* can store almost twice as much electrical energy as NiCad rechargeable batteries. Today larger groups of lithium-ion batteries are connected electrically to make a battery with a larger amount of current.

When a lithium-ion battery is being charged from a power supply or solar panel, the battery starts its chemical activity. It does this through a chemical reaction that shunts lithium ions (lithium atoms that have lost an electron to become positively charged) from one part of the battery to another. When you unplug the power and use your laptop or phone, the battery switches into reverse: the ions move the opposite way and the battery gradually loses its charge.

Most lithium-ion batteries have a circuit built into them that can interrupt charging and discharging. This circuit isolates the battery from the source of power to prevent overcharging and overheating and from the circuit to prevent the load from discharging completely. If the lithium-ion battery is allowed to completely discharge, the battery will become harder to charge up again.

Advantages

- The lithium-ion battery provides a large number of hours of operation for a given weight.
- The battery is very successful for mobile applications such as phones and notebook computers.

Disadvantages

- Lithium is more expensive than lead as a raw ingredient in the battery.
- Lithium costs more than other materials.
- There is no established comprehensive system for recycling lithium-ion batteries as there is for lead batteries.

9.8.5 Nickel-Cadmium Batteries

Nickel-cadmium (NiCad or NiCd) batteries are rechargeable batteries that have been used for many years. The current major application of nickel-cadmium batteries is for portable power tools. The nickel-cadmium batteries have higher specific energy and a better life cycle than lead-acid batteries. They were originally not considered for solar energy and hybrid electrical vehicle (HEV) applications because they did not deliver sufficient power over longer periods of time. New research has created NiCad batteries that can be used for solar applications and electrical vehicles.

A NiCad cell contains a positive electrode plate that is made of a nickel III oxide-hydroxide and a negative electrode plate that is made of cadmium. The electrolyte is concentrated in the separator for the NiCad battery, and it is a form of potassium hydroxide. NiCad batteries usually have a metal case with a sealing plate equipped with a self-sealing safety valve. During manufacturing the positive and negative electrode plates are isolated from each other by a separator sheet that is on top of the positive plate, between the two plates, and on the bottom of the negative plate. This configuration ensures that the two plates are isolated from each other and sandwiched between the separator sheets that are saturated with the electrolyte. After the separator sheets are in place, the plates in the separator sheets are rolled in a spiral shape and placed inside the case. This design makes the final assembly of the separator sheets and positive and negative plates looked like a jelly roll. Since the sheets of material are very thin and have a large area, this design allows a NiCad cell to deliver a much higher maximum current than an equivalent-size alkaline cell. Currently research and development efforts on the NiCad batteries are ongoing to improve their performance and life-cycle costs. Recently, newer technologies for batteries have been using metals such as nickel metal hydride in the batteries to replace NiCad, and these new metals have better energy characteristics and do not contain toxic cadmium.

Advantages

- NiCad batteries have a much higher energy density than lead-acid batteries.
- NiCad batteries tolerate deep discharge for long periods and work best when they have been deeply discharged.
- NiCad batteries can operate in a wide range of temperatures.

- NiCad batteries can typically withstand a large number of charge/discharge cycles better than other rechargeable batteries.

Disadvantages

- Cadmium is a heavy metal that is toxic and its recycling must be controlled.
- NiCad batteries are three to five times more expensive than lead-acid batteries of the same size.
- NiCad batteries may develop a memory if they are not completely and deeply discharged each time. The memory is a point of partial discharge, which means that the battery does not deliver all of its power.

9.8.6 Nickel Metal Hydride and Nickel Zinc Batteries

Other types of batteries that uses nickel include *nickel metal hydride battery* and the nickel zinc battery. Nickel metal hydride batteries are currently used in computers and medical equipment, and they have a specific energy and specific power capabilities that can do this job. Since this type of battery does not have cadmium in it, it is easier to recycle. Nickel metal hydride batteries have a much longer life cycle than lead-acid batteries.

A nickel metal hydride battery operates similarly to the NiCad battery in that it can be charged and discharged many times, but it is not as vulnerable to the memory effect. This means that nickel metal hydride batteries can be partially discharged and recharged, and during the next discharge, they are able to discharge completely and deeply without a problem.

Nickel zinc (NZ) batteries are similar to the NiCad battery, but they have been designed specifically for larger power systems including hybrid electrical vehicles. Since the nickel zinc battery has the ability to store larger amounts of power, it is in the initial stages of research for use in solar energy applications. The main problem with using the nickel zinc battery extensively for these applications is that now they have a short service life due to the zinc electrode. Newer technology is using a nickel foam material rather than the metal ceramic material, which will eventually cut the higher cost of manufacturing and possibly extend the life of the battery.

The first nickel zinc batteries were created around 1900 by Thomas Edison, and the newest versions use nanotechnology. Research is continually advancing developments to extend the life of the battery and make it withstand multiple discharge and recharge cycles.

Advantages

- Nickel metal hydride and nickel zinc batteries are reliable and lightweight.
- New types of nickel zinc batteries are projected to have very long cycle lives, equal to 100,000 miles when used in automobiles and multiple years of life when used with solar applications.

Disadvantages

- The nickel and zinc used in the battery are 25 times more expensive than lead.
- Some nickel zinc batteries generate large amounts of heat during charging and have some self-discharge problems.
- Nickel in some forms has been identified as a carcinogen.
- Actual applications in electrical vehicles and solar system electrical storage applications are relatively new and detailed historical data are not available.
- Recycling capability for zinc and nickel is not widespread at this time.

9.9 BATTERIES DESIGNED AND USED SPECIFICALLY FOR SOLAR ENERGY STORAGE APPLICATIONS

In recent years as the number of solar energy installations that need battery storage potential has increased, battery manufacturers have begun to design batteries and battery systems specifically for these applications. The requirements for these types of applications have become more stringent, and expectations have caused manufacturers to create systems that will last longer and operate with a larger number of charge and discharge cycles. At the present time the battery storage system in many solar photovoltaic energy systems and electronic controls for these systems are nearly as large of an expense as the photovoltaic panels. Currently the life expectancy of these battery systems is far less than the life expectancy of the photovoltaic panels. This basically means that the batteries used for the electrical storage may need to be changed completely once or twice during the life of the solar photovoltaic system. This causes the annual expense of operation to nearly nullify the advantage of getting electricity from sunlight for free. When residential and commercial users evaluate the total cost of the solar photovoltaic system, they must include any costs of replacing batteries or solar panels over the life of the project. Research today is trying to improve the life cycle for batteries so that they do not wear out and need to be changed as often.

In smaller stand-alone residential photovoltaic systems, the batteries that are used are either deep-cycle lead-acid types or shallower cycle maintenance-free batteries. Deep-cycle lead-acid batteries are not maintenance-free, and they must be checked quite frequently to ensure that the electrolyte does not boil off completely and leave the plates uncovered. The absorbed gas mat (AGM) type battery, which is also called the captive electrolyte battery, requires less frequent checks and is considered to be more maintenance-free. Some of the AGM batteries have been designed specifically as special shallow-cycle maintenance-free batteries. These batteries are designed for applications that have a less frequent discharging cycle, such as roadside signs and other systems that do not completely discharge when sunlight is not available. Typically these types of PV applications are designed with the battery storage capability being large enough that the battery bank does not have a depth of discharge (DOD) of more than 25%. This design is accomplished by utilizing a battery that is two to three times larger than necessary, which ensures that the battery does not discharge more than approximately 25%. The advantage of this type of design is that it provides a long-life battery system. This battery system used in photovoltaic applications with correct maintenance can last up to 15 year. In comparison, a poorly designed battery system for a photovoltaic application can cause the batteries to fail prematurely in less than a few years.

This section will provide examples of several applications of batteries specifically designed for solar photovoltaic applications. It is important to understand that larger solar photovoltaic systems for larger commercial and industrial systems generally do not include a battery storage system, since the cost would be exorbitant and the space to house the batteries would have to be very large and multiple locations might be needed. For this reason most of the larger commercial and industrial solar photovoltaic systems are grid connected and do not have a battery storage component.

One type of battery specifically designed for solar photovoltaic panels electrical storage is the advanced gel battery. Figure 9-7 shows a typical gel battery. The advantage of this type of battery when used with solar photovoltaic panels is that the battery is virtually

FIGURE 9-7 Advanced gel battery. (Courtesy Power Battery)

maintenance-free and does not need constant inspection, because the electrolyte is in the gel form. The battery is sealed, so the electrolyte cannot boil off as it might with a lead-acid battery. The drawback of this type of battery system is its slightly higher expense. The advanced gel battery is available as a 6, 12, or 24-V battery, so you may have to connect additional batteries in series or parallel for your application. Typically batteries are added in series to increase the voltage, and they are connected in parallel to provide additional current.

9.10 SOLAR HIGH-POWERED FLOODED BATTERY

Another type of battery system specifically designed for solar photovoltaic storage applications is the solar high-powered flooded battery. This type of battery is used for deep-discharge applications typical in solar photovoltaic panels. Figure 9-8 shows an example of a solar high-powered flooded battery. This type of battery contains one of the thickest plates available and has a typical life of more than 10 years. The battery usually does not need to have water added for 2 to 6 months unless the battery is overcharged. The specific

gravity of the electrolyte of this battery is 1.270 at 77°F when fully charged.

9.11 ADVANCED GEL-SEALED BATTERY

The advanced gel-sealed battery is available with top and front electrical terminals. The advanced gel-sealed batteries will provide the user with a long-life product when used for solar applications. Newer advanced gel technology provided in the new batteries combines the best features of absorbed glass mat (AGM) and gel into one battery design. The addition of an advanced high-density deep-cycle paste mix as an electrolyte provides a battery that has a wide range of capacities that will endure the rigors of cycling in high temperatures better than traditional AGM or gel batteries. The deep-cycle batteries are the ideal choice for any solar energy application for both off-grid and grid-tied systems.

9.12 MODULAR RACKS FOR SOLAR BATTERIES

One of the problems with setting up a battery storage area for a solar photovoltaic system is how to store a large number of batteries so that they can safely be charged every day and be accessible for maintenance and inspection. Figure 9-9 shows a typical battery rack and you can see all of the batteries positioned so that they can be wired together to create the proper amount of voltage and current. The batteries are also positioned so they can be inspected periodically. Lead-acid batteries will need to be inspected for the electrolyte level on a frequent basis. An absorbed glass mat (AGM) battery will not need to be inspected for electrolyte level as often because it is a sealed battery, but you will still need to have access so that you can connect the battery electrical connections at the battery terminals.

FIGURE 9-8   Solar high-powered flooded (HPF) battery. (Courtesy Power Battery)

FIGURE 9-9   An example of a battery rack to store batteries for solar energy system.   (Courtesy NREL and U.S. Department of Energy)

TABLE 9-1 Number of batteries connected to create strings of four batteries to produce 48 V. Additional strings are connected together in parallel to provide additional amounts of current.

48-V System

Model	System Cap. @ 100hr rate		Dimensions				Weight	Battery Model	No.of Batteries	No.of Strings	No.of EZ1 Racks	No.of EZ2 Racks	Bus-Bar Assy.
	AH	kWh	W	D	H1*	H2*	lbs						
EZS-48105	105	5.116	20.3	23.7	12		348	PSG-12105FT	4	1		1	Optional
EZS-48210	210	10.231	20.3	23.7	22.5		693	PSG-12105FT	8	2	1		Optional
EZS-48255	255	12.424	20.3	23.7	22.5		709	PSG-12255FT	4	1	1		Optional
EZS-48315	315	15.347	20.3	23.7	34.5	42.5	1041	PSG-12105FT	12	3	1	1	Induded
EZS-48420	420	20.463	20.3	23.7	45	53	1386	PSG-12105FT	16	4	2		Induded
EZS-48510	510	24.847	20.3	23.7	45	53	1418	PSG-12255FT	8	2	2		Induded
EZS-48765	765	37.271	20.3	23.7	67.5	75.5	2127	PSG-12255FT	12	3	3		Induded
EZS-481020	1020	49.694	2 × 20.3	23.7	45	53	2836	PSG-12255FT	16	4	4		Induded
EZS-481275	1275	62.118	2 × 20.3	23.7	67.5	67.5	3545	PSG-12255FT	20	5	5		Induded
EZS-481530	1530	74.542	2 × 20.3	23.7	67.5	75.5	4254	PSG-12255FT	24	6	6		Induded
EZS-482040	2040	99.389	4 × 20.3	23.7	67.5	67.5	5672	PSG-12255FT	32	8	8		Induded
EZS-482550	2550	124.336	4 × 20.3	23.7	67.5	67.5	7090	PSG-12255FT	40	10	10		Induded
EZS-483060	3060	149.083	4 × 20.3	23.7	67.5	75.5	8508	PSG-12255FT	48	12	12		Induded

(Courtesy Power Battery)

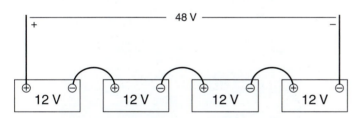

FIGURE 9-10 Four 12-V batteries connected in series to give 48 V total.

9.13 CONNECTING BATTERIES IN SERIES AND PARALLEL FOR SOLAR BANKS

When batteries are put together for a solar photovoltaic electrical storage system, individual batteries will need to be connected in series to increase the total voltage potential of the system and additional batteries will need to be connected in parallel to increase the current capacity of the system. Table 9-1 shows the data for connecting 12-V batteries in series to create a 48-V system. Since each battery provides 12 V, you can see in the top row that four of the 12-V batteries must be connected in series to create the 48-V power system. The first column in the top row shows that these four batteries can provide 105 ampere-hours and 5.116 kilowatt-hours.

If you need more current capacity (in ampere-hours), you would select eight batteries and connect four in series to make two strings. Then you would connect the two strings of four batteries that are in series in parallel to increase their amperage capacity. Figure 9-10 shows the four batteries connected in series to create the 48-V string. Figure 9-11 shows the two strings of four batteries (total of eight batteries) together to double the current capacity for the batteries. The data on the second row of the table show the number of batteries for this configuration is eight batteries, and the total number of ampere-hours doubles to 210 hours and produces 10.231 kilowatt-hours.

You can see that batteries can continue to be added until 48 batteries are connected in series and parallel to give 48 V and a total of 3,060 ampere-hours, which produce 149.083 kilowatt-hours. The batteries are placed on the battery racks in such a way that metal-conducting bus bars can easily be attached to each battery's terminals to connect them into strings of four batteries and then connect the strings in parallel to provide additional current.

Figure 9-12 shows a bank of the more expensive gel or absorbed glass mat (AGM) batteries. Since these batteries do not need to be refilled and do not normally generate explosive gasses, they will fit in a much tighter battery rack. This type of rack provides a large bank of batteries that are mounted close together in a vertical steel rack. Since this application uses AGM batteries, a vapor-proof enclosure or vent pipe system is not needed because the batteries do not vent gases of any kind.

Figure 9-13 shows a less expensive battery rack for a solar energy system. This rack allows a variety of batteries to be placed close to one another where they can be joined with bus connectors. This type of rack provides substantial strength to hold a number of batteries safely and to keep them off the ground and secured closely together.

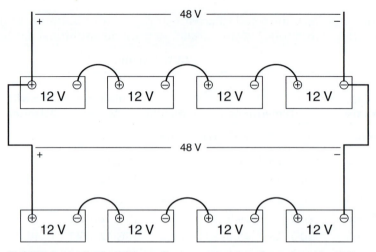

FIGURE 9-11 Four 12-V batteries connected in series to provide 48 V. A second set of four 12-V batteries is connected in parallel to double the amount of current.

FIGURE 9-12 A battery rack for AGM batteries allows the batteries to be placed closer together. (Courtesy NREL and the U.S. Department of Energy.)

9.14 PERIODIC MAINTENANCE FOR SOLAR ENERGY STORAGE BATTERIES

Maintenance for solar energy electrical storage batteries can be broken down into several sections, which include checking the level of the electrolyte, the specific gravity of the electrolyte, and the tightness of the terminal connections, as well as corrosion prevention and repair. When the battery is inspected for corrosion, each terminal connection and all exposed metal are checked to ensure that the acid fumes have not started the corrosion process. If corrosion is detected, the corrosive areas on thte battery should be cleaned using a baking soda and water solution, a couple of tablespoons to a pint of water. Baking soda will neutralize the acid and any damage caused by acid fumes. The major problems will occur where battery cable connections have exposed metal and near any other exposed metal.

After the battery terminals have been cleaned, they should be checked to ensure that they are tight.

FIGURE 9-13 Battery racks for solar batteries. (Courtesy NREL and U.S. Department of Energy.)

Many battery problems can be caused by terminal connections that become dirty or loose. A wet cell battery will need to have its electrolyte fluid level checked. If the electrolyte fluid level is low, you can bring it back to its proper level by adding distilled water that has all impurities removed to avoid contaminating the cells of the battery with minerals that may be in some water. When refilling electrolyte it is important to not overfill battery cells, since the fluid will expand slightly when the battery is being charged and the fluid becomes warm. If the battery cell is overfilled, it will push excess electrolyte fluid out of the cell through the vent cap, and this solution will be extremely corrosive since is nearly pure acid.

Corrosion on exposed metal parts and terminals can be eliminated by coating the terminals with grease. Other ways to control the corrosion is to use a felt battery washer that is specifically designed to prevent corrosion on the terminals. Coating the terminals and any other exposed metal with grease will keep out both liquid acid and any acid fumes that could cause corrosion.

9.15 TROUBLESHOOTING BATTERY PROBLEMS IN SOLAR ENERGY SYSTEMS

At times you will encounter problems with a battery that is part of the electrical storage system for a solar energy system. When the problems occur, it is important to determine where the problem is and what the appropriate steps are to fix it. This section will explain many of the basic problems you will find with the battery and charging system for solar energy systems.

9.15.1 Battery Will Not Take a Charge

One of the failure modes of a battery occurs when it will not take a charge. When this happens, you should suspect that the cables and connections to the battery are faulty; the power source providing the voltage for the charge is the problem in that the voltage is insufficient (below the current level of the battery voltage) or the problem can be a defect in the battery.

One way to determine what is causing the problem is to use a voltmeter to check the voltage level of the battery and compare it to the voltage specification for the battery. If the voltage level of the battery is lower than the voltage specification, the battery should take a charge. For example if the battery is a 12-V battery and you measure the battery voltage as 11.2 V, the battery should take a charge. If the voltage level in the 12-V battery is 12.4 V, the battery is nearly fully charged and it will not take much more charge.

To determine if the battery has problems with its cables or connections, you can visually inspect the cables and clean and tighten them if necessary. Over time the battery terminal connections become corroded and create a high resistance path for electrical power to flow to the battery. Once the terminals are cleaned and tightened, you can check the battery and see if it will now take a charge.

If the battery still does not take a charge, you can move to the next step and check to see if the power source (the photovoltaic panels) is delivering sufficient voltage. Remember that the voltage from the PV panels must be higher than the voltage in the battery or the current will not flow to the battery. You can check the open circuit voltage from the PV panels with a voltmeter to ensure that the voltage is high enough. If the voltage from the PV panel is higher than the battery voltage, the battery should accept the current from the PV panels and begin to charge.

The final problem that may keep a battery from accepting a charge is that one or more of the battery cells are damaged. The biggest problem with a battery that will not accept a charge is a damaged cell. The cell can be damaged in a number of ways, including repeated overcharging, low levels of electrolyte, or a cracked or damaged cell. You can check the electrolyte level and specific gravity by using an electronic battery tester such as the one shown in Figure 9-14, or a hydrometer such as the one shown in Figure 9-15. Figure 9-16 shows a battery refractometer, which can measure the specific gravity of the electrolyte.

FIGURE 9-14   Electronic battery tester.

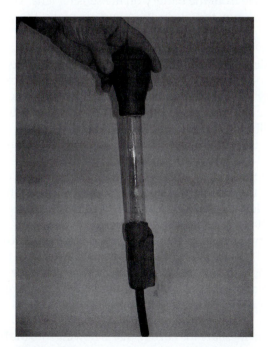

FIGURE 9-15   Glass tube and bulb hydrometer.

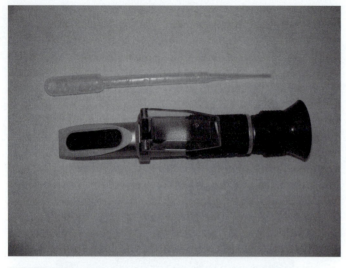

FIGURE 9-16   Battery refractometer shows the specific gravity of battery electrolyte.

TABLE 9-2	Specific gravity for lead-acid batteries to indicate state of charge.		

State of Charge	Specific Gravity	Voltage	
		12 V	6 V
100%	1.265	12.7	6.3
75%	1.225	12.4	6.2
50%	1.190	12.2	6.1
25%	1.155	12.0	6.0
Discharged	1.120	11.9	6.0

This type of tester operates by providing a small drop of electrolyte from the battery on the sensing lens. The tester breaks down the electrolyte and displays the specific gravity in a spectrum, so you check the amount of electrolyte. The specific gravity of lead-acid batteries in all conditions is shown in Table 9-2. Specific gravity is the ratio of the density of a unit volume of a substance to the density of a given reference material such as water. Specific gravity may be referred to as "relative density." Typically for a battery the specific gravity of acid used as the electrolyte in a lead-acid battery ranges between a low of 1.120 and a high of 1.265.

If the electrolyte level is low, you must refill the battery to the correct level with distilled water. If the electrolyte level indicates the battery is in a state of discharge, the battery should accept a charge.

You can use a battery charger designed for automotive wet cell batteries to provide additional checks on the battery. You can disconnect the battery from the solar energy system and test it with an automotive battery charger that has the ability to check the battery for problems as well as charge. The automotive battery charger will also have a means to measure the current that is flowing into the battery to determine if it is charging successfully. If the battery is not

taking a charge after you have made the other tests, you can suspect that it has a broken plate or other internal problem that is not fixable, and the battery must be replaced.

9.15.2 Battery Will Not Keep a Charge

Another frequent problem with batteries is that they will not remain charged after they have been completely charged up. This problem is usually due to one of several problems. First, check to see if the load circuit that is connected to the battery is using excessive current or if the current draw is always present due to a short circuit or load that is left on in the load circuit. An easy way to check to see if the problem is caused by the battery itself or if it is caused by the load is to disconnect the battery from its load circuit. Then see if it keeps its charge for several days. It is normal for a battery to self-discharge slightly, but the amount of discharge should be very low (less than 0.1 V per week). If the battery is able to keep its charge when it is isolated (disconnected) from its load, but it loses its charge rather quickly when it is connected to its load, you should now suspect the problem is in the load. You will have to use an ammeter or ohmmeter to test the circuit while you disconnect sections or segments of the load circuit. If you are using an ohmmeter, you should have the voltage turned off to the circuit and you are looking for the resistance to increase when the problem load is removed. Remember that this type of problem that causes the battery to discharge and not keep a load is caused by a low resistance load that remains connected to the circuit. When you disconnect the section of the circuit where the problem occurs, the resistance will increase.

If you are using an ammeter to troubleshoot the circuit, the total current from the battery will go down when the bad part of the circuit is disconnected. After you find the section of the circuit that is causing the problem, you will need to inspect every load in that part of the circuit and change out the part of the circuit that is causing the problem.

If the battery is failing, it will not be able to keep its charge when it is disconnected from its circuit, and you should suspect that the battery is beginning to fail because of sulfate buildup on the plates. This can occur when the battery is continually overcharged or when the electrolyte level remains low for a period of time. When the plates in a lead-acid battery are allowed to become uncovered because of low electrolyte level, they will begin to grow a layer of sulfate. Sometimes you can add a chemical to the battery electrolyte to help remove some of the sulfate, but normally the battery is beginning to fail and it will eventually need to be changed out.

Batteries will begin to build up sulfates on their plates or the plates will begin to harden when their specific gravity falls below 1.225 or voltage measures less than 12.4 for a 12-V battery, or 6.2 for a 6-V battery. If the battery plates begin to harden due to sulfation, it will reduce the ability to keep a charge and may totally destroy the ability of the battery to generate electrical power.

9.15.3 Ways to Test the Battery

Testing the battery can be done by using the test for specific gravity or by a battery load test. One of the ways to determine the condition of the battery charge is to measure the specific gravity and battery voltage. The specific gravity is measured with a temperature-compensating hydrometer (tube type or electronic type) and the voltage is measure with a digital DC voltmeter. You will be able to compare the hydrometer readings with the readings in Table 9-2. You will not be able to use a hydrometer with a sealed battery, so you must use a load tester that measures the voltage and the current draw over a short period of time.

A battery load test consists of removing current from the battery under a small load and then measuring the amount of voltage loss and the amount of current that is flowing during the test. If the battery is bad or has problems, it will not be able to sustain the rated current flow for 15 seconds without losing a substantial amount of voltage. The current flow for the battery load test is generally half of the rated cold cranking amps. When you are using a load test for a battery, you should begin by ensuring that the battery is fully charged. After the battery is fully charged, you will need to remove the surface charge by discharging the battery for 2 to 3 minutes with an external load that will draw 1 to 2 A. After you have removed the surface charge, you can disconnect the external load and then begin the load test.

During the load test the battery is connected to a load that causes it to draw half of the cold cranking amperage (CCA) for 15 seconds. During the load test the voltage will not drop too much if the battery is good. If the battery has problems, the voltage will drop substantially, usually less than 10.5 V for a 12-V battery. Another way to test the battery during the load test is to test its specific gravity, and the specific gravity should not vary more that 0.05 between each of the cells.

Another way to test the battery if you have time is to put a fixed load on the battery over 2 hours and allow the battery to discharge with a load that draws 10% of its rated current over the 2-hour period and then measure the battery voltage. You can determine the amount of load by using Ohm's law and calculating the amount of resistance or you can use an ammeter to measure load. At the end of the test you can measure the specific gravity of each cell and you should not see a difference of more than 0.05 between cells. The object of this test is to ensure that all of the cells are equal, and that none of the cells is going bad. Typically if a battery has one or more bad cells, it will show up during the load test. The result is very simple; a battery will begin to present problems in not holding a charge or not recharging correctly, and the end result is

the battery will need to be replaced. The main objective is to be sure that the battery is causing the problem rather than the charging equipment or dirty terminals, which will not fix the problem if you change out the battery.

9.15.4 Battery Testing During Periodic Maintenance

During periodic maintenance you can test the batteries with a digital voltmeter and the reading for a 12-V lead-acid battery should read higher than 12 V when it is fully charged. The sealed AGM and gel-cell battery full-charge voltage should be approximately 12.8 to 12.9 V. If the voltage reading for any of the types of fully charged 12-V batteries is less than 10.5 V, it indicates one or more of the cells in the battery are bad, and the battery will need to be changed.

9.16 THE EFFECTS OF TEMPERATURE ON BATTERIES

When the temperature drops to around the freezing point, the number of amperes that battery can hold will decrease. This creates a problem with battery storage systems for solar energy systems that are in areas of the country where the temperature in the winter goes down near freezing. If the temperature drops near freezing, the batteries' ability to store a charge may be reduced by 20%. The main point of understanding this problem is that you will need to increase the size of the battery storage by 20% if the batteries will be subjected to cold temperatures, or the system will not function well during the winter and the system will not be able to store and provide enough electrical power.

Another way around this problem is to use a battery storage building that can be heated during the coldest months to maintain a temperature above 40°F. The only problem with this is that the cost of heating the building may be more than the cost of the solar energy that you are producing.

The opposite condition occurs in the summer when the temperature increases, and the battery will be able to produce more current than its rating. This condition works well with solar energy systems, as the solar PV panels will produce more electrical power during the summer months when the sun is shining longer hours and the location of the sun is more favorable to produce maximum electrical power from the panels.

9.16.1 Selecting a Battery for a Solar Energy Application

When you are selecting a battery for a solar energy application, you must keep in mind that the solar energy application requires a deep-discharge battery because the battery charges for extended periods of time during the daytime and then discharges for extended periods of time through the evening. The number of batteries that are needed will depend on the voltage of the system, and the amount of current the system will need is rated in ampere-hours. The number of batteries that are connected in series to create the voltage will depend on the amount of voltage the system needs. You can select 6- 12- or 24-V batteries to combine in series for these applications. Additional batteries can be added in parallel to increase the amount of current capacity the system needs.

The next step in the process after you've selected the batteries is to select a battery rack that is capable of holding all the batteries while providing access to the terminals, and access to the battery caps if the system includes wet cell batteries that need to be tested periodically. Access to the battery terminal connections is also important, so that they can be tightened and cleaned periodically. If the battery system is housed in an enclosed room, a vent system may be required to ensure that the gases from the battery are safely vented to the atmosphere.

When you are selecting the size of batteries for your system, you may have to refer to a sizing chart to ensure that you're getting a battery that has the proper amount of ampere-hours and the capability of storing enough electrical power for your needs.

Questions

1. Explain the difference between a primary cell and a secondary cell.
2. Explain what is meant by a deep-discharge battery.
3. Explain the advantages and disadvantages of an absorbed glass mat (AGM) battery.
4. Explain the advantages and disadvantages of a lead-acid battery.
5. Explained what an electrolyte is for a battery.

Multiple Choice

1. When the specific gravity of an electrolyte in a battery is high, _____
   a. the charge in the battery is low.
   b. the charge in the battery is high.
   c. it is impossible to tell the amount of charge the battery has from measuring its specific gravity.
2. When batteries are connected in series _____
   a. the voltage of the group of batteries increases.
   b. the current of the group of batteries increases.
   c. both the voltage and current of the group of batteries increases.
3. When batteries are connected in parallel, _____
   a. the voltage of the group of batteries increases.
   b. the current of the group of batteries increases.
   c. both the voltage and current of the group of batteries increases.
4. A lead-acid battery _____
   a. has a paste electrolyte.
   b. has a porous material mat wrapped in a tight spiral that absorbs the electrolyte.
   c. uses a liquid electrolyte.
5. An absorbed glass mat (AGM) battery _____
   a. has a paste electrolyte.
   b. has a porous material mat wrapped in a tight spiral that absorbs the electrolyte.
   c. uses a liquid electrolyte.
6. Temperature affects a battery by _____
   a. increasing the amount of voltage a battery can store in cold weather.
   b. increasing the amount of voltage a battery can store in hot weather.
   c. has no effect on the amount of voltage a battery can store in hot or cold weather.

10

The Grid and Integration of Solar-Generated Electricity into the Grid

OBJECTIVES FOR THIS CHAPTER

When you have completed this chapter, you will be able to

- Identify the three main sections of the electrical grid.
- Explain what the grid is and how it works to distribute electrical power to customers.
- Explain what the smart grid is and how it is different from the grid.
- Explain how a transformer works.
- Identify the two windings of a transformer.
- Explain what a substation is and why it is important to the grid.
- Identify three things that the IEEE 1547-2003 standard addresses for technical requirements for power that is connected to the grid.
- Explain what a net meter is.
- Explain what power quality issues are.

TERMS FOR THIS CHAPTER

Apparent Power
Brown Out
Capacitive Load
Flicker Power
Grid
Grid Companies (Gridco's)
Grounding
Induce Current
Induction
Institute of Electrical and Electronics
  Engineers (IEEE)
Kilo Volt Ampere (kVa)
Low Voltage Ride Through (LVRT)
National Electrical Code (NEC)
Net Meter
Power Factor

Power Quality
Primary Winding
Public Utility Regulatory Policies Act (PURPA)
Reactive Power
Resistive Load
Secondary Winding
Service Drop
Smart Grid
Step Down Transformer
Step Up Transformer
Substation
Transformer
Transmission Companies (Transco's
True Power
Turns Ratios
Volt Ampere (VA)

10.0 OVERVIEW OF THIS CHAPTER

This chapter will discuss how electrical output from a solar photovoltaic system is connected to the electrical grid. It is important to understand that if you are hired to work on the electrical system of the solar photovoltaic system or the electrical portion of the grid and any power lines, you will need to have a complete working knowledge of electricity and receive specialty training to be able to work around voltages above 200 volts. The information in this chapter is provided to give you a solid background in understanding what the grid is and how solar photovoltaic systems are connected to it.

It will begin by explaining what the grid is in its present condition in respect to size and potential for growth. Information about the new Smart grid will also be identified, and the features of the Smart grid will be explained. Information about the National Electrical Code and other rules and regulations pertaining to the grid will be identified and described.

Another part of this chapter will discuss how electricity from a solar photovoltaic system is used to supply AC power directly to a commercial building or residence by sending the DC voltage through an inverter. Information regarding the basic utility meter and the net meter will be presented. This section will also identify the switches and connections for power distribution for a grid-tied system. Additional information will be provided to explain how the electrical power from solar photovoltaic systems may need to be cleaned up through an inverter to match the voltage and frequency of the power on the grid. This can be accomplished by sending the DC voltage from the solar photovoltaic system to an electronic inverter to ensure that its frequency is the same as the grid's. Other information about voltage control and reactive power control will also be presented. Additional problems with connecting voltage produced from solar photovoltaic system to the grid will be discussed, including power quality issues and safety circuits in the solar photovoltaic system.

Information in this chapter will also explain how the solar photovoltaic electrical system is grounded, and how the power lines from the solar panels are connected to the substation or grid power source on site. This information will include why the cable installation goes underground and overhead and the use of transformers. If the solar photovoltaic system is part of a larger solar farm, it will be tied in to the substation, which will be explained at the end of this chapter.

10.1 UNDERSTANDING THE GRID

In order to understand the grid, this first section will provide basic information about it. *Grid* is the name that is applied to all the electrical power distribution systems in the United States and North America. If a solar photovoltaic system is designed to produce power for the grid, it must be connected to an existing section of the grid and its power must meet the specifications for that section. The grid is used to distribute electricity across the United States, and it is actually made up of miles of interconnected wires that are used to move electricity from where it is produced to where it is consumed. When electricity is produced from photovoltaic panels, it will be at values that are typically 600 V or 1,000 V DC, which will be converted by the inverter and transformer to 15kV (15,000 V) or 35kV (35,000 V) AC.

This voltage must be stepped up with transformers to higher voltages of 545 kV to 765 kV for long-distance transmission over the grid, and it may be stepped up to 345 kV to 545 kV to be transmitted for short distances. When the electricity reaches the areas where it will be used, it is transformed back down at a substation near the city or an industrial site to a level of about 12,470 V and is routed to where it is finally consumed. Service transformers on the power pole outside of the residence or at the smaller substations for an industrial application will step the voltage from 12,470 V to 120 V per line for residential customers and 240 V per line for industrial users. Figure 10–1 shows the distribution system from the generating site through transmission lines to a residential customer.

Subsections of the grid include generation sites, transmission systems, sub-transmission systems, substations, feeders, service systems, and customers. Generation

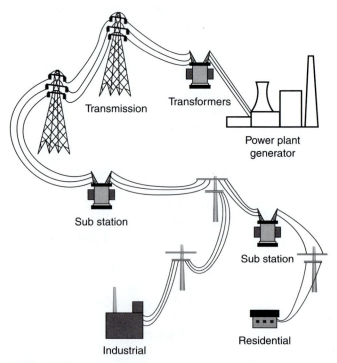

FIGURE 10–1 Electrical power distribution from generation station to residential customer.

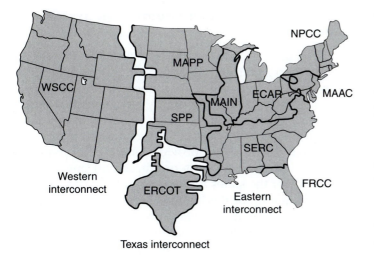

FIGURE 10–2 A map of the United States showing the Eastern Interconnect, the Western Interconnect, and the Texas Interconnect of the grid.

sites may include coal-fired steam plants, nuclear power plants, hydroelectric plants, wind turbines, and photovoltaic (solar) sites. The transmission and sub-transmission lines include the high-voltage and extra-high-voltage transmission lines that carry power across the country over longer distances. A *substation* is a large set of transformers located at the edge of a city or near an industrial site that steps down the electrical power transmitted from the generation site at high voltage. Feeder lines provide voltage throughout cities and residential areas. Service systems include the transformers in wiring to provide the service drop to industrial, commercial, and residential systems. The customer portion of the grid includes any wiring around the customer site, such as a large industrial application.

The map in Figure 10–2 illustrates how the grid is connected in each state. The grid is divided into three major sections: the Eastern Interconnect, the Western Interconnect, and the Texas Interconnect. The Eastern Interconnect is further divided into subsections: the Northeast Power Coordinating Council (NPCC), which includes Maine, Vermont, New Hampshire, New York, Massachusetts, and Rhode Island; the Mid-Atlantic Area Council (MAAC), which includes Maryland, Delaware, New Jersey, and parts of Pennsylvania; the East Central Area Reliability Coordination Agreement (ECAR), which

includes Michigan, Indiana, Ohio, West Virginia, and Kentucky; the Southeastern Electric Reliability Council (SERC), which includes the southern states except Florida; the Florida Reliability Coordinating Council (FRCC), which includes all of Florida; the Mid-America Interconnected Network (MAIN), which includes Wisconsin and Illinois; the Mid-Continent Area Power Pool (MAPP), which includes the Dakotas, Nebraska, Minnesota, and Iowa; and the Southwest Power Pool (SPP), which includes Kansas and Oklahoma. The Texas Interconnect includes the Electric Reliability Council of Texas (ERCOT), which includes all of Texas; the Western Interconnect includes the Western Systems Coordinating Council (WSCC), which includes all of the western states from New Mexico, Colorado, Wyoming, and Montana westward.

Table 10–1 shows the demand, supply, and capacity margin for the United States as of 2009. The capacity margin is the amount of electrical power that is produced compared to the amount of electrical power that is consumed. It is important that the amount that is available be more than the amount being used to meet industrial and residential needs. At that time, the capacity margin of the Eastern Interconnect Grid was 19.3%, for the Texas Interconnect Grid it was 13.6%, and for the Western Interconnect Grid it was 22%. The capacity margin for the entire United States was 19.4%. It is important to understand how much power is needed in each section of the grid, and how much power is available to supply the need in each section because that is the amount of electrical power that must be produced every hour of every day, and solar energy systems and other alternative energy must produce this amount if they are going to replace traditional energy sources. You will learn that energy produced by solar energy systems and other alternative energy sources provide only a small amount of this larger demand.

The grid provides electricity for each of the areas shown in the map, and if you are providing electricity to the grid from solar energy systems, you must know which part of the grid you will be connecting to and be familiar with the specifications for that part of the grid. When a solar project is connected to the grid, a permit to connect to the grid must be filed with the utility to begin the process.

When you are connecting a solar photovoltaic system to the grid, you are referring to the voltage

TABLE 10–1	U.S. Capacity Margins, 2009			
	Eastern Grid	**Texas Grid**	**Western Grid**	**U.S. Total**
Demand (MW)	554,412	63,376	136,441	754,229
Supply (MW)	688,783	72,204	174,978	935,965
Capacity margin (%)	19.3	13.6	22	19.4

Source: U.S. Energy Information Administration

transmission portion of the system. This is basically where voltage is created at the solar panels and connected to the grid where the voltage is moved across the power transmission system to its end point in an industrial, commercial, or residential area. The actual transmission lines, towers, transformers, and switch gear are owned by individual companies called *transmission companies (Transco's)* that manage this hardware. *Grid companies (Gridco's)* are sets of companies that manage the grid function, which is the interconnecting and routing of electricity through the hardware cables so that areas receive the electricity they need. Some parts of the grid, hardware, and grid management may be controlled by some local or regional utility companies.

The electricity that flows through this hardware is treated as a commodity and is bought and sold on short-term and long-term contracts. Some of the ownership of the equipment and the sales of the electricity that flows through the lines are regulated by the states where the equipment exists, and where the electricity is being consumed. Other parts of the system, in which electricity moves between one or more states, are covered by interstate regulations. This is why it is important to understand which part of the grid you will be connecting to with your solar photovoltaic system, and which laws apply in which states. For example, some states have laws that require utility companies that produce power and companies that transmit power to purchase power that is generated from solar photovoltaic systems. Other states have not reached this level of cooperation between solar photovoltaic system owners and the utility companies. You will find that in most areas, you will be required to fill out an application and make a request to be able to connect your system to the grid, and to get reimbursed for the power that the solar photovoltaic system produces.

The main voltages on the grid are separated into four different voltage levels for long-distance transmission. One of the highest voltages used to transmit voltage inside the U.S. grid is 765,000 V. Some higher transmission lines are used to transmit voltage from larger hydroelectric generation stations. Other lines transmit voltage at 500,000 V, 345,000 V, and 230,000 V. Typically the higher the voltage, the longer the distance the voltage is being transmitted.

The grid is divided into three distinct sections: the Western grid, the Eastern grid, and the Texas grid. The three major sectors are isolated to the extent that a problem in one grid sector can be contained and not affect the other two. There are large control centers in each of the subsectors that monitor the grid continually to ensure that a sufficient amount of electrical energy is available, and that blackouts are limited to a very small area if they do occur. In the remaining sections of this chapter, you will learn about the pros and cons of the grid system as it exists today. You will also see how the grid is being updated to something called the Smart grid.

10.1.1 Connecting Solar Photovoltaic Systems to the Grid

After you understand what the grid is and how it works, the next important item is understanding how to get a solar photovoltaic system connected to the grid. Approximately 42 states currently have laws that allow or require utilities to permit solar photovoltaic systems and other electricity-producing technology to be connected to the grid, and to provide some type of compensation for the energy that is produced and put into the grid. Some of the states require the utility to pay only a minimal amount for this energy, thus making the energy the solar photovoltaic system produces less valuable, whereas others require utilities to pay the same or nearly the same value as they charge for their electricity. A federal law known as the *Public Utility Regulatory Policies Act (PURPA)* was passed in 1978 as part of the National Energy Act. This law requires utility companies to purchase power from independent providers, but the law does not determined the rate that should be paid for the power.

One way of providing the connection between the solar photovoltaic system and the grid is with a utility meter that can measure the amount of power going into the grid or coming from the grid. Since this meter can measure the voltage that the solar photovoltaic system produces, or the voltage that is used at a residence, it is called a *net meter.* Some versions of the net meter allow the excess generated electricity to be banked, or credited to the customer's account, which basically means that the electricity that is generated by the photovoltaic panels is sold at the same rate as electricity that is used. The net meter allows the electric meter to spin forward and backward, which simplifies the process of measuring how much electricity was produced or consumed from the grid.

10.2 THE SMART GRID

The *Smart grid* is basically an update of the present-day grid. Currently information and energy in the grid go in one direction: from the point where it is generated to the point where it is consumed. The power company that generates electricity or the solar photovoltaic system that produces electrical power basically provides that power to the grid, where it flows with other electrical current to the end point where it is used by consumers. Any data or information that will assist energy producers currently flows in one direction from the energy producer to the end user, the consumer.

The Smart grid is based on the idea that information should flow in both directions on the grid. For example, it would be useful if the electric meter in the residence or commercial location had the ability to report back to the generating company or the transmission company the details of how much energy was being consumed at any point in time through out the day from that location.

These data could be recorded and analyzed so that predictions could be made about the times energy would be consumed in the future. Additional information could be provided through this meter to let consumers know the cost of the energy that they were consuming at any given time, and if the energy would be more expensive or less expensive at a different time during the day. For example, there is an excess amount of electrical power available on the grid late at night, which is a low-peak or off-peak time. Electrical generating companies are able to provide electricity at a cheaper rate during these times, if consumers can use the power then. For example, it may be possible to heat water or to heat bricks in a furnace storage unit during this off-peak time, saving the consumer a considerable amount of money.

Basically the Smart grid identifies each step in the grid transmission process and each end user with a computer address that allows all the equipment at that location to send and receive data about the amount of energy being used, the time of day, and other information such as temperature or other environmental conditions. Since each point in the grid will have its own specific address, it can be accessed much like a large computer network in which information can flow to and from each node in the network. Since the information is similar to that on a computer network, it can be shown on a small display monitor or on the computer in the home or commercial establishment. This display of information will help consumers understand how much energy they are using at any given point, and if there are any alternatives available at a later date or at a lower price. For example, if you have an electric vehicle that needs recharging from time to time, the Smart grid would be able to indicate what time of day or what day of the week the electricity would be at the minimum cost to complete this recharging process.

Since the electrical meter and any switch gear to each end user can be monitored, they can also be controlled. For example, the individual loads in a home could be connected to a controller to be individually monitored as well as turned on or off for a small period of time to shed power during a peak power period. The controller would send and receive control information to the air-conditioning system, the electrical hot water heater, electric clothes dryer, and the electric range. The Smart grid could monitor these components and begin to determine how often they were used and at what time they were being used. The utility company could then offer the homeowner several pricing packages that would fit with the peak and non-peak times the power company has. The power company would price the electrical power consumed during the peak times the most expensive, and the power consumed during the off-peak times the least expensive. The reason power is less expensive during the off-peak times is that there is more generating capacity available than the amount of power that is being consumed at that time. This usually occurs during the late-night hours because customers are asleep, and the number of factories and commercial establishments that are using large amounts of energy is minimal. The homeowner could then program some of the electrical loads such as heating hot water, recharging an electric car, or heating thermal bank electric furnaces at that time to ensure that they were using power when the lowest-priced electricity was available. This would allow the electrical energy consumers to better match the times when they choose to consume electricity for some loads and match it to the time when the least expensive electrical energy is available.

Today, power-producing utilities try to predict the amount of load at any given period and then burn the energy such as coal or nuclear power to produce the electricity that will be consumed during the next several hours. Since this process of consuming the energy to create electricity is time delayed, some of the energy may be wasted; for example, if coal is burned to produce steam to make electricity and it was not used 2 hours later. You can see that since the current system does not monitor in real time the usage patterns and the changes that occur to the system due to weather and other external conditions, it is very inefficient.

The Smart grid with residential commercial controllers will also allow end users to enter into contracts with energy producers to agree to have some of their largest loads disconnected or turned off for short periods of time when the grid is becoming overloaded, or if a brown out is occurring. A *brown out* is a period of time when more power from the grid is being used than what is being produced, and a low-voltage condition occurs where the voltage drops 10% to 20%. If the brown out continues for a longer period of time, large electrical loads that have motors, such as air-conditioning compressors and pumps, may be severely damaged and overheated. One method of fixing a brown out condition is to allow the distribution control for the grid to begin to shed larger loads and heavy users until the voltage level comes back up. The Smart grid will allow large industrial users to make agreements with the power company to have large sections of their electrical equipment disconnected for short periods of time, in return for a better rate for the electricity they purchase. This system has been in existence for more than 10 years with large industrial users, and with the Smart grid this pricing arrangement will be available to residential users as well. A residential user may agree to have the air-conditioning shut off during high-peak periods or in the event of a brown out, and the power company would agree to offer a reduced rate for the electricity for that consumer.

The Smart grid would also allow the controller to lock certain loads out and allow them to be energized only during the lowest-peak periods of the day or week, depending on the need of the electrical load.

Since the Smart grid will need to send and receive information like a computer network, it will need to connect to homes through existing computer networks, or the signals can be sent in packets over the high-voltage lines. Computer communications and networking equipment would be able to filter this information off the single-phase and three-phase high-voltage lines so that the computer data could be sent and received through the high-voltage lines.

On the transmission side, the Smart grid would be able to determine which areas of the country and which areas of the grid sector are using the most power. This would help the utility distribution companies understand more quickly how to redistribute electrical power at any hour to the sectors that need it most. It would also allow the interconnection system to identify problems and disconnect and isolate small sections faster, to ensure that they do not affect the entire grid. You might remember a problem in August 2003, when lightning struck a power line in the Cleveland, Ohio, area, and the series of events that followed caused large sections of Ohio, Pennsylvania, New York, Michigan, and Ontario in Canada to be blacked out. The blackout occurred because the grid was not able to disconnect the intermediate sectors fast enough to prevent the blackout from spreading across several states. The new technical equipment that would be used in the Smart grid would be able to limit the blackout area to a much smaller region, possibly within several blocks of where the lightning strike occurred, instead of allowing it to spread across several states.

Another problem with the present-day grid is that it is inefficient in its transmission capability. Some experts believe that as much as 30% of the power in the grid may be lost either through heat loss due to older wiring or poor equipment and connections. The Smart grid would be able to identify the most efficient means of transmitting the power from where it is produced to where it is consumed.

Some modern larger solar photovoltaic systems may have a programmable logic controller (PLC) for control or dedicated computer control, so they will be able to integrate with the Smart grid easily. The amount of voltage and current the solar photovoltaic system is able to produce can be closely monitored at all times, and the amount available to send into the grid will be measureable at all times.

Some of the largest energy companies in the United States, such as General Electric, IBM, and Siemens Electric, are working to improve the Smart grid and get it integrated as quickly as possible. One of the problems with the integration of the Smart grid is the high cost of converting residential and commercial users to new equipment as well as changing out the transmission equipment and switch gear. The Smart grid is currently being created and integrated, and this integration will last for many years. Many experts compare the creation and integration of the Smart grid to a time when electricity was first introduced to all parts of the United States. At that time, it took many years to get the electricity to all the points where it was needed.

10.3 TRANSFORMERS, TRANSMISSION, AND DISTRIBUTION INFRASTRUCTURES

Before you can begin to understand the transmission and distribution infrastructures for electricity, it is important to understand the concept and theory of a transformer, and the groups of transformers that make substations. A *transformer* is an electrical component that takes voltage from one level such as 900 V AC and increases that voltage to 12,470 V for transmission. The transformer consists of two coils of wire that are located close to each other inside a metal case, but the two windings are not connected to each other in any way. The two coils are called the primary coil and the secondary coil. AC voltage is applied to the primary coil, and it is induced into the secondary coil. Since the voltage applied to the primary coil is AC, it has a characteristic sine wave that starts at zero, goes to a positive peak, returns to zero, goes to a negative peak, and returns to zero again. Each time the voltage increases from zero to its peak, it will saturate current into the primary winding, which creates a strong magnetic field around the wires in the winding. When the voltage decreases from its peak back to zero, the magnetic field and its flux lines will collapse and the flux lines in the magnetic field will cut across the wires in the secondary coil. Any time flux lines cut across the wires of the any coil, an electrical current is induced into that coil. This means that the primary winding of the transformer will receive the AC voltage, and its voltage increases to the positive peak, returns to zero, decreases to the negative peak, and returns to zero, a corresponding AC waveform will be created in the secondary winding of the transformer. The waveform in the secondary winding will be out of phase with the voltage in the primary winding by 180°. Since the primary winding and the secondary winding are not connected, the voltage in the secondary winding is induced because the collapsing flux lines of the magnetic field from the voltage in the primary winding cut across the wires in the winding of the secondary winding.

The amount of voltage in the secondary winding will be determined by the ratio of the number of coils and the primary winding to the number of coils in the secondary winding. If the secondary winding has twice the number of coils as a primary winding, the secondary voltage will be twice as large as the primary voltage, and the secondary current will be half the amount of the primary current. The amount of power that goes through the transformer is measured in *volt amperes* (VA) the amount is usually in thousands of volt amperes (kVA) could be changed to the amount is usually in kilo volt amperes (kVA). This means that if you know the amount of voltage for a 10-kVA-transformer, you can divide kVA rating by the voltage and determine the amount of current.

If the amount of secondary voltage for a transformer is larger than the primary voltage, the transformer is called a *step up transformer*, and if the secondary voltage is smaller than the primary voltage, the transformer is called a *step down transformer*. A number of transformers that are used together at the edge of the community make up a substation.

10.3.1 Overview of Transformers

Transformers are needed to step up or step down voltage levels. When voltage is generated, it needs to be stepped up to several thousand volts, and in some cases several hundred thousand volts, so that its current will be lower when it is transmitted. When the current is lower, the size of wire used to transmit the power can be smaller. Smaller wire weighs less, so smaller transmission towers can be used. After voltage is transmitted on the grid over hundreds of miles at thousands of volts, it is then sent on to a city where it is consumed. The high voltage needs to be stepped down to approximately 12,470 V as it flows around the city so that it is less dangerous. This level of voltage is sufficient to transmit power throughout a city. When voltage arrives at a factory, it must be further stepped down to 480, 240, or 208 V and 120 V for all the power circuits in the factory, including the offices. When voltage is transmitted to residential areas, the voltage will be 240 V line to line and 120 V line to neutral. Transformers provide a means of stepping up or stepping down this voltage.

10.3.2 Operation of a Transformer and Basic Magnetic Theory

The transformer consists of two windings (coils of wire) that are wrapped around a laminated steel core. The winding where voltage is supplied to a winding transformer is called the *primary winding*. The winding where voltage comes out of the transformer is called the *secondary winding*. Figure 10–3 shows a typical transformer with the primary coil and secondary coil identified.

A transformer works on a principle called *induction*. Induction occurs when current flows through the primary winding and creates a magnetic field. The magnetic field produces flux lines that emanate from the wire in the coil as current flows through it. When this current is interrupted or stopped, the flux lines collapse, and the action of the collapsing flux lines causes them to pass through the winding of the secondary coil that is placed adjacent

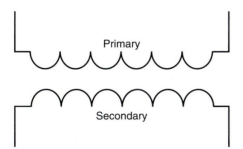

FIGURE 10–3 Example of primary and secondary windings of a transformer.

to the primary coil. You should remember that the AC sine wave continually starts at 0 V and increases to a peak value and then returns to 0 V and repeats the wave form in the negative direction. The process of increasing to peak and returning to zero provides a means of creating flux lines in the wire as current passes through it and then allowing the flux lines to collapse when the voltage returns to zero. Because the AC voltage follows this pattern naturally 60 times a second, it makes the perfect type of voltage to operate the transformer.

When the AC voltage returns to zero during each half-cycle, the flux lines that were created in the primary winding begin to collapse and start to cross the wire that forms the secondary coil of the transformer. When the flux lines from the primary winding collapse and cross the wire in the coil of the secondary winding, the electrons in the secondary winding begin to move, which creates a current. Because the current in the secondary winding begins to flow without any physical connection to the primary winding, the current in the secondary winding is called an *induced current*.

The following information provides a more technical explanation of the relationship between the AC voltage and the windings of the transformer. The magnetic field in the primary winding of the transformer builds up when AC voltage is applied during the first half-cycle of the sine wave (0° to 180°). When the sine voltage reaches its peak at 90°, the voltage has peaked in the positive direction and it begins to return to 0 V by moving from the 90° point to the 180° point. When the voltage reaches the 180° point, the voltage is at 0 V, and it creates the interruption of current flow. When the sine wave voltage is 0 V, the flux lines that have been built up in the primary winding collapse and cross the secondary coils, which creates a current flow in the secondary winding.

The sine wave continues from 180° to 360°, and the transformer winding is energized with the negative half-cycle of the sine wave. When the sine wave is between 180° and 270°, the flux lines are building again; when the sine wave reaches the 360° point, it returns to 0 V, which again interrupts current and causes flux lines to cross the secondary winding. This means that the transformer primarily energizes a magnetic field and collapses it once in the positive direction and again in the negative direction during each sine wave. Because the sine wave in the secondary winding is not created until the voltage in the primary moves from 0° to 180°, the sine wave in the secondary is "out of phase" by 180° to the sine wave in the primary that created it. The voltage in the secondary winding is called induced voltage because it is created by induction.

The induced voltage is developed even though there is total isolation between the primary and secondary coils of the transformer. Figure 10–4 shows the four stages of voltage buildup in the transformer and the flux lines building and collapsing at each point as the sine wave flows through the primary coil.

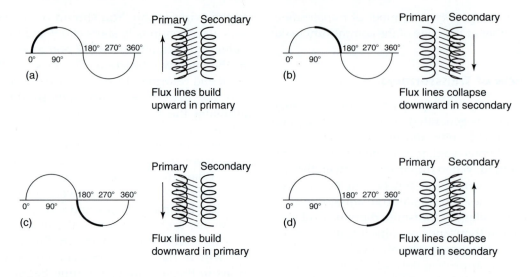

FIGURE 10-4 (a) Flux lines build upward in primary winding. (b) Flux lines collapse downward in secondary. (c) Flux lines build downward in primary. (d) Flux lines collapse upward in secondary.

10.3.3 Transformer Voltage, Current, and Turns Ratios

The amount of voltage a transformer will produce at its secondary winding for a given amount of voltage supplied to its primary is determined by the ratio of the number of turns in the primary winding compared to the number of turns in the secondary. This ratio is called the *turns ratio.* The amount of primary current and secondary current in a transformer also depends on the turns ratio.

Figure 10-5 shows the primary voltage (E_p), the primary current (I_p), the number of turns in the primary winding (T_p), the secondary voltage (E_s), the secondary current (I_s), and the number of turns in the secondary winding (T_s).

The primary and secondary voltage, primary and secondary current, and the turns ratio can all be calculated from formulas. The turns ratio for a transformer is calculated from the following formulas:

$$\text{Turns Ratio} = \frac{T_s}{T_p} \text{ or } \frac{V_s}{V_p} \text{ or } \frac{I_p}{I_s}$$

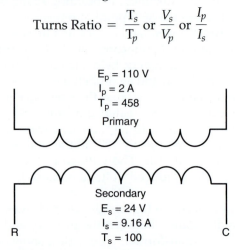

$E_p = 110 \text{ V}$
$I_p = 2 \text{ A}$
$T_p = 458$
Primary

Secondary
$E_s = 24 \text{ V}$
$I_s = 9.16 \text{ A}$
$T_s = 100$

R   C

FIGURE 10-5 Transformer with primary voltage at 110 volts and secondary voltage at 24 volts.

The ratio of primary voltage, secondary voltage, primary turns, and secondary turns is

$$\frac{E_p}{E_s} = \frac{T_p}{T_s}$$

From this ratio, you can calculate the secondary voltage with the formula

$$E_s = \frac{E_p \times T_s}{T_p}$$

The ratio of primary voltage, secondary voltage, primary current, and secondary current is

$$\frac{E_p}{E_s} = \frac{I_s}{I_p}$$

(Notice that the ratio of voltage to current is an inverse ratio.) Using this ratio you can calculate the secondary current:

$$I_s = \frac{E_p \times I_p}{E_s}$$

If you know the primary voltage is 240 V, the primary current is 2 amperes, the primary turns are 458, and the secondary turns are 229, you can easily calculate the secondary voltage and the secondary current using these formulas:

$$E_s = \frac{E_p \times T_s}{T_p}$$

$$\frac{240 \text{ V} \times 229 \text{ turns}}{458 \text{ turns}} = 120 \text{ V}$$

$$I_s = \frac{E_p \times I_p}{E_s}$$

$$\frac{240 \text{ V} \times 2 \text{ A}}{120 \text{ V}} = 4 \text{ A}$$

10.3.4 Step Up and Step Down Transformers

If the secondary voltage is larger than the primary voltage, the transformer is a step up transformer. If the secondary voltage is smaller than the primary voltage, the transformer is a step down transformer. Typically on mid-size and large–size solar photovoltaic systems, the generator creates voltage at approximately 900 V, and a step up transformer is used to get the voltage to12,470 V so it is compatible with the grid. The reason the voltage is increased through the step up transformer is that when the voltage is increased, it lowers the secondary current so that more power can be transferred on smaller-size wire, which weighs less so the transmission towers do not need to be large. A step down transformer is used when the voltage arrives at the substation in a community at 12,470 V and must be stepped back down to 480 V for use in factories and 240 V for residential areas.

10.3.5 VA Ratings for Transformers

The VA (volt-ampere) rating for a transformer is calculated for either the primary side or the secondary side. The VA rating calculation can be accomplished by multiplying the primary voltage and the primary current, or by multiplying secondary voltage and secondary current. If the primary voltage is 240 V and the primary current is 4 A, the VA rating for this transformer is 960 VA. The VA rating indicates how much power the transformer can provide or handle. As a technician you can say that the primary VA is equal to the secondary VA, and not worry about the losses. If you were a design engineer, you would need to be more precise and you would find that the secondary VA is slightly less than the primary VA because of transformer losses. It is important that the VA rating of the transformer be large enough for the application you are using it for. If the VA rating is too small, the transformer will be damaged and fail prematurely because the extra current will cause it to overheat. Any time you need to replace a transformer, you must be sure the voltage ratings match and that the VA size of the replacement transformer is equal to or larger than that of the original transformer.

On large transformers that are used to transmit large amounts of power from a solar photovoltaic system that produces 1.8 megawatts to the grid, the secondary voltage from the step up transformer will be 12,470 V and the current could be as high as 140 A. The transformer for this application would have to be approximately 1,800 kVA.

10.3.6 Troubleshooting a Transformer

It is important to understand that you will need special training and special personal protective equipment to work on transformers with voltages above 200 V. Larger voltages create an electrical hazard called *arc flash*, which has the potential to cause injuries as well as death if an arc occurs when power is turned on or off or at other times. The information provided in this section about troubleshooting applies to small transformers that are disconnected from an electrical circuit. The information is provided for background so you will better understand the workings of a typical transformer. If the transformer you are testing is not connected to power, you can test each of the windings with an ohmmeter for continuity. Each of the coils should have some amount of resistance that indicates the amount of resistance in the wire in each coil. A measure of infinite resistance (∞) indicates that the winding has an open; a measure of 0 ohms indicates that one of the windings is shorted.

Another way that you can test a small transformer is by applying power to its primary winding. It is important that you complete the continuity test first to detect any shorts before you apply power. If the transformer has a short, do not apply power. It is also important to understand the safety rules when working around live voltage and use the rules during this test. If the transformer windings are not shorted, you can apply any amount of AC voltage that is equal to or less than the primary rating of the transformer. If the transformer is working correctly and is operational, a voltage will be present at the secondary terminals of the transformer. If no voltage is present at the secondary, be sure to check all the connections to ensure that they are correct. If voltage is still not present at the secondary, the transformer is defective and should be replaced. It is important to remember that the transformer has a primary winding and a secondary winding that are placed in close proximity; when AC voltage is applied to one winding, induction will cause voltage to be available at the other winding.

10.3.7 Common Voltages Use in Power Distribution Systems in Cities

Some city and suburban distribution systems use transformers to create a range of voltages, such as 7,200/12,470, 7,620/13,200, 14,400/24,940, and 19,920/34,500 V. The higher voltages are needed to keep the wire size as small as possible as power is distributed around and through a city. The transmission lines within a city will be a mixture of overhead lines and construction utilizing traditional utility poles and wires. In some newer installations, there might be some locations that include underground construction with cables and indoor or cabinet substations. However, underground distribution is significantly more expensive than overhead construction. Typically overhead power lines and power poles are used to distribute power in rural areas.

10.3.8 Substation and Power Distribution Around the City

A substation is a group of large transformers that are located in one area where power is brought into the city. The high voltage from the grid is transmitted across country through overhead wires that are supported with large steel towers. Figure 10–6 shows a set of high-voltage towers where power is transferred across the country and eventually is connected to a substation. Notice the size of the towers, and how they are arranged in straight lines to provide the shortest path between where power is generated and where it reaches a substation.

The power from the high-voltage lines feeds the input of a large group of transformers at the first substation on the outside of the city. Figure 10–7 shows the large number of transformers mounted on the ground and the structural steel that provides support for all of the cables that interconnect the transformers. You will also notice lightning protection at the very top of each tower. The function of the substation is to break the large volume of high voltage and current down into subsections that feed different parts of industrial, commercial, and residential areas of the city.

After the large voltage and current are brought into the large substation, they are broken down and sent to areas of the city that have large industrial applications, commercial applications such as malls and large stores, and then to residential areas. Figures 10–8 and 10–9 show the smaller substations that are found around a city closer to the industrial applications, commercial applications,

and residential applications. The figures show a trailer that has several portable transformers mounted on it inside the substation connected to the grid. The portable transformers can be quickly installed in case of transformer failure or for additional load capability. Figure 10–10 shows a set of three transformers that provides three-phase voltage to a

FIGURE 10–8 Substation for small commercial or industrial application.

FIGURE 10–9 A smaller substation for residential applications. Notice the portable transformers on a trailer on the far right side of the picture.

FIGURE 10–6 High-voltage power lines that feed voltage from where it is generated to substations just outside a city.

FIGURE 10–7 A large substation at the edge of a city. Notice the transformers are located on the ground, and a large steel structure overhead is used to ensure that the wires feeding the transformers do not touch each other.

FIGURE 10–10 Three transformers provide three-phase voltage to a small commercial application.

small commercial application, such as several stores located on the same property. In residential areas, one power pole may have a transformer that services 8–10 homes. In some residential areas, the power lines are buried, and the transformer sets in the backyard of one home but services multiple homes.

10.4 THE GRID CODE RULES AND REGULATIONS

A set of rules and regulations applies to any connection that is made between a producer of electricity such as a solar photovoltaic system and the grid. These rules and regulations include nontechnical issues that address the exchange price of power and purchase agreements between the owner of the solar photovoltaic system and the operators of the electrical grid in that area. Technical issues concern safety and power quality, and other conditions concerning electrical connections and isolation between the solar photovoltaic system and the grid.

These rules and regulations address electrical safety and electrical power quality issues. The rules and regulations may fall under any number of major code and safety organizations that publish interconnection codes and standards, including the National Fire Protection Association (NFPA), which publishes the National Electrical Code (NEC); the Underwriters Laboratory (UL); and Institute of Electrical and Electronics Engineers (IEEE) (pronounced I triple E). Two federal labs are NREL (National Renewable Energy Laboratory), part of the U.S. Department of Energy, Office of Energy Efficiency and Renewable Energy that is operated by the Alliance for Sustainable Energy LLC and Sandia National Laboratories, which also work closely with the NFPA, UL, IEEE, and the distributed generation (DG) community on code issues and equipment testing. The labs are not responsible for issuing or enforcing codes, but they do serve as valuable sources of information on PV and interconnection issues. The technical and safety issues are part of codes and standards for the interconnection process, and they provide some consistency and standards that ensure safety in any grid connection.

10.5 NATIONAL ELECTRICAL CODE AND OTHER REQUIREMENTS FOR THE GRID

The National Fire Protection Association (NFPA) writes the *National Electrical Code (NEC)*. The latest codebook was published in 2008, and it is republished every 3 years. The National Electrical Code specifies code requirements for electric wiring to ensure that it does not cause a fire or electrocution hazard. It also establishes sizes for current carrying conductors and identifies requirements for grounding and other safety issues. Since the generator for the solar photovoltaic system produces electrical power, all connections, interconnections, and installation and sizing of the wiring must be completed according to the National Electrical Code. Other important issues such as grounding, installation practices for the wiring in and around the solar panels must also meet code requirements. You can purchase a copy of the National Electrical Code and should have it with you to ensure that you are following code.

10.5.1 The Institute of Electrical and Electronics Engineers (IEEE)

IEEE has written a standard that addresses all grid-connected distributed generation, including renewable (wind) energy systems. IEEE standard 1547-2003 was established to provide technical requirements and tests for grid-connected operation, regardless of where the electrical power originates. This standard was written in cooperation with NREL, which is part of the Department of Energy (DOE). This standard include interconnection systems and interconnection test requirements for interconnecting distributed resources (DR) with Electric Power Systems (EPS). This ensures that when a solar photovoltaic system generator is connected to the grid, it follows the standards and regulations that have been written. Writing and implementing the standards and regulations help move all sectors that connect electrical power into the grid closer to the new standards for a Smart grid. The standards are written for all distributed energy resources (DER), which are defined as small-scale electrical generation that is connected to distributed generator (DG) systems, which would include solar photovoltaic system generators. This includes the solar photovoltaic system farms that connect directly to the grid and those that do not connect to the grid but connect directly to a residence or small commercial establishment.

The interconnection technical specifications and requirements include

- Voltage regulation
- Integration with area EPS grounding
- Synchronization (to ensure the phase is close to the phase of the existing voltage on the grid)
- DR on secondary grid spot networks
- Inadvertent energizing of the area EPS
- Monitoring provisions
- Isolation device (interconnecting switch gear)
- Interconnect integrity

Interconnection technical specifications and requirements also cover

- Area EPS faults
- Area EPS reclosing coordination
- Voltage levels
- Frequency
- Loss of synchronism
- Reconnection to area EPS (electrical power system)

10.6 SUPPLYING POWER FOR A BUILDING OR RESIDENCE

When a solar photovoltaic system is designed to provide power directly to a building or residence, it may need to follow some of the same standards and regulations that are provided for the grid connection. But since the power being fed to the building does not need to comply with other existing voltage, it may be able to withstand some slight variations in voltage and frequency, without causing damage to the equipment in the building. If the power from a solar photovoltaic system is used mainly for resistive loads such as heating and lighting, the issues of voltage levels and frequency control are not as important. If the solar photovoltaic system provides part of the power for the building, and the remainder of the power comes from the grid, then all connections, equipment, and the power itself must meet the same standards as though it were connected directly to the grid.

In some small applications, these issues can be taken care of by a single piece of equipment that includes an inverter and safety switch. The inverter will take care of AC voltage and frequency issues, and the safety switch will take care of interconnect and disconnect issues.

10.7 SWITCHES AND CONNECTIONS FOR POWER DISTRIBUTION

The voltage from the solar photovoltaic systems has to have a point at which it is connected to the grid. Figure 10–11 shows a connection where the power from two solar photovoltaic systems connects to the grid. You can see the interconnection and the switch at the top of

FIGURE 10–11 Electrical connection between the solar photovoltaic equipment and the grid. The switch gear for the connection is near the top of the pole.

the power pole. Power is fed underground from the solar photovoltaic system through a conduit that protects the wire as it is routed to the top of the pole.

When a solar photovoltaic system is connected to the grid, it must use safety equipment that protects the solar photovoltaic system equipment from being damaged from the electricity from the grid. One of the major components is the safety disconnect switch. The safety disconnect automatically or manually protects the wiring and components of solar photovoltaic generation system from power surges and other equipment malfunctions on the grid by disconnecting and isolating the equipment from the grid. The disconnect safety switch also ensures that the solar photovoltaic system can be isolated from the grid for maintenance and repair. The safety disconnect also ensures that solar photovoltaic system equipment can be isolated from the grid so that it does not send any power back into the grid when the grid needs to be isolated. This is a very important to ensure that power from the photovoltaic panels does not enter the grid when the grid needs to be isolated and powered down for maintenance, such as after severe high winds, ice storms, or hurricanes.

The safety equipment must include grounding equipment. Grounding includes using grounding rods that are inserted directly into the earth to a depth of about 8 feet. The grounding rods provide a low-resistance path from the generation system to the earth ground to protect the system against current surges from lightning strikes or equipment malfunctions. The grounding system also protects the metal services around the solar photovoltaic system from causing electrical shock hazards to any humans working and touching metal surfaces. If a live wire touches any metal surface and the surface is not grounded, anyone who touches the metal surface could receive a severe electrical shock. If the metal surface is connected to the ground system, as soon as the live wire touches the metal, the ground circuit will cause the current in the live wire to increase quickly to a point where it causes a fuse or circuit breaker in the circuit to open and turn off voltage to the circuit. In this way, the ground circuit ensures that any unwanted or unsafe voltage creates a short circuit that causes the fuse or circuit breaker to blow.

Other types of safety equipment include surge protectors. This type of equipment prevents electrical power surges from reaching the generation equipment. The power surge can occur when the solar photovoltaic system generator becomes out of phase with the power in the grid or from lightening strikes. The surge protection can open and clear the surge, or it can divert the excess energy directly to the ground.

The solar photovoltaic system must have a switch that connects and disconnects its circuits to the grid. When the solar photovoltaic system is stopped for maintenance or needs to be disconnected from the grid for any reason, the switch is placed in the open position, which disconnects all grid voltage from the solar photovoltaic system generator. When the solar photovoltaic system is producing voltage at the correct voltage level and frequency, the switch will be closed and the solar photovoltaic system output will be fed to the grid. If the amount of sunlight is too low to produce voltage that is the right amount in the right frequency, the solar photovoltaic system is electrically disconnected from the grid, or in some cases the lower voltage is send to batteries.

If the solar photovoltaic system is allowed to remain connected to the grid when it is not producing the correct amount of output voltage, the voltage from the grid could feed back into the panels. The electronic controls disconnect the solar photovoltaic system from the grid when this occurs until it is producing power that has the correct voltage and frequency and then the system is connected to the grid again. One last important factor called "voltage phase" must be monitored, so that the connection to the grid takes place when the phase of the voltage in the solar photovoltaic system is nearly identical to the phase of the voltage on the grid. If the electrical circuit from the solar panels is connected to the grid, and the phase of the generated voltage is not close to the phase of the voltage on the grid, a large surge current will be sent through the solar panels and can cause severe damage to them.

The switches that are used to connect the solar photovoltaic system circuit to the grid are similar to the switches used by utility companies to connect and disconnect the power their generators produce, and the switches have been utilized for many years. This means that most of the problems with the switches have been corrected, and the switches have been modified over the years so that they perform the switching process seamlessly. The only problem with voltage that is produced by the solar photovoltaic system is that the voltage tends to have more irregular levels of voltage and changes to the frequency than do voltage and frequency that are produced by larger energy producers such as nuclear power and coal power energy producers.

Another time that the disconnect switches are used is when sections of the grid need to be powered down and isolated so linemen can work on them. For example, after a severe ice storm or a hurricane, large portions of the electrical distribution wiring for the grid can become damaged in a geographical area. When this occurs, that section of the grid is disconnected from the rest of the grid, and all voltage is removed so the line workers can work safely on any downed wires. This type of disconnection is called "power isolation," and any electrical energy producing equipment such as the solar photovoltaic system generator must be disconnected and isolated from the grid to ensure that voltage is not fed into a section of the grid to prevent a line worker from being injured or killed.

Other safety equipment includes a protection system that lets the solar photovoltaic system ride through grid faults such as low-voltage problems without tripping or adding to the problem. Safety equipment may also include systems that check for the frequency that drops below or raises above the standard 60 Hz.

10.8 UTILITY GRID-TIED NET METERING

Net metering is a method of measuring the energy consumed or produced when a solar photovoltaic system is connected to a home or business. When a net meter is used to measure power, it will record the amount of energy the home or business uses from the its power company (grid) when the solar photovoltaic system is not operating and all of the electrical power is coming from the grid. This is similar to a typical household or business that does not have a solar photovoltaic system and uses only electrical power from the grid. The electric meter measures the amount of power used in kilowatts (kW) and the customer is billed for that amount. When the solar photovoltaic system begins to produce electrical power, the residence or business will begin to use only the power from the solar photovoltaic system, and the electric utility meter will not record this power and the customer is not charged for it. If the home or business uses less power than the solar photovoltaic system is producing, the excess power will be fed back into the grid and the net meter will spin in the opposite direction and record the amount of power the solar photovoltaic system is putting into the grid.

Some utility companies pay a lower rate for the power that solar photovoltaic systems put back into the grid. Other utilities basically allow the net meter to spin backward when the solar photovoltaic system is producing power, which will provide the customer with full retail value for all the electricity the solar photovoltaic system produces.

If a state or location does not allow net metering, under existing federal law (PURPA, Section 210) the utility customers can use the electricity they generate with a solar photovoltaic system for their own needs. This basically offsets any electricity they would otherwise purchase from the utility at the utility's retail price. If the solar photovoltaic system produces any excess electricity that is more than the customer needs, the net meter does not run backward; rather the electrical utility purchases that excess electricity at the wholesale or "avoided cost" price, which is generally much lower than the retail price. The excess energy is metered on a separate meter, rather than the net meter. Today, nearly 30 states require the utility companies under their jurisdiction to offer some form of net metering for small solar energy systems. It is important to understand that these requirements vary broadly from state to state. Electrical power from the photovoltaic system can also store any excess power into a bank of batteries and it can be used at a later time.

In most states, net metering rules were enacted by state utility regulators, and these rules apply only to utilities whose rates and services are regulated at the state level. In most of the states with net metering statutes, all utilities are required to offer net metering for some of the photovoltaic systems, although many states limit eligibility to small systems. If a net meter or an auxiliary meter is used, it must be installed at the customer's expense. Net metering simplifies this billing arrangement by allowing

FIGURE 10–12 This net meter will measure the power used from the electric utility or the power fed back to the utility.

the customer to use any excess electricity to offset electricity typically used from the utility company. In other words, the customer is billed only for the net energy consumed during the billing period. Figure 10–12 shows a net meter.

10.9 OVERVIEW OF POWER QUALITY ISSUES

The term *power quality* refers to issues dealing with the frequency of voltage and current that is produced by the inverter for the solar photovoltaic system. Power quality issues also refer to several items that filtering keeps out of the system such as electrical noise, DC injection, and harmonics. Electrical noise and harmonics can be put into the electrical power from other equipment that is connected to the grid and is using power. The ideal power quality exists when voltage and current have a sinusoidal waveform with a frequency of exactly 60 Hz. The voltage of ideal power stays relatively constant on the grid regardless of changing conditions from the solar photovoltaic system. The ideal power can be obtained from a solar photovoltaic system by using electronic (inverter) and other types of controls to ensure that the power has the highest quality. Problems with the amount of voltage or frequency must be detected and be corrected before the voltage from the photovoltaic panels is allowed to connect to the grid.

10.10 FREQUENCY AND VOLTAGE CONTROL

Frequency and voltage on the grid are controlled by the individual power producers that put large amounts of electrical power into the grid. The grid receives voltage from large energy producers such as nuclear-powered generators, coal-fired generation systems and hydroelectric generators, as well as from solar photovoltaic systems

and other forms of alternative energy. The large energy producers have an easier time controlling voltage and frequency than the solar photovoltaic systems does because the energy source that they use to make steam that turns their generators is more consistent, whereas the energy from the sunlight that causes the solar photovoltaic system to produce electrical power tends to be more variable on cloudy days, and the amount of sunlight changes from summer to winter, which can also affect the amount of voltage that is produced.

Since the majority of the electrical energy on the grid has constant voltage and constant frequency from the larger energy producers, the grid can withstand some small amount of variation of voltage and frequency from the energy produced by the solar photovoltaic systems. The solar photovoltaic systems have a variety of methods to control the frequency of the voltage they produce. Larger solar photovoltaic systems use various types of inverters to control the frequency. The inverter can control the frequency that is at exactly 60 Hz and the amount of voltage to match the voltage and frequency on the grid.

On small and midsize solar photovoltaic systems, an electronic inverter can be used to create an output voltage that has exactly 60 Hz and at the rated voltage required for the grid connection. In these applications, the inverter must be sized large enough to handle all the voltage and current the photovoltaic panels produce. In the inverter, the pure DC voltage from the photovoltaic panels is sent through a set of transistors, such as insulated gate bipolar transistors (IGBTs), where the transistors are turned on and off so that their output looks similar to an AC sine wave. The frequency of the voltage output and the voltage are controlled by the electronic control circuit in the inverter that controls the IGBTs. The electronic inverter does an excellent job of producing AC voltage at exactly 60 Hz and the voltage required for the grid, but it must also be large enough to handle all of the voltage and all of the current the solar photovoltaic system produces and the load or grid consumes. This means that size of the inverter for this type of system must be as large as possible.

10.11 VOLTAGE, TRUE POWER, AND REACTIVE POWER

As you know, voltage is the force that causes electrons (electrical current) to flow. Voltage is also called electromotive force (EMF), and the stronger the voltage, the more force moves the current to flow through the system. When you multiply the amount of DC voltage in a system by the amount of DC current, the result is electrical power that is measured in watts. In an AC circuit, the impedance and reactance must be taken into account to measure power. The current in the AC circuit can be used by several different types of loads. A pure *resistive load* includes loads such as an electrical hot water heater element, electrical resistance heating element in a furnace, and incandescent light bulbs. Another type of load is called an *inductive load* and includes current consumed by motors, magnetic coils, and other inductors (coils). The third type of load is called a *capacitive load,* and it includes current that flows through capacitors, which are used to correct power factor and are also used in some switch mode power supplies used in some computer power supplies.

Real power is also called *true power* (TP) and it is the voltage and current that flow only through the resistive loads. In these loads, the voltage and current waveforms are in phase with each other and there is no wasted energy. When voltage and current go through capacitive or inductive loads, the voltage and current waveforms are out of phase. In an inductive circuit, current waveform lags behind the voltage waveform. In a capacitive circuit, the voltage waveform lags behind the current waveform. The power that flows through the inductive loads and the capacitive loads is called *apparent power* (AP). The ratio of true power (TP) to apparent power (AP) is called the *power factor* (PF). The formula for power factor is TP/AP and since true power is always less than apparent power, the power factor is always less than 1.00. The closer the PF is to 1.00 the more efficient the circuit that uses the power is and the less the losses are.

Any time the voltage and current are out of phase, power losses begin to occur. If a grid circuit is connected to a factory that has a large number of large motors, the circuit will have losses due to the inductive loads. The circuit losses can be corrected by adding an amount of capacitance so that the capacitive reactance matches the amount of inductive reactance. When capacitance is added to a circuit to correct the inductive losses, it is called power factor correction. The capacitance for correcting the power factor can be added at the end user who is causing the large inductive loads (the factories); when the power factor is corrected, the industrial end user will have a better rate. The power factor can also be corrected where the power is produced. This means that if the inductive loads are scattered throughout the grid circuit, such as a large number of air-conditioning compressor motors in residential areas, capacitance cannot be added at each home, so it is added by the electric utility where the electrical power is produce. This means that any electrical power that is produced by solar photovoltaic systems must have the ability to add capacitance in its circuit to correct the power factor of the area of the grid it is connected to. This is called power factor correction and each solar photovoltaic system must meet the standards required by the grid to correct the power factor for the power it is putting into the grid.

10.12 LOW VOLTAGE RIDE THROUGH (LVRT)

Low-voltage problems in the grid may occur for several AC voltage cycles, or they may last for several hours. They may also be caused in one phase of a three-phase

system or in all three phases. Low voltage is a condition that exists any time the voltage on the grid drops 10% or more below the rated voltage level.

When low voltage occurs in the grid system where the solar photovoltaic systems are connected, the solar photovoltaic system must have the capability to "ride through" this condition. It does not matter if the solar photovoltaic system is causing the low voltage or if the low voltage exists in the grid where the solar photovoltaic system is connected. The safety system for this problem is called *low voltage ride through* (LVRT). This means that when the low-voltage fault occurs in the grid, electrical instrumentation will measure the severity and the duration of the problem. The solar photovoltaic system must have a protection system to ride through the low-voltage condition and not trip off.

If the solar photovoltaic system is causing the low-voltage condition, such as on a cloudy day, it must be able to correct the low-voltage condition quickly or disconnect the panels from the grid. The choices the solar photovoltaic system has during the low-voltage conditions may include completely disconnecting from the grid and automatically reconnecting after the low-voltage dip has cleared, or it can remain connected, stay in operation, not disconnect from the grid, and ride through the low-voltage condition. The third choice is to stay connected to the grid and help correct the problem by adding *reactive power* from its capacitive banks, which will help correct the low-voltage condition. The method the solar photovoltaic system is programmed to use will depend on the type of problems the grid it is connected to has.

10.13 FLICKER AND POWER QUALITY

Flicker power is defined as a short-lived voltage variation in the electrical grid that might cause a load such as an incandescent light to flicker. This problem is more noticeable when a solar photovoltaic system is connected directly to a residential or commercial application as the sole source of power, rather than having a battery storage system or getting power from a combination of the solar photovoltaic panels and the grid. When the sun shines brighter for short durations, the solar photovoltaic panels will produce slightly more voltage, which will cause a light bulb to glow brighter during that duration. Eventually the controls on the solar photovoltaic system will realize that the voltage is increased due to the increase in sunlight, and control circuits inside the inverter will control the voltage and bring it back to the normal level. The sequence to identify the excess voltage and bring it under control may require several seconds, during which the increased brightness of the light will be noticeable. This problem can be controlled in residential applications by sending voltage produced by the solar photovoltaic panels directly to a battery and then pulling the voltage from the battery and sending it to an inverter to control the voltage and frequency directly. If the batteries are used, a battery charging control circuit must be used to ensure that the batteries are not allowed to over- or undercharge. If the solar photovoltaic system uses an inverter and it is connected directly to a grid, controls will have to be used to ensure that problems with the voltage are kept to a minimum in both duration and amount.

10.13.1 Power Quality

Several factors in the production of electrical power affect power quality. These include voltage level, frequency, phase shift, and power factor correction. The solar photovoltaic electrical system must have controls in place to correct deficiencies in each of these factors. For example, an electronic voltage regulator circuit may be used with the inverter to be sure the voltage from the inverter matches the voltage on the grid. Phase shift problems occur when the power from the inverter is connected to the grid and there is a slight difference of phase angles between the two voltages. This means that sine waves for one of the voltages start a few microseconds before the other, and the sine waves from both voltages do not start at zero degrees at the same exact instant. Complex electronic circuits monitor the phase of the active voltage on the grid continually and ensure that the phase of the voltage being created by the solar photovoltaic system is similar to that of the grid before the switch connecting the solar photovoltaic system voltage coming from its inverter to the grid is closed. If the phase difference between the voltage from the solar photovoltaic system inverter and the voltage on the grid is too large, adjustments will be made to the voltage coming from the inverter of the solar photovoltaic system to bring its phase relationship closer to the voltage at the grid so that the sine wave of the inverter voltage and the grid voltage both start at their zero degree angle point at exactly the same time. When the phase relationship of the solar photovoltaic system voltage is identical to the voltage on the grid, the connection switch will be closed, allowing the voltage from the inverter of the solar photovoltaic system to be connected to the grid.

Electronic circuits are also used in the solar photovoltaic system to measure the power factor for the voltage from the solar photovoltaic system and the voltage on the grid where the connection is made. Any time the power factor meters indicate the circuit has too much inductive reactance, capacitance from the banks of capacitors will begin to become connected to the circuit, which will correct the power factor. The banks of capacitors will be constantly connected and disconnected from the grid as required to correct the changing power factor. It is important to understand that the inductive motors at the factories that are using voltage from the grid are causing the inductive reactance to occur in the circuit, and these motors may be switched on and off a number of times during the day, which will cause the amount of reactance

in the circuit to continually vary. The electronic circuit that constantly monitors the power factor will become aware of the changes in the inductive reactance in the circuit and immediately adjust the amount of capacitance from the capacitor banks to ensure that the power factor on that sector the grid is continually corrected.

10.13.2 Islanding

If a small section of the electrical grid becomes disconnected from the main grid, this condition is referred to as *islanding*. Islanding may occur when lightning strikes or a short circuit occurs in one of the areas of the grid, and the circuit breakers trip in one or more sections of the grid and cause the isolation. This may also occur if an automobile strikes a power pole, or some other event occurs that causes the circuit breakers to trip. In some cases, it is possible for a solar photovoltaic system that is connected to the isolated area to continue to produce enough voltage to support consumers and loads in the isolated section of the grid. The problem that arises during this time is that the phase of the voltage from the solar photovoltaic system may drift out of phase from the voltage in the remainder of the grid. If the condition that caused the disconnection is cleared, the small section of grid could be connected back to the main grid while the voltage is out of phase. The phase differential problem can cause a strong current surge in the solar photovoltaic system inverter and severely damage it. For this reason, it is important that an electronic detection system is used to continually monitor the main grid voltage and ensure that the phase differential between it and the solar photovoltaic system voltage is minimal when the reconnection is made. The phase detection circuits continually monitor the phase differential between the solar photovoltaic system and the main grid. If the phase difference becomes too large, the voltage from the solar photovoltaic system inverter is disconnected from the grid through its disconnect switch. When the phase in the voltage from the solar photovoltaic system inverter is adjusted and the differential between the solar photovoltaic system voltage and the grid voltage becomes small again, the voltage from the solar photovoltaic system inverter is reconnected to the grid.

10.14 SYSTEM GROUNDING

The *grounding* system for a solar photovoltaic system consists of one or more copper grounding rods that have been inserted into the earth to a depth of 8 feet. Heavy copper conductor is used to connect the metal frame of the solar panels to the rod. When the metal frame is connected to the rod that is in the ground, the metal of the solar photovoltaic system is at the same electrical potential as the earth ground. Any metal in the system will be bonded to the metal frame, which means it may be mechanically connected with nuts and bolts, or it may have a wire conductor connecting to the frame. Bonding ensures that all of the metal parts of the metal frame and mounting hardware have a low resistance path to the earth ground. If more than one rod is used, a copper conductor will connect each of copper ground rods.

The ground system is used for two reasons. First, the ground will provide a low-resistance electrical path to the earth between any metal in the solar photovoltaic system frame and the ground rods. If any electrical wiring that has voltage in it become chafed and allows electrical voltage to come into contact with the metal of the solar panel frames or other parts of the solar photovoltaic system, this voltage would immediately be taken to the earth ground, which would cause an extremely high current to occur. The high current would then cause a circuit breaker or fuse protecting the high-voltage lines to open and interrupt the circuit. If the system were not grounded, and the circuit breaker did not trip, anyone who came into contact or touched any metal parts could receive severe electrical shock or be electrocuted. If you come upon a circuit breaker that is open, you must test the system to see if a short circuit has occurred. When you close the circuit breaker and it immediately opens again, you should suspect a short circuit somewhere in the system and you will need to begin isolating wires until you find the wire or section of the system that is the problem.

The second reason the system is grounded is to help protect against lightning strikes. If the solar photovoltaic panels or any other metal part of the solar photovoltaic system such as the pole on which a system is mounted is struck by lightning, the large amount of electrical energy in the lightning bolt must be quickly dissipated into the earth where it will not cause damage. If the solar photovoltaic system or the pole is not grounded properly, the location where the lightning strike enters the equipment will receive severe damage because the electrical energy in the lightning has not been diverted to the earth. The ground system must be checked periodically to ensure that it is connected correctly and that all connections provide a low-resistance path to the earth. In some installations of solar panels, additional lightening arrestors may need to be added to protect the system from lightening strikes.

10.15 UNDERGROUND FEEDER CIRCUITS

On larger solar photovoltaic farms, the voltage lines from the solar photovoltaic arrays are run underground through conduits to the point where they are connected to the combiner box or large inverters and the grid. These wires are run underground to help protect them from the elements, and this also makes the solar photovoltaic system site more pleasant to view. If the wires were run above ground, power poles and other hardware would have to be used that might become a problem during high winds or ice storms that can bring them down.

The underground conduits are generally poured into any concrete that the poles are mounted in, and additional conduit is run between the solar photovoltaic panels and the point where their voltage is connected to the grid on the power pole or transformer. In Figure 10–13, you can see where three separate conduits are used to bring the power out of the ground and up to the top of the pole where the connection to the grid is made. Today, it is possible to run the underground conduits for thousands of feet with a piece of equipment that tunnels through the ground at approximately 3 feet or deeper to go below any surface obstructions. This piece of equipment is called a horizontal directional drill, and it uses a hydraulic system to power a drill bit on a horizontal path. The horizontal directional drill uses 10-foot sections of pipe to push the drill farther and farther toward its target. The technician checks the location, direction, and depth of the drill head every 10 feet

FIGURE 10–13 Electrical connection between the solar energy equipment and the grid. The switch gear for the connection is near the top of the pole. The interconnection wires between the solar panels and the power pole are buried underground. A conduit protects these wires as they come above ground at the base of the pole and rise to the top of the pole.

as another section of pipe is added. The operator uses a small joystick to adjust the head of the drill bit to move left, right, up, or down to ensure that the drill and pipe continue in the exact direction and depth. A small hole approximately 18 inches in diameter and 3 feet deep or deeper is dug at the target location, and the pipe and drill are pushed to the ground until they emerge in this hole. Once the drill head reaches the target hole, a large spool of plastic conduit is connected to the drill head and is pulled back through the tunnel the drill created. The pipe for the horizontal directional drill is pulled back and each time a 10-foot section becomes free, it is unthreaded from the long section and placed back on the magazine for the directional drill to be used for the next section. When the last section of pipe is pulled from the tunnel, the plastic conduit will be inserted in place from point to point, waiting for final connection. The plastic conduit will have a pull cord inserted into its complete length when it is manufactured, and this cord will be used to pull the wires through the conduit to complete the installation of the underground service. The directional drill can tunnel under driveways, trees, or other objects to bury conduit and wire underground. Underground installations for wiring are more expensive than those above ground, but they are able to withstand problems such as ice storms and tornadoes that may bring down power poles.

10.16 CABLE INSTALLATION

The cables that connect the solar photovoltaic system to the grid conductors can be run underground or overhead. If these cables are run underground, connections are made at service points along the distance of the conduit, where they can be opened and inspected later if there is a problem. If the wire is run overhead, the connections are made at power poles and transformers. Utility companies and companies that provide service installation of high-voltage lines have been in business for many years and are very good at this job. If they are making service connections overhead, they will use trucks with hydraulic lifts and buckets to lift the service technicians to the level where the connections are made. They also have specialized service equipment to help make underground installations simpler. Many times this type of cable installation work is contracted out to companies that have specialized equipment and technicians with specialized training.

10.17 OVERHEAD FEEDER CIRCUITS

Wooden power poles and steel electrical transmission towers are used to bring high-voltage power lines into and out of the substations. Figure 10–14 shows an example of this mixture of power poles and transmission towers. The overhead feeder circuits must be high enough above the ground to allow service vehicles to

FIGURE 10–14 Wooden power poles and metal electrical transmission towers used to bring power into and out of electrical substations.

move freely underneath without touching them. These overhead power lines must also be supported between the poles and towers so that the lines do not group or touch each other in strong winds. Another reason the lines are run overhead is that the air surrounding the wires will help keep them cool, which allows them to be loaded to the maximum extent.

Overhead feeder circuits consist of electrical wire that is supported by power poles. The power poles typically run close to streets and highways, where the utility company owns a small strip of land called the right away. Any trees that are located in this right-of-way under the power poles must be trimmed and maintained to ensure that they do not bring down power lines in a storm or in heavy ice and snow. Figure 10–15 shows power lines connected to the power poles for support. You can also see how close the power poles are positioned to the highway. Each residence or commercial

FIGURE 10–15 Overhead power lines are run along streets and highways where utility companies own right-of-way strips of land.

application that needs electrical power will have a connecting line run from the main wires to the building. The wire that connects from the building to the main power lines is called a *service drop*. You can see a service drop connection right below the three transformers on the power pole. Originally these poles were called telephone poles because telephone lines were connected to them. Today, electrical power lines, service drops, telephone lines, and cable television lines may be all mounted on the same pole at different heights. The feeder circuits run electrical wire from the substation throughout the city and area so that each residence, commercial establishment, and industrial site can be connected to the electrical grid.

10.18 SOLAR FARM SYSTEM SUBSTATION AND MAIN COMPONENTS

When a large number of solar photovoltaic systems are located on a large solar farm, an electrical substation must be provided for the number of step up transformers needed to boost the voltage so it is ready for long-distance delivery. A substation is a smaller group of transformers that is located close to where solar photovoltaic systems create electrical power. The transformers at these substations step up the voltage to a level where it is large enough to be transmitted to where it will be used. Figure 10–16 shows a substation on a solar farm. All the electrical wiring interconnections between the solar photovoltaic systems and the substation are made through underground conduits. The high-voltage lines leaving the substation make interconnections to the grid above ground with overhead connections. The high-voltage lines then use existing high-voltage transmission towers and lines to distribute the electrical power to the cities and industries that need the power. The substation at the solar farm may also contain all the electrical disconnecting switches,

safety fuses, and circuitry to be compliant with connecting to the grid.

10.19 CONNECTING TO RESIDENTIAL OR COMMERCIAL SINGLE-SOURCE POWER SYSTEMS

A solar photovoltaic system connected directly to a residence or commercial location is a single-source power system. Figure 10–17 shows the electrical hardware for a single-source power system. The power lines from the solar photovoltaic system are installed in underground conduits and come up into the switch and metering boxes at the bottom. If backup power from the grid is provided, it will come in overhead, as at a normal residential or commercial location. You can see the switch gear, metering boxes, and overhead service from the grid in this figure. The switch gear and net meter must be provided when the solar photovoltaic system is installed, and they need to be included in the cost of installation for the system. The net meter will run one direction when the solar photovoltaic system is providing power to the grid, and it will run in the opposite direction when the commercial application uses electrical power from the grid rather than from the solar photovoltaic system.

FIGURE 10–17 Electrical disconnect and metering components are shown for this commercial application. The power from the solar photovoltaic panels comes in from underground into these components, and electrical service from the grid comes in from overhead.

FIGURE 10–16 Substation for a solar farm.

Questions

1. Explain what the grid is and how it works to distribute electrical power to customers.
2. Explain what the Smart grid is and how it is different from the grid.
3. Identify the three main sections of the electrical grid.
4. Identify the federal law that requires utility companies to purchase power from solar photovoltaic system generators and what year the law was passed.
5. Explain how a transformer works.
6. Identify the two windings of a transformer.
7. Explain what a substation is and why it is important to the grid
8. Identify three factors that the IEEE 1547-2003 standard addresses for technical requirements for power that is connected to the grid.
9. Explain what a net meter is.
10. Identify two factors that are addressed in power quality issues.

Multiple Choice

1. The grid is _____
   a. All of the electrical power distribution systems in the United States and North America.
   b. The electrical connection between the solar photovoltaic system generator and the solar photovoltaic system tower.
   c. The electrical connection that exists only between solar photovoltaic panels in a solar farm.
2. The smart grid _____
   a. Is exactly like the regular grid and can deliver only electrical power.
   b. Has the ability to send and receive information to and from each customer along with electrical power.
   c. Filters and purifies the voltage as it is delivered.
3. Transformer windings include _____
   a. The first and second winding.
   b. The input and the output winding.
   c. The primary and the secondary winding.
   d. All the above.
4. The turns ratio for the transformer effects _____
   a. The number of turns of wire in the primary and secondary windings.
   b. The ratio of voltage in the primary and the secondary windings.
   c. The ratio of the current primary and secondary windings.
   d. All the above.
5. Which is most true about the step up and step down transformers?
   a. The voltage in the step up transformer is larger in the primary than in the secondary winding.
   b. The voltage in the step down transformer is smaller in the secondary than in the primary.
   c. The voltage in the step down transformer is larger in the secondary than in the primary.
6. How do power factor, true power, and apparent power relate to one another?
   a. Power factor is the true power divided by the apparent power.
   b. Power factor is the apparent power divided by the true power.
   c. Power factor is the true power multiplied by the apparent power.
   d. All the above are true.
7. Low voltage ride through (LVRT) is _____
   a. The ability of an electrical generation system like a solar photovoltaic system to not fault out when the voltage on the grid gets too low.
   b. A special circuit in a solar photovoltaic system that ensures that the proper voltage is supplied when the solar photovoltaic system is not connected to the grid.
   c. A special circuit in a solar photovoltaic system that causes it to trip off when voltage gets too high.
   d. All the above are true.
8. Flicker is a problem that occurs when _____
   a. A solar photovoltaic system increases or decreases voltage.
   b. The voltage in the grid causes the solar photovoltaic panel to create more voltage.
   c. The power switch that connects the solar photovoltaic system to the grid turns on and off quickly.
9. Grounding is _____
   a. The insertion of a copper rod into the earth down to 8 feet.
   b. An electrical connection of all metal parts of a solar photovoltaic system to a system of copper conductors that terminates at copper rods that are inserted into the earth.
   c. A safety circuit that helps protect the solar photovoltaic system from lightening strikes and short circuits.
   d. All the above are true.
10. The disconnect switch for the solar photovoltaic system _____
    a. Connects or disconnects the electrical circuit of the solar photovoltaic system to the grid.
    b. Places a cover over the photovoltaic panel to make it stop producing voltage.
    c. Is installed on each solar panel when the panel is manufactured.

Installing, Troubleshooting, and Maintaining Solar Energy Systems

OBJECTIVES FOR THIS CHAPTER

When you have completed this chapter, you will be able to

- Explain the steps involved in installing a residential solar PV system.
- Explain the steps involved in installing commercial solar photovoltaic panels on a rooftop.
- Explain the steps involved in installing commercial solar photovoltaic ground panels.
- Identify the electrical components in a commercial photovoltaic panel installation.
- Explain the parts of a closed loop control system.
- Explain the parts of the single-axis and double-axis solar panel tracking systems.
- Identify the tests to troubleshoot problems with solar photovoltaic panels.

TERMS FOR THIS CHAPTER

Backup Generator	Process Variable (PV)
Ball Screw	Rotary Actuator
Bypass Diode	Servo System
Closed Loop Control	Set Point (SP)
Combiner Box	Single-Axis Solar Tracking System
Error	Stepper Motor
Feedback Signal	Summing Junction
Inverter	Two-Axis Solar Tracking System
Linear Actuator	Uninterruptible Power Source (UPS)
Open Loop Control	Voltage Fading

11.0 OVERVIEW OF THIS CHAPTER

This chapter will show the steps to install solar panels for residential commercial and industrial systems that are grid tied, grid tied with battery backup, and smaller solar panel installations that use using only battery backup. The chapter will explain how the project is designed and installed on single-story as well as two-story homes. The second part of this chapter will show the steps involved in planning and installing solar panels on the roof of a large commercial installation. This section will show how the hardware and mounting rack are installed, and how the electrical controls and panels are installed and connected.

This chapter will not cover the installation of solar products that are embedded in shingles or other roofing products; this is a more complex type of installation that you will learn about in the field. This chapter will provide enough detail so that you understand that installing solar panels requires some mechanical skills, because panels are pulled together and bolted onto hardware that is mounted directly onto the rooftop, and it will include learning skills in reading electrical diagrams, connecting conduits, and installing converters and other boxes into the main electrical system.

The final section of this chapter will explain how to troubleshoot and perform maintenance correctly on installed solar photovoltaic systems. You will learn basic steps in troubleshooting and identifying a problem and being able to test each of the suspected parts of the system from the solar panels to the wiring and switches. Some of the more complex electrical troubleshooting will be covered in Chapter 12.

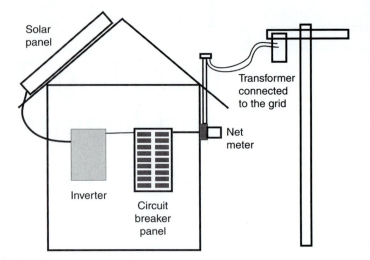

FIGURE 11–1 Grid-tied residential solar energy system without batteries.

11.1 INSTALLING SOLAR PANELS AT A RESIDENTIAL LOCATION

Installing solar panels at a residential location involves several steps. The first step in this process is a planning meeting with the homeowner. During this step the project is discussed to determine the type of system that the homeowner wants to install. It is also important to discuss any federal or state funding that the homeowner might be eligible for. It is important to remember that for the homeowner to be eligible for this funding, the installer for the project must be certified by the North American Board of Certified Energy Practitioners (NABCEP) or be electrical contractors that are licensed for solar installations. It is important to double-check to ensure that the person doing the installing is certified and whether the project is large enough to qualify for federal and state funding, if that is desired.

The types of solar energy systems that can be installed include solar panels that are mounted on the roof, or in the backyard, and connected to the electrical system of the house. If the house is connected to the grid, and the solar panels are being used to generate additional power, a decision needs to be made about whether the power will be stored in batteries or if the system will operate without batteries. Figure 11–1 shows the electrical diagram of solar panels being connected to a residential electrical system that does not use batteries. Since the residence is connected to the grid, the system is a grid-tied solar energy system. This type of system requires an *inverter* to change the DC electrical power produced by the solar panels to AC power at 60 Hz. Since some of power is returned to the grid, a net meter is also required, which measures the amount electrical power consumed from the grid and also the amount of electrical power that is returned to the grid. One advantage of this type of system is that a battery charge controller is not needed since batteries are not being used to store the energy. The disadvantage of this type of system is that if the solar

panels produce more electrical power than the home requires to operate or is using at any particular time, this additional electrical power will be returned to the grid. In some locations across the United States, the electric utility will purchase this power, but at a lower price than the power it is selling to the residence.

The second type of system is shown in Figure 11–2. It is similar to the first system but uses batteries to store the excess electrical power that is produced by the solar panels. Since this residence is tied to the grid, the electrical controls can be designed so that any excess power that the solar panels produce that is not used by the residents at any given time can be stored in the batteries. This type of system requires that an inverter and a battery charge controller be used to ensure that the power going to the batteries is regulated and controlled so the batteries do not overcharge. Also the charge controller ensures that the batteries do not go completely dead when they are being discharged. If the system has enough solar panels to produce more electrical power than the residents can use, and enough power to charge the batteries, additional power can be passed back to the grid and sold back to the electrical utility. It is important to understand that the rate that is paid for the power that is sold back to the grid may be substantially lower than the rate the homeowner pays for electrical power that is supplied by the grid. This type of system provides the homeowner with the ability to produce electrical power from the solar panels and use it as needed during the daytime, charge its batteries, and then use extra power from the batteries at night or at times when the solar panels are not producing.

The main problem with this type of system is that the batteries are an additional expense, and they must be maintained on a regular basis to ensure that the level of liquid in the batteries is sufficient, and the terminals must be checked, tightened, and cleaned periodically. This means that the homeowner must be capable of performing these duties or additional costs will be incurred to

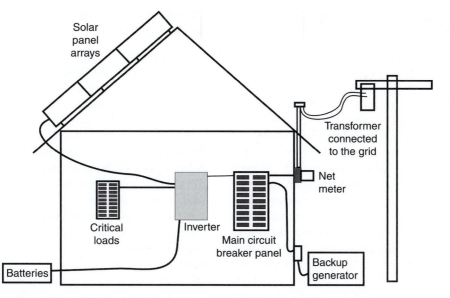

FIGURE 11–2 Grid-tied residential system with backup batteries and generator.

hire trained technicians. In the diagram you can see that this type of system can also be supplemented by a small generator. The generator can provide electrical power any time power is interrupted from the grid, such as in the severe thunderstorm, or if the area's power lines are damaged by a hurricane or ice storm. In this type of application any time the residents use more power than the batteries can provide, the generator can add additional power if the grid cannot supply it. It is important to remember that a switch must be used with the generator that is specifically designed to isolate the generator from the grid power any time the generator is used. If the generator and the grid are both on at the same time, and the grid goes down due to a problem such as downed power lines, it is important that the generator does not continue to produce electrical power and send it back into portions of the grid where electric utility workers may be working since they will presume all power has been disconnected. The switch is specifically designed to isolate the generator in such a way that it provides power to the home or residence but does not allow that power to get it back into the grid.

The third type of solar installation is basically for residential applications where the home is in a remote area and electrical power from the grid is not available. In this type of application the solar panels provide electrical power for the residents, and excess power is stored in batteries. This type of application can also use a generator to provide additional power. Several things must be considered in this type of application, such as how much fuel can be stored for the generator operation, and how many batteries can be brought to the remote location to store the backup electrical power. Another point to consider with this type of application is that the residents may not use the home on a full-time basis; rather it may only be used for short periods of time for vacations, which would mean they would

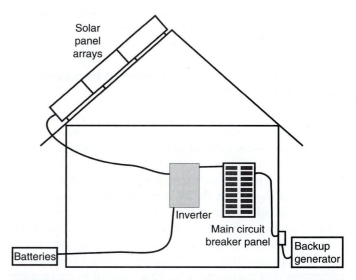

FIGURE 11–3 Diagram of electrical residential solar system that is completely off the grid and uses only solar and backup batteries with a generator.

need less electrical energy. Figure 11–3 is a diagram of this type of system. The solar panels are connected to the main power for the house through an inverter. The inverter converts the DC electrical power to AC power to operate appliances such as a refrigerator and lighting. It is also possible in these types of applications to purchase appliances and lighting that operate with DC power, and then an inverter will not be required. Since the solar panels are used to charge the battery system, a charge controller must be used to ensure that the batteries are not overcharged or allowed to drain completely during the discharge cycle. If a generator is used, it must be connected to the system with a switch that protects the system any time power is supplied by the generator.

11.1.1 Designing the System and Applying for Grants and Permits

The next step in the installation process is to complete the design diagram and get a signed contract from the homeowner. The second step in this part of the process is to apply for any required state and local permits. In some states this process must be completed by the homeowner, and in other states the installer must apply for these permits. If the homeowner intends to receive tax rebates or funding from the local, state, or federal government, the appropriate forms must be submitted and approved. This step is vitally important to ensure that the homeowner is eligible and can receive the maximum amount of grants that are available. If the instructions in the grant or tax refund applications are not followed completely, it is possible that the homeowner will not receive the tax credit or refund for the project. The final piece for the permits requires that the completed project be inspected, and the project must pass inspection before electrical power can be energized. When the system is using a grid-tied system, the electric utility must also be involved, and the project, equipment, and installation must pass its inspection as well.

11.1.2 Determining the Equipment Required to Install the Project

The next step in the process of installation is to determine how the solar panels are going to be mounted onto the roof or if they are going to be mounted at the ground level. The type of panel that has been selected for the project will also determine the type of mounting hardware required. During this stage of the project the location where the panels will be mounted is selected and the hardware that supports the panels will be anchored in place. For example, if the solar panels are to be mounted at ground level either on rails or on a pole, preparation must be made, such as pouring concrete slabs and burying or installing conduit to connect the solar panels to the electrical system. The conduit ensures that the system wiring is protected from outside elements at all points between the solar panels and the electrical boxes. State and local electrical codes must be followed precisely for the project to be approved and pass inspection. This is why it is so important to use a certified installer and electrical company that understand state and local codes.

After the equipment for the project has been identified, and the residence owner has signed a contract, all the electrical equipment including the solar panels, inverters, and meters can be purchased and delivered to the site. If the solar panels are being installed on a new home that is not occupied, it is important that the material is not delivered and left unattended before it is permanently installed on the home site, as it can be easily stolen. This means that the material must be delivered in the amounts that will be used each day, or special arrangements must be provided to lock up all this equipment before it is installed.

11.1.3 Installing the Hardware to Mount the Solar Panels

After the material and equipment for the project have been delivered, the installation can start. Typically solar panels require some type of metal hardware for mounting. This may include rails or poles with brackets so that the panels can be mounted directly onto a rooftop or a pole. It is important to remember that solar panels will remain installed for many years, and they must be securely held in place so that they cannot be moved or damaged by the winds or other weather conditions. When the hardware is mounted onto the roof of a dwelling, it is important that the watertight integrity of the roof is maintained. This means that the roofing material, such as shingles or coatings, cannot be damaged or have a large number of holes created in them, else the roof will leak any time it rains or snows. Special hardware components have been designed for these types of applications that will cause the minimal amount of damage to the roof yet hold the solar panels securely. In some cases if solar panels are being installed on an existing home, it may be important to include any roof repairs or possibly a complete new roof while the solar installation is in progress. It is important to understand that even though the solar installation works perfectly, the homeowner will be unhappy with the installation if there are any leaks.

It is important to select the proper installation hardware for the solar panels you are using and ensure that it matches the type of roof that you are working on. For example, the roof may be nearly flat or have a slight slope, or it may be a very steep slope, and each of these applications requires different types of mounting hardware. If the panels are to lie flat, or if they are to be raised to some angle, they will also require specific hardware. If the roof is made of metal or tile, special brackets and hardware must be used in special processes to secure the hardware to the roof.

If the solar panels are to be mounted at ground level, it is important that the concrete pad is prepared correctly, and far enough in advance to allow the concrete to cure. It is also important to remember that any electrical conduits that are being used in conjunction with a concrete pad must be secured in place prior to the time when the concrete is poured. If the solar panels are mounted on the roof, electrical conduits may be added at the time the panels are installed.

11.1.4 Electrical Hardware for the Solar Panel Project

When the solar panel project is installed, it is important to ensure that the electrical hardware is installed simultaneously. A place must be determined to mount the inverter and any other electrical switches involved for the solar panels. Typically this location must be near the entrance panel for the residence. The entrance panel is sometimes called the distribution panel, and it is the place where all the circuit breakers for the electrical system are installed. The circuit breakers provide electrical protection against overcurrent for each of the circuits in the house. The distribution or entrance panel is typically located in a utility room or laundry room in the residence, where the circuit breakers can be accessed from inside the home. Typically the electrical power from the electric utility (the grid) is brought into the home through an electrical service entrance that will include a meter panel that is on the outside of the home. An electrical conduit is provided to bring electrical wiring from the metering panel into the circuit breaker box. Any electrical switches and the inverter for the solar panel must be located near the electrical service entrance. If a solar electrical meter is added to the system, it may be mounted near the electric utility meter. If the application is designed to sell power back to the grid, a net meter must be purchased. The net meter is a special electrical meter that can measure the amount of electricity coming into or going from the house. In this way it is possible to measure the power that is sold by the electric utility to the residence, as electricity flows into the house, and also measure any electricity that is generated by the solar panels and returned to the grid.

11.1.5 Installing the Solar Panels on the Roof of the Residence

The next step in the project is to physically install the solar panels on the mounting hardware that has previously been mounted on the roof. Usually the solar panels are bolted together while they are on the ground to create a solar array that will produce the amount of voltage that is desired. Figure 11–4 shows two workers moving solar panels that have been pulled together onto the roof where they will be secured to the hardware rails that were previously mounted on the roof. During this process the workers must be careful not to drop or damage the solar panels while they are being moved. It is also important that they are aware of the electrical wiring for

each panel so that is not damaged and is free to be connected once the panels are secured to the mounting rails. Notice that the workers used ladders and scaffolding to move the panels to the roof level. The scaffolding provides a secure platform for the workers to work on while they are above the ground. You should also notice in this image that the workers must wear protective harnesses any time they are working above the ground. The harnesses are connected to a safety system that will prevent the worker from falling from the roof and being injured. After all the panels are secured on the mounting rails, and electrical power is connected at the roof, the workers can continue the installation of the electrical system and controls on the ground. It is important to understand that any time the solar panels are exposed to light, they will begin generating electrical power, which may present an electrical shock hazard. In some installations covers are provided for all of the panels to ensure that they do not generate voltage during the installation process.

Figure 11–5 shows technicians installing solar panels on a roof that has a slight pitch. The advantage of this type of roof is that it is nearly flat and allows the workers to move about more safely than they could on a roof with a high pitch. Typically ladders are used to get onto the roof and to move the solar panels safely from the ground to the roof. Figure 11–6 shows two workers working on a complete assembled solar array on the roof. Since this house is a single-story house and it has a deck, the distance the solar array must be lifted is not too high. It is also important to understand that different types of solar panels and solar arrays will weigh more or less, which may require additional helpers to lift the panels to the roof or to be on the roof to securely move the panels to the mounting rails. Figure 11–7 shows a small lifting crane that is used to lift larger solar panels to the second-story roof of a residence. The lifting crane can be used any time the weight of the solar panels becomes too large for individual workers to

FIGURE 11–4 Workers moving assembled solar panels onto the roof where they will be attached to the mounting hardware rails. (Courtesy NREL and the U.S. Department of Energy.)

FIGURE 11–5 Workers installing solar panels on the roof with a slight pitch. (Courtesy NREL and U.S. Department of Energy.)

FIGURE 11–6 Workers lifting an assembled solar panel onto the roof. (Courtesy NREL and U.S. Department of Energy.)

handle manually, and it may also be used any time the height of the roof is more than can safely be reached from extension ladders. The cost of the crane will more than offset problems that result in trying to lift extremely large panels to a second-story roof. During the planning stages of the project these issues must be addressed and accounted for in the cost analysis. If a crane must be used, the cost of renting it and scheduling the crane operator to be available on the appropriate day are important details to include in the initial considerations. If a crane is being used, it is important to ensure that the company that owns the crane has the correct insurance to protect the installer and homeowner during the installation process.

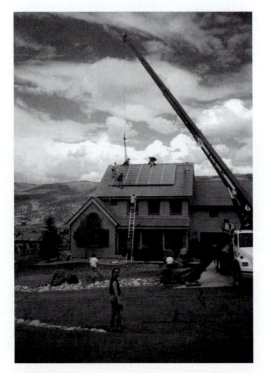

FIGURE 11–7 A small crane is used to lift larger solar panels onto the roof of the two-story residence. (Courtesy NREL and U.S. Department of Energy.)

11.1.6 BRINGING THE SOLAR PANELS ONLINE AND TESTING THE FINAL INSTALLATION

After the solar panels are installed and all the electrical connections are completed, the entire system must be brought online and tested. If the system is grid tied, the electric utility company must supervise this test. It is also a time when the final inspection can be completed and signed off on by the local authorities. During this start-up time, the charging system and inverter must pass the inspection and work as designed. The test may be conducted during the latter part of the day, when the system charging can be inspected, and then later the battery system can be tested to ensure that it is providing power when the solar panels are not producing electrical power.

During this time the electrical utility company will ensure that the installation meets all the requirements for safety and operation that are stated in the utility company policies. Since electrical power from the solar panels is run through an inverter, it is important that the inverter be synchronized with the electrical power from the grid. The entire system must also be grounded properly to ensure that it meets all electrical codes. After the entire system has been tested and approved, the project can be signed off and the customer can be billed.

At the time that the project is being approved, the customer must be told about his or her responsibilities and any maintenance required by the system, such as battery maintenance or maintenance of the switch gear and inverter. A homeowner who is not capable of these maintenance checks must be aware that he or she will need to employ someone to provide these checks periodically.

If the solar panels or the switch gear show any problems during the start-up procedures, these problems must be identified and fixed immediately. These kinds of problems may include loose wiring or wires that are not connected correctly to the proper terminals. If the wiring is not completed correctly, the inverter will not operate correctly and cannot provide electrical power to the system for the household electricity, to the grid, or for charging batteries. Each of the sections of the system must be checked off to ensure that it is working properly. When the project is completed and signed off by the electric utility and the customer, the project is considered complete.

11.2 COMMERCIAL INSTALLATION OF SOLAR PANELS

The installation of large number of solar panels on a commercial or industrial application is more complex than installing three or four solar panels on a residence. The major differences include the larger number of panels that may need to be installed, and the integration of the electrical power from the solar energy system to the three-phase electricity from the grid that is used to provide power for the commercial or industrial site. When solar panels are installed on the roof of an industrial building, they must coexist with the large amount of heating, ventilating, and air-conditioning (HVAC) equipment and any vents that are mounted on the rooftop. The company that is installing the solar equipment on the roof of industrial building must spend hours with the plant maintenance facilities manager to ensure that all of the details are worked out. When the solar installation is ongoing, workers must have access to the roof and be able to move their equipment in close proximity to the building, which may cause problems with the normal day-to-day operation of the building. When the solar installation is complete, and it is connected to the existing electrical power to the building and to the grid, the building's electrical power may need to be shut off for a short period of time while these connections are made. This section of the chapter will describe how to plan and install photovoltaic panels on a rooftop of a large company building. The images in this section will follow the installation process from beginning to end.

11.2.1 Planning, Sizing, and Designing the Solar Photovoltaic Installation

The first step in a large project to install solar photovoltaic panels on an industrial building includes a planning meeting that includes the company that will be installing the panels and the maintenance facilities manager for the building where the panels will be installed. During this meeting the company that is installing the solar panels will describe the scope of the project and present a detailed financial plan including a return on investment (ROI) projection and an installation calendar to begin the process. At this point the project is still tentative; financial details and contracts must be drawn up, and an agreement from all sides must be reached to get the proper approval to proceed with the project. Once the company that owns the building accepts the financial contract and the projected calendar for installation, the documents are signed by both parties and the project will proceed to the next step.

Before the company that is installing solar panels submits a proposal, it must make one or more site visits to determine the available space on the roof for the number of panels that can fit with existing HVAC and other rooftop equipment. After the space requirement is identified, the number and types of solar panels that can fit in the rooftop space will be identified, and the output of these panels will be determined. The solar panel installing company may develop several plans with different types of solar panels that will provide a variety of cost and return on investment details. By having a variety of proposals, the solar panel installer is giving the building owner several options on the amount of investment needed to make the return on investment acceptable.

Another detail that must be completed during the site inspection is a complete evaluation of the condition of the roof to determine if it can support the weight of the solar panels that will be mounted on it. In some cases the roof of the building will be more than 20 years old, and in need of some repairs. Many solar installation companies also have the ability to do roof repairs and roof sealing at the same time they are installing the solar panel system. If roof repairs are part of the process, it must be pointed out in the proposal, so the proposal can be compared to those that may be submitted by other solar panel installers.

11.2.2 Preparing the Roof and Installing the Solar Panel Mounting Hardware

Figure 11–8 shows a rooftop of a large building that is going to receive a large number of solar panels. You can see the existing HVAC equipment, and that the roof repair and

FIGURE 11–8 Overhead shot of roof of building prior to starting solar panel installation. (Courtesy Ron Swenson, www.SolarSchools.com)

FIGURE 11–9 The original roofing material is stripped away so that the roof can be repaired and waterproof flashing can be installed. (Courtesy Ron Swenson, www.SolarSchools.com)

sealing process has begun on the right side of the building. The old roof material has been stripped away during this process. It is important that the process of stripping and resealing the roof is scheduled in advance, and it is also important to check the weather forecast to ensure that you have dry weather for 1 to 2 days that the roof is exposed. Typically the roof is never exposed more than 24 hours.

When the roofing material has been removed, repairs and patches can be made to ensure that the new roof will support the solar panel hardware, and technicians can safely walk about on the rooftop. This may include removing any rotted material and providing a patch and waterproof flashings around the existing HVAC units and roof vents. Figure 11–9 shows the waterproof flashings applied around each HVAC unit, and you can see where patches have been completed on the bare roof.

Figure 11–10 shows the short post for mounting the solar panel rails screwed into the deck of the roof. Prior to the sealing process the location for mounting each photovoltaic panel is marked on the roof deck, and the short posts are screwed into the roof at precise locations. After the short posts are secured to the roof, foam is sprayed around each of the posts and allowed to accumulate to provide a watertight seal. You can see that a number of support blocks have also been secured to the roof to hold the drain lines for the HVAC systems at the correct pitch so that condensate water running from the HVAC units will move through the copper pipes across the roof to a common drain. The mounting rails for the solar photovoltaic panels will go over the top of these copper pipes.

11.2.3 Mounting the Hardware Rails for the Solar Panels onto the Post

After all the posts have been screwed to the roof deck, and foam has been applied, the entire roof will have several coats of foam applied to create a waterproof barrier. The foam must be allowed to dry and then technicians can begin to install the rails on which the photovoltaic panels will mount. The metal rails are secured by bolts to the short posts that are already screwed directly to the roof. When the rails are bolted to these posts, the solar panels can be bolted to the rails and held securely in place on the roof so that they will not be damaged by high winds or other weather.

Figure 11–11 shows the metal rails after they have been bolted to the post. You should notice that the rails make a metal frame to hold the solar photovoltaic panels securely in place. After the rails have been bolted to the post, technicians can begin to deliver the solar panels to the roof. Notice that the panels are stacked flat, and they are placed crossways on the rails to keep the boxes off the roof. When the solar panels are stacked in this manner, they will protect the new roof so that it is not perforated or damaged. Notice the solar panels are stacked completely across the roof in such a manner that they will be near the location where they will be installed.

During the time that the solar panels are being installed on the roof, technicians and other workers will constantly be walking across the roof. They must be careful not to damage existing HVAC equipment or electrical and drain lines that have been previously installed.

FIGURE 11–10 The short post (stanchions) that the rails for the solar panels are mounted to are precisely located and screwed into the roof deck. Foam is sprayed on each post to ensure that it has a watertight seal with the roof.   (Courtesy Ron Swenson, www.SolarSchools.com)

Figure 11–12 shows sheets of plywood carefully placed over the electrical and plumbing lines so that workers can walk across without damaging them. These plywood ramps also make it possible to use small handcarts to carry one or more solar panels from place to place throughout the roof. When the construction has been completed and the panels have been installed, the sheets of plywood will be removed.

FIGURE 11–11 Solar panels in their shipping cartons are brought to the roof and stacked crossways across the metal rails so that they do not damage the new roof coating.   (Courtesy Ron Swenson, www.SolarSchools.com)

FIGURE 11–12 Sheets of plywood are carefully positioned to create ramps and walking paths across the surface of the roof. The ramps allow two-wheel carts to be used to move solar panels from one side to the other on the roof where they will be installed. (Courtesy Ron Swenson, www.SolarSchools.com)

11.2.4 Working from Blueprint to Locate and Wire Solar Panels

During the design portion of the project the exact location of the solar panel mounting hardware is measured out and identified on a blueprint. Another set of blueprints shows how the solar panels are electrically connected into arrays and how the arrays are wired into the combiner boxes and junction boxes. The wiring diagram for the panels' electrical connections determines the exact number of panels to be connected in series and parallel so that the voltage the panels produce is at the correct level.

The blueprint shows all the parts to use, the hardware to connect the parts, and the exact dimensions for the installation. Technicians use the blueprints to determine the dimensions and then measure the exact location where the rails are installed. The electrical blueprint shows where all the electrical conduits are installed, and where the combiner boxes and junction boxes should be located. Figure 11–13 shows an installation blueprint. Figure 11–14 shows technicians measuring the locations and positions for mounting the metal rails that the solar panels will be bolted to. You can see that the metal rails for the solar panels are mounted just above and over the top of the copper condensate lines for the HVAC system. It is difficult to ensure that the panels and the rails are mounted at exact locations so they do not conflict with existing equipment and plumbing on the roof.

11.2.5 Bolting Multiple Solar Panels Together to Form an Array

After the hardware and mounting rails are secured in place, the technicians will begin removing the solar panels from

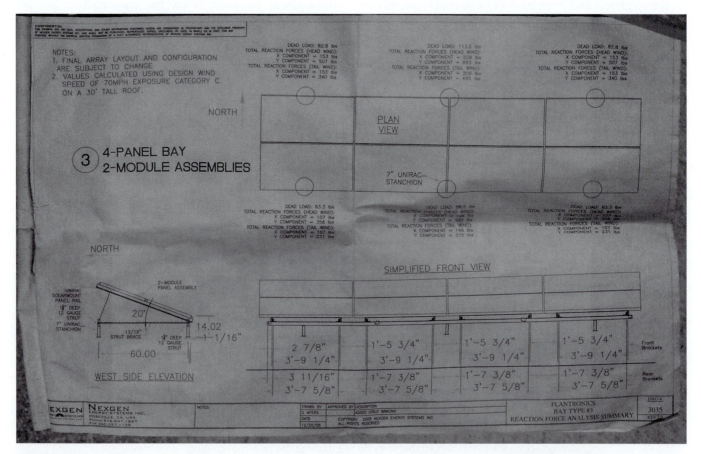

FIGURE 11–13 A blueprint that shows the dimensions for properly locating and installing the metal support rails for the solar panels. (Courtesy Ron Swenson, www.SolarSchools.com)

FIGURE 11–14   Measuring the position and location of the mounting rails for the solar panels. (Courtesy Ron Swenson, www.SolarSchools.com)

the cardboard shipping boxes. The technicians must be careful not to scratch or break the glass on the panels while handling and removing them from their shipping boxes. The solar panels are carefully laid on the roof and positioned so that two or more can be bolted together mechanically to form an array. On this job the technicians will bolt three panels together to form an array. Figure 11–15 shows the technicians positioning the panels and connecting them with nuts and bolts. The technician uses a socket wrench to tighten the nuts and bolts to securely hold

FIGURE 11–15   Three solar panels are bolted together to form an array.   (Courtesy Ron Swenson, www.SolarSchools.com)

FIGURE 11–16 Technicians complete tightening the nuts and bolts from the back of the solar panel. Notice the electrical wiring for each panel has a termination connection at one end on the back of the panel.   (Courtesy Ron Swenson, www.SolarSchools.com)

the three panels together in an array. Two sets of three panels are bolted together to make an array of six panels. You can see in the background panels that have been mounted together and are already installed. Six panels are bolted together and then bolted to the mounting rail that was previously installed. The wiring for each panel is carefully positioned for final installation.

Figure 11–16 shows the technicians completing the final assembly of the panels from the back side. For this process, the panels are carefully positioned so that technicians can work from the front and the back to ensure that the nuts and bolts are tightened correctly. You can also see the wire termination for each panel on the back side. These wires are connected and then joined to wiring that goes through the panel.

11.2.6 Mounting the Solar Panels on the Rails and Bolting Them into Place

The next step in the installation process is to place the solar panels on the rails and bolt them in place. Figure 11–17 shows a close-up of the hardware that is used to connect the solar panel securely to the mounting rails. You can see in this picture that solar panels are bolted to the cross rails and the cross rails are bolted to the main rails, which are bolted to the top of the mounting posts (lower left). The mounting posts are secured to the roof deck with screws and then foam is applied to make a water-tight seal. The nuts and bolts that hold the solar panels securely in place

must be tightened so that the panels cannot be lifted when the wind blows. You can see that the mounting system allows a small amount of movement at the rails and at each fixture that ensures that each panel will align properly with its mounting rails.

FIGURE 11–17 Close-up view of hardware connections showing how solar panels are bolted to the cross rails and the cross rails are bolted to the main rails, which are bolted to the top of the mounting posts (lower left). The mounting posts are secured to the roof deck with screws and foam.   (Courtesy Ron Swenson, www.SolarSchools.com)

11.2.7 Raising the Panels to the Proper Tilt Angle

When the panels are first mounted to the rails, they are laid flat so all the final hardware connections can easily be made. Figure 11–18 shows an array that has just been installed and connected to the hardware rails lying flat. After the panel is connected to all hardware, the array can be lifted at the back side to tilt the panels to the proper angle to receive the maximum amount of sunlight. In this picture you can see that the panel in front of the one lying flat has been tilted and locked into place. The amount of tilt is adjustable; when the panel is tilted to just the right angle, the final hardware nuts and bolts are locked down to ensure that the panel is secure. Up to this point the hardware that allows the panel to tilt has not been locked down tight, so that the panel can be moved up or down to the tilt angle. Once the panel has been moved to the proper tilt angle, the remaining nuts and bolts are all tightened. Vertical braces are used on the back side of the panel to better secure the panels and adjusted to keep the tilt angle at exactly the correct angle to receive the maximum amount of sunlight.

11.2.8 Connecting the Electrical Wiring Between the Solar Panel Arrays

The next step in the process involves connecting the wires from the solar panel array to the other solar panel arrays on the rooftop that are in the same row. Figure 11–19 shows an array of solar panels that have been locked in a proper tilt position, and the wiring from the panels is hanging down underneath them. These wires are interconnected among the individual panels in the array so that only two wires for each array remain to be connected to other arrays.

The wires for each panel in the array are interconnected so that only two wires make the final electrical connection between the array and the combiner box. The *combiner box* is a special junction box that provides a place to interconnect the wires from the solar panels and keep their polarity straight.

Figure 11–19 shows the interconnecting wires from each solar panel in the array hanging down from the bottom of each panel. The wires include one positive wire and one negative wire from each panel, and they are connected to the positive and negative wires of the panel next to it. The exact connections for the panels has been determined during the design phase of the project, and a blueprint has been created for the technicians to follow to ensure that the panels are connected correctly in series or parallel. After the wires are connected, they are routed inside the metal hardware rails so they are protected and out of sight. When the wires are routed through the metal rails, they are protected from blowing freely in the wind and they are protected from the elements such as direct rain or other adversities that may cause that wires to stretch or become damaged.

FIGURE 11–18 When the panel array is first installed, it is placed on the hardware rails in the flat position. The next step in the process is to lift the back side of the panels in the array so they are tilted to the proper angle to receive maximum sunlight.   (Courtesy Ron Swenson, www.SolarSchools.com)

FIGURE 11–19   Wires are shown hanging down from the back of each solar panel in the array. The wiring from the solar panels is interconnected so that only two wires remain to make the final connection.   (Courtesy Ron Swenson, www.SolarSchools.com)

Figure 11–20 shows a close-up picture of the wires inside the metal hardware rails. This picture also shows the wires being routed from the hardware rails into the electrical metal conduit where they will be routed back to combiner boxes or junction boxes that are mounted on the back of the solar panels. Since the wires from several

FIGURE 11–20   A positive and negative wire from each array are routed through the mounting rail; at the end of a row they are pulled into a metal conduit.   (Courtesy Ron Swenson, www. SolarSchools.com)

arrays are pulled into the same combiner box, the combiner boxes can be spaced out so that one box is mounted to each group of arrays.

11.2.9 Pulling Wires into the Combiner Box and Junction Box and Making Connections

Figure 11–21 shows the wires from the solar panel arrays that have been pulled into combiner boxes and are ready for final connection. The combiner boxes are mounted on the back side of the panel arrays. When the wires are first pulled into the combiner boxes, a small amount of excess wire will stick out of the front of the box. Eventually the solar technician or electrician will make the proper termination connections and put lug terminals on the end of each wire. Notice in this picture that the wires that have been pulled into the combiner boxes and junction boxes are getting larger as they get farther away from the solar panel arrays. The reason the wires are getting larger is that the amount of current the wires must carry is getting larger as more and more arrays are being connected into these wires.

Figure 11–22 shows the wires with their final connections made in the box on the left and wires ready for connection in the boxes on the right. These junction boxes are mounted on the roof right above where the inverters are mounted on the ground. The reason this installation has two sets of boxes is that two different types of solar

FIGURE 11–21 Combiner boxes are special junction boxes that are mounted on the back of sets of arrays. Electrical connections from the arrays are made in the combiner boxes. Additional electrical conduits connect the combiner boxes to the inverters. (Courtesy Ron Swenson, www.SolarSchools.com)

panels were used in this project due to problems of getting enough panels of one kind to complete the project on time. This project required more than 800 panels; they had to use 165-watt panels and 175-watt panels to get all of the panels within the project's schedule. Since two types of panels were used, the voltage from each type of panel string was slightly different and had to be isolated from each other in the combiner boxes, junction boxes, and all the way to the inverters. The 175-watt panels used the Xantrex 225 kW inverter and the 165-watt panels used two smaller Xantrex 30 kW inverters. It is important to understand that on large projects, the availability of panels, inverters, and switch gear may dictate the types, numbers, and sizes of equipment that end up in the installation.

A large electrical conduit comes out of the back of the box and down the wall to the boxes where the inverter is mounted on a pad on the ground. You can see that all of the positive wires are connected together at the top of the box, and all of the negative wires are connected at the bottom of the box. Each combiner box and junction box is purchased with a specific number of terminal connectors in the box. All the positive terminals in the box are electrically connected, and they are separate from all the negative terminals that are also connected. Each terminal has a threaded bolt called a stud, and each of the wires has a terminal connector lug placed on their ends. Each of the lugs has a hole that allows the lug to be placed over the stud and a nut and washer are tightened down on the lug to ensure a good electrical connection. It is very important that the terminal connectors are tightly connected on the lugs so that electrical current can flow through them easily. If any of the lugs is not tightened correctly or becomes loose, a high electrical resistance will occur and the terminals will begin to heat up as current flows through loose connections. At some point the temperature will be so high that the insulation on the wire and the wire itself will become damaged. For this reason it is very important to ensure that the terminals and the hardware on the terminal lugs is tightened.

Figure 11–23 is a close-up view of the internal electrical connections inside another combiner box. The smaller positive wires from the solar panels are connected to individual terminals on a main block on the left side of the box. All of the smaller negative wires from the solar panels are connected individually to the electrical terminals on the large block on the right side of the combiner box. One large wire is connected to the positive terminal and one large wire is connected to the negative terminal; these two larger wires go out the

FIGURE 11–22 Wire connections are shown terminated at the combiner box on the left. The wires are connected to the terminal lugs according to their (+) and (−) polarities. Notice the wires enter the combiner box through metal conduits.   (Courtesy Ron Swenson, www. SolarSchools.com)

FIGURE 11–23 A close-up view of the internal connections inside a combiner box. The smaller positive wires from the solar panels are connected to a main block on the left side, and one large wire for the positive power comes out of this combiner box. On the right side the negative wires from the solar panels are connected individually to the large block, and one large wire for the negative power comes out of the combiner box. (Courtesy Ron Swenson, www.SolarSchools.com)

bottom of the combiner box through a conduit and finally end up at the inverter. The reason these wires are larger is that they have to carry all the current from the individual solar panels that are wired into this box.

11.2.10 Connecting Wires from the Combiner Boxes to the Inverter, Meter, and Disconnects

The final junction box on the roof is mounted against the outside wall directly above where the inverter and switch gear are mounted on a pad on the ground. Figure 11–24 shows a picture of several inverters, their switch gear, and junction boxes all mounted on a concrete pad. You can see conduits that go up the wall and through the outside wall where they connect into the junction boxes on the roof. These conduits provide a passage for the wires that are connecting the junction box and combiner box on the roof with the inverter on the ground. The conduits ensure that the wires are protected from rain and other elements that could damage them. Electricians or solar technicians must pull wires through these conduits and then make the final connections at each end. The main reason the inverters and their switches are mounted on a concrete pad is because of their weight. If the inverter were placed on the roof, the roof would need to be reinforced so that it could support the weight of this equipment.

Additional electrical equipment is mounted on the pad for this solar installation. This equipment includes

FIGURE 11–24 Inverters, transformers, disconnects, metering, and junction boxes are mounted on a pad directly below the roof. These hardware components are mounted on a concrete pad instead of on the roof because of their weight. (Courtesy Ron Swenson, www. SolarSchools.com)

FIGURE 11–25   (a) Solar electrical system disconnect; (b) solar electrical meter; (c) the disconnect switches for two small inverters; (d) solar electrical system splice box; (e) two isolation transformers; (f) two smaller inverters.   (Courtesy Ron Swenson, www. SolarSchools.com)

FIGURE 11–26   (a) DC disconnect switch; (b) DC enclosure; (c) DC disconnect; (d) the three-phase inverter; (e) the three-phase isolation transformer for the system.   (Courtesy Ron Swenson, www.SolarSchools.com)

(step-up) transformers, disconnect switches, a solar electric meter, and various junction boxes as well as several inverters. The next images will identify the equipment in detail. Figure 11–25 shows the front view of the equipment that is pictured in the outside row in Figure 11–24 where you can only see the backs of the panels. In the front view you can see the main disconnect at the far left. All the power from the combiner boxes on the roof enters the main disconnect from a conduit at the bottom of the box. The conduit was buried in the pad when the concrete was poured. The main disconnect is mounted directly over this conduit. The main disconnect consists of a large switch that can be turned on and off to disconnect all the electrical power from the solar panels on the roof if necessary for maintenance. The next box to the right of the disconnect houses the electrical power meter that measures all the electrical power the solar panels produce. Two smaller disconnect switches are mounted side-by-side to the right of the meter panel, and each of these disconnect switches controls power to a 30 kW Xantrex inverter and an isolation transformer. The inverters are the white boxes in the picture, and the transformers are the smaller gray components between the inverters.

The inverters convert the DC electrical power from the solar panels to three-phase AC power, and the transformer steps the three-phase AC voltage up to match the three-phase AC grid voltage that is supplied to the building by the electric utility company. In this installation two smaller Xantrex 30 kW inverters are provided for the 165-watt solar panels, and the larger Xantrex 225 kW inverter is used with the 175-watt solar panels. The two smaller isolation transformers are provided for the smaller inverters, and one larger isolation transformer is connected to the larger 225 kW inverter. In this application the transformers step up the three-phase AC

voltage from the inverters and also provide electrical isolation for the system since the primary and secondary windings of the transformers are not connected in any way.

Figure 11–26 shows the front view of the electrical equipment on the pad that is mounted directly against the building. DC electrical power is fed from the rooftop combiner box through a large conduit that is mounted directly to the wall. This conduit feeds DC power into the top of an electrical box called an enclosure. The enclosure has a front door that opens and allows access to the electrical connections inside the box. When the door is closed, the box becomes sealed so that rain cannot enter it. The box immediately to the left of the DC enclosure is a DC disconnect switch, and it has a switch handle on the front of the box that can turn DC power on and off. The left side the DC disconnect is connected directly to the 225 kW Xantrex three-phase inverter. This larger inverter carries the majority of the power load for the system, and it converts the DC electrical power from solar panels to three-phase AC electrical power.

The AC electrical power from the inverter is fed to an isolation transformer that is mounted directly to the left of the inverter. The transformer steps up the AC voltage to match the three-phase AC voltage that is supplied to the building from the electrical power grid that the electrical utility provides.

Figure 11–27 shows the inside of the three-phase disconnect switch. You can see that wires enter the box through the conduits on the lower right side. One set of wires is bringing power into the switch, and the other is taking power out of the switch. After all the electrical connections are made, the box door is closed, and a manual handle on the right side of the box allows the switch to be turned on or off with the door closed. The

FIGURE 11–27 The wires coming into and going out of the disconnect switch are ready to be connected to the terminals in the switch disconnect. (Courtesy Ron Swenson, www. SolarSchools.com)

FIGURE 11–28 The large inverter with its door open so the DC wires and the AC wires can be connected. Notice the large cooling fans and heat sinks at the top of the inverter. (Courtesy Ron Swenson, www.SolarSchools.com)

disconnect usually has a feature that allows the switch to be locked in the off position, which is used for a lockout procedure that ensures that the power will remain off during maintenance.

Figure 11–28 shows the larger 225 kW inverter with its door open so technicians can make the final electrical connections. The wires from the DC disconnect enter the large inverter on its right side and send three-phase AC power on the left side. You can see the large heat sinks and cooling fans in the top of the inverter. The cooling system is needed because the inverter creates a large amount of heat when it is running. Even though the wiring in the inverter looks very complex, the wiring for the solar electrical system requires connecting only the positive and negative wires on the DC side of the inverter and the three wires for the three-phase AC power. The rest of the electrical circuits inside the inverter are prewired and do not need any connections made during the installation process. After the two DC wires and the three AC wires are connected, the door to the inverter is closed, creating a sealed enclosure except for the air flow needed to keep the inverter cool.

11.2.12 Connecting the Three-Phase AC Voltage from the Inverter into the Building's Main Electrical Entrance Panel

The next step in the installation process is to connect the electrical power from the inverter and transformer into the main electrical panel where the power from the electric utility is connected. During this process the electrical

power from the utility company must be turned off, so that electricians or other technicians are not exposed to high-voltage electrical power when the connections are made. The first step in this connection process is to remove the front cover from the main electrical switch panel. Figure 11–29 shows technicians removing the front cover to this panel, and they are adding a switch to the main panel where the three-phase electrical power from the inverter and transformer will be connected. This switch will provide the terminal connections for the wires that come from the inverter and transformer for the solar electrical system. The switch also provides a simple way to disconnect the electrical power from the solar panels from the main power that is provided from the electric utility at a later date any time that power from the solar panels needs to be turned off. The output side of this switch is connected to electrical bus bars in the back of the main entrance panel. The electrical connections on the bus bars also provide terminals where the main power lines from electric utility are connected. Electrical conduit is installed between the transformer, which is mounted outside the building on a pad with the inverter, and the main entrance switch, which is inside the building in the main electrical entrance panel. The conduit is buried in the concrete and comes in underneath the electrical entrance panel. The three-phase electrical wires that connect the transformer and the inverter to the main

FIGURE 11–29 Technicians are installing a disconnect breaker switch for the solar electrical power into the main electrical panel for the building.   (Courtesy Ron Swenson, www. SolarSchools.com)

switch in the electrical entrance panel are run through the electrical conduit that connects the two panels. A pull cable or rope is run through the conduit and connected at one end to the wires that need to be pulled into the conduit. When the pull cable or rope is connected, the electrical wiring is pulled into the conduit and the final connections on each end are made. It is very important that the three-phase wires are clearly identified individually so that the correct phase orientation is provided and the A, B, and C phase wires from the solar panel match up with the A, B, and C phase wires from the utility company. It is important that the phase relationship between the electrical power from the utility company matches the phase relationship of the wires coming from the inverter for the solar panels before the switch is turned on and the power from the solar panels inverter is connected with the power coming from the electric utility.

11.2.13 Synchronizing and Utilizing Backup Power from a Diesel Generator with Power from the Solar Panels

Many larger companies require a means to provide backup power if power from the electric utility is interrupted. For example, if there is a lightning storm or other condition that causes the utility power to be disabled, typically all electrical power to the company building would be turned off for a short time. If the company has a large number of computers or other vital fire and safety equipment that must receive power at all times, a backup power system must be integrated with the power from the utility company. Many times this type of system will

have batteries to provide what is called an *uninterruptible power source (UPS)* or they may use a *backup generator.* In the solar installation we have been studying, the uninterruptible power source was installed and operating since the building was built. The UPS system was enlarged several times as more and more computers were added to the company. When the solar electrical system was added to the roof, special switch gear had to be added to make sure that when the electrical power from the electric utility went down, and electrical power from the solar panels was being generated, that the electrical power from the solar panels did not cause a disruption in the system and was not fed back into the grid system. Since power from the solar panels comes into the electrical entrance panel just like the power from the electric utility, it merges with the three-phase AC power when the solar disconnect switch is closed. Any time there is a problem with the electric utility power and the utility goes down, the electric power from the solar panels can continue to be used by the company, but it must be isolated from the utility grid.

When the power from the utility company is lost for any reason, the electrical control in the building recognizes this and the electrical system goes into a mode where all of the larger HVAC loads and some of the lighting loads are turned off to reduce the load for the building. The computer loads will switch to the uninterruptible power source that provides electrical power from a bank of batteries. The 100 kW backup diesel generator immediately comes on and begins to provide additional power to run the safety circuits, fire control systems, and security doors. If the sun is shining and the solar panels are producing, the solar

FIGURE 11–30 A 100 kW backup generator. The backup generator is integrated into the electrical system for the building. (Courtesy Ron Swenson, www.SolarSchools.com)

power can be used to offset some of the electrical power needed in the building. The sensors will determine how much power is available from the solar panels and adjust the loads inside the building accordingly. If the power outage occurs at night, all of the larger loads are immediately shut off and only the most important loads are allowed to remain on.

Figure 11–30 shows the backup generator for this system. The backup generator can provide enough electricity for the safety circuits in the building. The most important parts of this system is the series of switches that can isolate the electrical power provided by the solar panels and the electrical power provided by the backup generator from the each other and from the power from the grid.

The main electrical meter for the building is a net meter that can measure the power the building uses from the electric utility (the grid). On weekends and during the daytime, the solar panels produce enough power to sell some back to the grid. The net meter can measure the power the solar panels produce and send back to the grid, and it can measure the electrical power the company uses from the grid. The company gets a predetermined rate for all the power it sells back to the utility.

11.2.14 The Completed Project

After the main switch is installed in the service entrance panel, the connection between the solar panels and the

main electrical disconnect from the utility is completed. When this switch is closed, and the sun is shining, the solar panels are able to produce up to 267 kW from 860 solar panels. The panels cover 61,960 square foot of rooftop. Figures 11–31 and 11–32 show the two types of solar panels used on this project. Figure 11–31 shows the Schott SAPC-175 high-efficiency 175-watt mid-sized photovoltaic module. These modules use 125 mm single-crystal silicon solar cells that have 16.4% encapsulated cell efficiency and 13.45% module efficiency, which is the maximum usable power per square foot of solar array.

This module has a nominal 24 VDC output, that is perfect for creating arrays for grid-connected systems, and each module uses a *bypass diode* that minimizes the power drop caused by shade. The module has superb durability to withstand rigorous operating conditions, and it is ideal for grid-connected as well as stand-alone applications. It has a textured cell surface to reduce sunlight reflection and uses tempered glass, EVA resin, weatherproof film, and an anodized aluminum frame for extended outdoor use. The module has a 25-year warranty on power output.

Figure 11–32 shows a Schott SAPC-165 panel. This panel is a high-power 165-watt panel that uses 125 mm square multicrystal silicon solar cells. The solar panel has 12.7% conversion efficiency. The panel has a textured cell surface that reduces the sunlight reflection and it has a back surface field structure that improves cell conversion

FIGURE 11–31   Schott SAPC-175 high-efficiency 175-watt mid-sized solar modules. These are monocrystalline solar panels.   (Courtesy Ron Swenson, www.SolarSchools.com)

efficiency to 14.6%. This panel also has tempered glass, EVA resin, a weatherproof film, and a frame that is made from anodized aluminum. This panel has a 20-year limited warranty for power output. Schott solar photovoltaic panels are made by the Schott Applied Power Corporation in Rocklin, California.

Figure 11–33 shows an aerial view of the finished project. The roof has been sealed with foam and the solar panels have been fully integrated around the existing HVAC equipment and other vents in the roof. The completed project consists of 860 photovoltaic panels and the roof and covers 61,960 square feet. The panels can produce up to 267 kW of DC electrical power. The project was completed in 2005 and has produced approximately 20% of the building's electrical power needs on a continual basis throughout the year.

11.3 INSTALLATION OF GROUND PANELS ON A LARGE SOLAR FARM

Installing solar panels on a solar farm or solar installation where the panels are mounted on hardware that is on the ground, rather than on a roof, uses a process that is similar to mounting panels on the roof. This section will explain the process of creating a solar photovoltaic panel installation where the panels are mounted on the ground. Figure 11–34 shows technicians working on installing solar panels on a metal frame that has been mounted solidly into the ground.

When solar panels are mounted on the ground, the process starts with selection of a location that allows a sufficient number of panels for the application. If this location is not near an access road, an access road has to be designed and created prior to the beginning of the panel installation. The access road will allow various types of equipment to get to the area to level land; install the electrical conduits underground; pour concrete pads; and deliver electrical switch gear, inverters, transformers,

FIGURE 11–32   Schott SAPC-165 solar modules that produce 165 watts. These are polycrystalline solar panels.   (Courtesy Ron Swenson, www.SolarSchools.com)

FIGURE 11–33 Aerial view of the finished solar project. (Courtesy Ron Swenson, www. SolarSchools.com)

FIGURE 11–34 Technicians installing solar panels at the PSEG solar farm at Hackettstown, N.J. (Courtesy PSEG.)

and other electrical equipment required for connection to the grid. At the same time, it is important to identify the location of the nearest electrical power lines, where the solar panels can be connected into the grid.

The next steps in the process after an access road has been provided are to prepare the land for installing the underground electrical conduits that run between the panels and to set up the forms for the concrete pads that will support the hardware on which the panels will be mounted. In some installations larger concrete pads are used to mount the solar panel hardware, and in others like the one shown in the figure, the smaller feet of the frames are mounted directly onto the ground. This process of installing conduits underground will be very similar to installing conduits in the foundation of buildings. Once the location of each set of panels is identified and marked out, the process of installing the conduit can begin. The size of the conduit between each panel location and to the switch gear and inverter will be determined by the number of panels and the amount of electrical power they are designed to produce. Once the conduit sizes have been determined, the conduit can be installed in the trenches or with a mechanical system that drills into the ground and pulls the conduit underground without trenches. The ends of the conduit will be secured in the forms were concrete will be poured at a later time.

After the conduits have been installed, and the forms to hold the concrete for the pads for each set of panel hardware have been built, the pouring of the concrete can be scheduled. It is important to understand that all of the stanchions for mounting a solar panel frames to the concrete must be precisely located prior to pouring the concrete. Once the concrete is poured, the hardware that is poured in the form must be located so that the solar panel frames can be mounted to them and secured with fasteners such as nuts and washers. The concrete generally is poured in sections, since the number of forms to be poured may be large, and all the forms cannot be poured on the same day.

After the forms have been poured and the concrete is allowed to set for several days, the process of installing the solar panel frames can begin. This process starts with the delivery of the hardware for the solar panels. This hardware will be assembled to create the frame for the solar panels, and each section will be in mounted on the bolts that have been placed in the concrete form. After the frames are fully assembled and mounted securely into the ground, the technicians can begin to mount the solar panels to the frames as shown in the photo. The solar panels are secured to the frames with hardware to ensure that they will withstand any strong winds or other disturbances and remain in place without becoming damaged.

After all the solar panels have been mounted into the frames and secured, the electrical connections between the panels can be made. The electrical connections for ground-mounted panels are similar to those that were discussed for the roof-mounted panels. This means that combiner boxes and junction boxes are strategically placed between the strings and arrays of solar panels. The wires from each solar panel are interconnected according to the electrical blueprint that was provided for the project. The wires are then fed into the conduits where they are continually connected at junction boxes and finally end up at a DC disconnect switch that is mounted near the inverter and transformer. On large solar farms, multiple inverters and transformers may be required to provide the electrical connection for all the panels.

The inverters for the ground-mounted panels are similar to the ones used on the roof-mounted panels, and a concrete pad is provided to securely mount all the electrical equipment including disconnect switches, meters, inverters, and transformers. Electrical conduits from the last solar panels in the arrays are run underground and are terminated in the concrete pad that will hold all the electrical equipment. In this way the electrical wiring from the solar panels that go to the electrical equipment is buried underground and protected from damage and weather.

After all the wiring has been run in the conduits, and all the connections made in the combiner boxes and junction boxes, the final connections to the DC disconnect can be made. Wires from the DC disconnect run through a conduit to one or more inverters. The phase wires from the inverters are routed through additional conduits that allow them to be connected to the phase transformers. In some applications a single large (step-up) transformer is used to step up the AC voltage from the inverters so that it matches the voltage of the electrical grid where they will be connected. In other installations multiple smaller transformers may be used, and their outputs are connected to one set of terminals at the point where the connection to the electrical grid is made.

11.4 INSTALLING A SOLAR PANEL ON A POLE

One of the innovations in the past few years has been highway signs and other signs that need lighting. The electrical power for these signs is provided by a battery that is constantly charged during the daylight hours by a solar photovoltaic panel. Figure 11–35 shows a solar photovoltaic panel on a pole that is used to charge a battery that provides electrical power for a light that is used to illuminate a sign. This section of the chapter will go through the steps to install a photovoltaic panel on a post.

11.4.1 Installing the Pole

The first step in the process of installing a photovoltaic panel on a pole is to obtain a pole that is the proper height. Next the location where the pole will be mounted in the ground is identified. A large hole must be dug 3 to 4 feet deep with enough space to create a concrete pad at the base of the pole. The combination of a deeper hole and the concrete pad will provide enough counterbalance to keep the photovoltaic panel from moving once it is attached to the pole. The amount of concrete needed in the bottom of the hole must be calculated to ensure that it can handle the strain of wind blowing against the photovoltaic panel when it is attached to the pole.

FIGURE 11–35 A solar photovoltaic panel used to charge a battery that powers a display light.

After the hole is dug, the pole is mounted into the hole and supported so it does not move when concrete is poured. All the concrete can be poured at the base of the pole, or some of the concrete can be used to form a pad near the surface as well as a large amount being poured in the hole. There are several strategies to create a solid base for the pole to ensure that the pole does not move once the concrete is hard. After the concrete is poured and is allowed to cure, the supports that are holding the pole in place are removed.

11.4.2 Selecting the Proper Size of Photovoltaic Panel and Battery for the Job

The next step in the process is to select a photovoltaic panel for this job. The first step in this process is to determine the amount of light that is necessary to light the sign completely at night. In this application the light will be powered by DC electricity, and a battery will be used to store enough charge to power the light for several days. The reason the battery is large enough to power the light for several days is that there may be times when it is cloudy or other factors that prevent the panels from charging the batteries completely every day. By designing the system with a larger battery, the battery can be charged on sunny days and it can power the light for several days without being recharged. The battery is mounted on the pole below the solar photovoltaic panels. The battery is

covered in a weatherproof box, so it is not exposed to the harmful elements of the weather. The box is vented so any gas that is produced by overcharging can be safely released from the batteries and the battery box.

The next step in this process is to select the solar photovoltaic panels that are large enough to charge the battery in one day, which will include approximately 8 to 10 hours of sunlight. In this application two solar photovoltaic panels were selected to be mounted on the pole to charge the batteries.

11.4.3 Mounting the Solar Panel on the Pole

After the concrete is cured and the pole is secured in the ground, the solar photovoltaic panels can be mounted to the pole. Figure 11–36 shows the hardware used to mount the photovoltaic panels to their hardware rails, and the hardware that is used to mount the rails and the panel on the pole so that the panel is tilted at the proper location to receive the maximum amount of sunlight.

The bracket at the bottom of the solar panel is attached to the pole first, and it is secured tight enough to allow last-minute adjustments to be made to the panel. The amount of tilt the panel will have is determined by how high the top bracket is adjusted on the pole. The lower the top bracket is positioned, the flatter the panel will be, and the higher the top bracket is positioned, the more vertical the panel will be. Figure 11–37 shows

FIGURE 11–36 Side view of solar panel showing mounting hardware.

FIGURE 11–37 Back side of solar panel showing connection between panels and battery.

the back side of the solar panels. You can see where the electrical connections are made at each panel and how the wires are routed between the solar panels and the battery that is located in the electrical box.

11.5 TRACKING SYSTEMS FOR SOLAR PANELS

Since the earth moves around the sun to create seasonal change, and the earth rotates on its axis during each 24-hour period, the direction at which the rays of the sun strike a solar panel are always changing. The solar panel will produce maximum electrical power only when the sunlight is striking it at the optimum angle. One way to ensure that the solar panels receive the maximum amount of solar energy is to move the panels with the use of a tracking system so the face of the solar panel is receiving the maximum amount of sunlight and producing the maximum amount of electrical power. There are several types of tracking systems used with solar photovoltaic panels. These tracking systems can be manual or automated. The manual tracking system is designed so that the panels can be moved about one or two axes manually. This means that periodically someone must go to the site where the solar panel is mounted and loosen the tracking hardware and move the panels a precise distance to a new location. The direction, tilt positions, and

the times and dates the panels need to be moved are precisely calculated from data about the location where they are installed. Computer software is available to identify the information to allow the solar photovoltaic panels to capture the maximum amount of energy from the sun. The computer software identifies the proper direction and angle the solar panel must be moved to for each given day of the calendar year. To operate the software, the location coordinates where the panels are installed must be loaded into the software, and the software must be synchronized with the clock and calendar to give the precise data for each day.

The automatic tracking system is more sophisticated and more expensive. It utilizes a closed-loop servo system that will have sensors that determine the direction of the strongest sunlight and drive motors that change the position of the solar panels so that they receive the strongest amount of solar energy. Typically high-torque motor such as stepper motors or permanent magnet motors are used to drive the solar panel positional actuators.

11.5.1 Single-Axis Solar Panel Positioning Control

Solar panels can have single-axis or double-axis positioning capabilities. The simplest type of access control is *single-axis solar tracking system*. In some parts of the country solar panels may be mounted so that they always face south, and the tracking system will adjust the panels' horizontal position so that it can track the sun as it moves from east to west. In other parts of the country such as the

North, the sun changes position from a northerly direction in the summer to a southerly direction in the winter. For these applications critical position adjustments may be needed to continually harvest the maximum amount of solar energy.

Many varieties of solar panel tracking mounts are available. When tracking systems are adjusted manually, the single-axis solar panel tracking systems for vertical mounted panels stay at the same vertical angle but rotate around the post to follow the sun from sunrise to sunset. Horizontally mounted panels are adjusted so that their vertical angle is set to match the angle of the sun for each season, which means they will be adjusted to a nearly flat position for summertime, and at a steep angle for winter, with the exact angle depending on latitude, which is how far north or south of the equator the panels are located. One general way to adjust the panel if you do not have computerized system is to look on the map or use a GPS to find the latitude in which the panels are located. Then set the panels at an angle that is the number of your latitude minus 15° in summer, and the latitude plus 15° in the winter. This method should make the panels' position nearly flat through the summer and nearly vertical in the winter with the panels facing in a southerly direction.

11.5.2 Horizontal Axis Control of Solar Panels

One type of solar panel control changes the horizontal axis of the solar panel so that the panels' face moves about an arc and horizontal plane. Figure 11–38 shows several panels mounted in a horizontal position in such a way that their face tilts up toward the sun. Hardware for moving the panels in unison is part of the mounting system for the panels so that an actuator is the only part that needs to be adjusted and it automatically changes the position of all the panels mounted to the assembly. The position of these panels can be rotated over time at the center of their axis so that the face of the panels receives more sunlight. The arrow in the diagram shows how the face of the panel can be rotated so that it receives more sunlight. This type of single-axis adjustment can be made manually over a number of days to improve the solar panel's ability to harvest solar energy.

11.5.3 Vertical Axis Control of Solar Panels

Another type of solar panel axis control is vertical axis control. In some applications the solar panels are mounted on a base or pulled in such a way that the panel is located in the vertical or upright position. The vertical axis control system allows the solar panel to swivel on the base to change the position of the face of the solar panel so that it can collect more solar energy. Figure 11–39 shows three vertically mounted solar panels. The solar panel identified by the letter (a) shows the solar panel facing straight to the front; (b) shows the face of the solar panel rotated about 10° in the counterclockwise position; (c) shows the face of the solar panel adjusted another 10° in the counterclockwise position. In these three diagrams the solar panel is mounted to its base so that it will be secured from strong winds and other conditions that may damage it, but the panel has vertical axis control, which allows the face of the panel to be turned a number of degrees in a clockwise or counterclockwise direction so that the face of panel receives the maximum amount of solar energy.

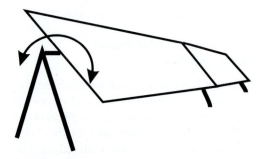

FIGURE 11–38 Single horizontal axis control of the solar panels.

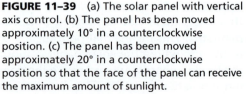

FIGURE 11–39 (a) The solar panel with vertical axis control. (b) The panel has been moved approximately 10° in a counterclockwise position. (c) The panel has been moved approximately 20° in a counterclockwise position so that the face of the panel can receive the maximum amount of sunlight.

11.6 THEORY OF SERVO SYSTEMS AND OPEN- AND CLOSED-LOOP FEEDBACK SYSTEMS

The axis control systems that are used to change the position of solar panels can use electrical motors or other types of actuators. These actuators can be controlled by open-loop control systems or closed-loop control systems. The *open-loop control system* does not use a sensor or other type of feedback signal. This type of system changes the actuator a small amount at various times based on calculations or tables that have been created to determine the proper position for the panel.

The *closed-loop control system* has a sensor that measures the strength of the sunlight, and the direction and angle it is coming from. The sensor changes the measured conditions into an electrical signal that is sent back to the controller that controls the motors that make the axis positioning hardware move. When a system uses a sensor, it becomes a closed-loop system, often referred to as a *servo system.*

It is important to understand that students can study open- and closed-loop control systems through several courses in order to fully understand them. This section is an abbreviated explanation of open- and closed-loop systems, and its goal is to provide a broad overview, including terms and basic operation so that students can begin to understand control loops. This means that many of the details of control systems will be covered lightly in this section, and students will need to refer to other sources or take additional classes to receive more details. Figure 11–40 shows a basic closed-loop system diagram with all the parts identified. It will be easier to understand these basic parts of a control system if they are applied to a specific solar panel axis positioning application. From the block diagram you can see that the system starts with a *set point* (SP) signal. The SP in this application can be any value between 0 and 100% and it represents the strongest amount of electrical energy the panel can produce and it is also called the desired value. For this example we will assign the amount of the

maximum electrical signal as 100%, so the SP would be 100% since we want the panel to be moved so it will produce the maximum amount of voltage.

The next part of the control system that we will examine is the *process variable* (PV), which is the signal that comes from the sensor that indicates the amount of electrical energy the sunlight is producing. In this system the PV is the signal from an electrical sensor that indicates the amount of sunlight that is being received and the signal is converted to 0–10 volts so it is compatible with the control system. The PV is also called the *feedback signal,* and it is the present value or actual value of the amount of sunlight that is being measured at the instant the sensor reading takes place. Since the sensor reading is continuous, the PV will change continually to indicate the changing amount of sunlight that is received when the panel is pointed in a certain direction.

The *summing junction* is the place in the control system where the set point (SP) is compared to the sensor (process variable) PV signal. Mathematically this means that the PV is subtracted from the SP (SP–PV) and the resulting answer is called the *error.* For example, if the SP is 100% and the PV signal indicates the actual amount of sunlight that is received is 90%, the difference is 10%. The summing junction is identified by the Greek letter sigma (Σ).

The solar panel axis control system in this application uses a stepper motor that rotates its shaft to cause the actuator to move the solar panel position. Stepper motors will be discussed in detail in Section 11.7. When this loop is programmed into the controller, the SP is the maximum amount of sunlight the panel can harvest if it is in the proper position. As the sun changes its position during the day, the amount of solar energy the sensor will receive will continually change. The closed-loop control system will make small adjustments to the solar panel axis control to move the face of the solar panel slightly so that it will continue to receive the maximum amount of sunlight. Constantly changing the position of the solar panel face so that it receives maximum sunlight is called tracking. The new position can be determined from a calculation that indicates where the best direction should be. The sensor

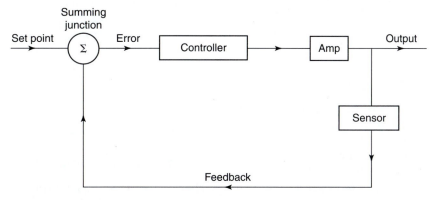

FIGURE 11–40 Closed loop (feedback) diagram that shows the set point, process variable, error, controller, and output.

for this system is a sensor that measures electrical energy converted from the sunlight it receives and its signal is also called the process variable. The SP is compared to the PV signal at the summing junction, which is inside the controller. When the loop control begins its operation, the computer does a mathematical calculation at the summing junction to determine the difference between the SP and PV, and this is called the error signal, which is sent to the controller part of the system.

The controller sends an output signal to an amplifier that controls the stepper motor. In this application the amplifier is contained directly in the stepper motor controller.

The stepper motor controller can cause the stepper motor shaft to change a number of degrees, which causes the solar panel axis control to move the face of the solar panel. The controller sends the output signal to the stepper motor controller, which changes the position of the shaft on the stepper motor. The shaft of the stepper motor can be rotated clockwise or counterclockwise depending on the conditions necessary to track the sunlight.

When the stepper motor changes the solar panel axis control hardware, the face of the solar panel is moved slightly and the amount of energy it receives from the sunlight will also change. If the amount of energy increases, the control system knows that it is moving the panel in the right direction and the right amount. The controller will continually increment small amounts of change and continually test to see if the amount of sunlight increased or decreased. If the amount of sunlight decreased, the controller will move the panel slightly back until the amount of sunlight being received reaches the peak amount. In this way the solar panel's face is continually being adjusted slightly to ensure that it receives the maximum amount of sunlight. It is important to understand that the maximum peak amount of sunlight changes over the entire day, and from week to week as seasonal changes occur.

This explanation tries to keep the control system as simple as possible for students to understand. In reality these closed-loop control systems are complex and require a large amount of technology to work properly. When the system fails to operate correctly in the field, technicians will need to ensure that the sensor is clean and receiving an unobstructed amount of sunlight. It is also important to ensure that the axis drive system is well lubricated and able to adjust the panel with as little energy as possible.

11.6.1 A Typical Open-Loop System

The block diagram for the control system shown in Figure 11–40 is called a "loop" because a sensor is used to sample the amount of sunlight and send its PV signal from this sensor back to the summing junction where it can be compared to the SP. The sensor signal is also called the feedback signal because it gets fed back to the summing junction. When this happens, it is closing the loop, and the system is called a closed-loop system.

In an open-loop system the feedback signal is not used. In this type of solar panel axis control, the amount of positional change is calculated ahead of time and the amount of change is based on the calendar day and the time of day. A table is created that determines the location the solar panel should be positioned to for that time of day and that day on the calendar. The table must use precise information about the location where the panels are installed, so that the overhead position of the sun is known and can be used in the calculations. Once the table is created, the information is fed to the controller as an incremental change, and it is based on a time. This means that the solar panel's face is adjusted small incremental amounts by the axis control system during a set number of time periods. For example, the time period can be 10 minutes, hourly, or daily and the signal is sent to the motor to change the position without regard to any feedback signal. The open-loop control system is less expensive and less complex, but it may need to be checked periodically to ensure that the motors are not moving the panels to positions that do not optimize the amount of solar energy they can receive.

11.7 STEPPER MOTOR USED TO DRIVE SOLAR PANEL AXIS CONTROLS

Stepper motors are used as the drive mechanism for many other types of rotary and linear solar panel axis controls. Stepper motors provide a means for precise positioning and speed control without the use of feedback sensors. The basic operation of a stepper motor allows the shaft to move a precise number of degrees each time a pulse of electricity is sent to the motor from the stepper motor controller. Since the shaft of the motor moves only the number of degrees that it was designed for when each pulse is delivered, you can control the pulses that are sent and the positioning and speed of the shaft with a stepper motor controller. The rotor of the motor produces torque from the interaction between the magnetic field in the stator and rotor. The strength of the magnetic fields is proportional to the amount of current sent to the stator and the number of turns in the windings. Stepper motors are available for a variety of motion-control applications, such as single- and multiple-axis control for solar panel axis control systems. It is important to understand that the stepper motor requires a special stepper motor controller that sends electrical signals as pulses to the motor to cause it to change its position. The polarity of the electrical pulse will determine the direction the motor's shaft will turn, and the number of pulses will determine the distance the shaft will rotate.

11.7.1 Types of Stepper Motors

There are three basic types of stepper motors: the *permanent magnet motor*; the *variable reluctance motor*; and the *hybrid motor*, which is a combination of the previous two. The rotor for the permanent magnet motor is called a canstack rotor. The permanent magnet stepper motor can have multiple rotor windings, which means that the shaft can control the movement of the rotor to fewer degrees as each pulse of current is received at the stator. For example, if the rotor has 50 teeth and the stator has 8 poles with 5 teeth each (total of 40 teeth), the stepper motor is able to move 200 distinct steps to make one complete revolution. This means that the shaft of the motor will turn 1.8° per step. The main feature of the permanent magnet motor is that a permanent magnet is used for the rotor, which means that no brushes are required. This type of motor has relatively low torque and is used for low-speed applications such as solar panel axis drives.

The variable-reluctance motor does not use permanent magnets, so the field strength can be varied. The amount of torque for this type of motor is still small, so it is generally used for small positioning tables and other small positioning loads. Since this type of motor does not have permanent magnets, it cannot use the same type of stepper controller as other types of stepper motors. The hybrid stepper motor is the most widely used and combines the principles of the permanent magnet and the variable reluctance motors.

11.7.2 Stepper Motor Theory of Operation

The theory of operation for a stepper motor uses magnets to make the motor shaft turn a precise distance when a pulse of electricity is provided. You learned previously that like poles of a magnet repel and unlike poles attract. The stator (stationary winding) has four poles, and the rotor has six poles (three complete magnets). The rotor will require 12 pulses of electricity to move the 12 steps to make one complete revolution. Another way to say this is that the rotor will move precisely 30° for each pulse of electricity that the motor receives. The number of degrees the rotor will turn when a pulse of electricity is delivered to the motor can be calculated by dividing the number of degrees in one revolution of the shaft (360°) by the number of poles (north and south) in the rotor. In this stepper motor 360° is divided by 12 to get 30°. The smaller the number of degrees the rotor moves with each pulse of electricity, the more precise it is when it is used to position a solar panel.

When no power is applied to the motor, the residual magnetism in the rotor magnets will cause the rotor to detent or align one set of its magnetic poles with the magnetic poles of one of the stator magnets. This means that the rotor will have 12 possible detent positions. When the rotor is in a detent position, it will have enough magnetic force to keep the shaft from moving to the next position. This is what makes the rotor feel like it is clicking from one position to the next as you rotate the rotor by hand with no power applied. This also means the stepper motor is very good at keeping its load (the solar panel) at the precise position it has moved to until an electrical

pulse is sent to the motor again to make it change its position. When the stepper motor receives an electrical pulse, it will move the solar panel, and when the electrical pulses are not being received, the magnetic field in the stepper motor will be strong enough to hold the shaft in position so the solar panel position does not drift from where it was moved.

When power is applied, it is directed to only one of the stator pairs of windings, which will cause that winding pair to become a magnet. One of the coils for the pair will become the north pole, and the other will become the south pole. When this occurs, the stator coil that is the north pole will attract the closest rotor tooth that has the opposite polarity, and the stator coil that is the south pole will attract the closest rotor tooth that has the opposite polarity. When current is flowing through these poles, the rotor will now have a much stronger attraction to the stator winding, and the increased torque is called "holding torque."

By changing the current flow to the next stator winding, the magnetic field will be changed 90°. The rotor will move only 30° before its magnetic fields will again align with the change in the stator field. The magnetic field in the stator is continually changed as the rotor moves through the 12 steps for a total of 360°.

11.7.3 Linear and Rotary Actuators for Solar Panel Axis Controllers

Solar panel position controllers can be moved by *rotary actuators* (motors) or *linear actuators* that use a ball screw and slide. Figure 11–41 shows a ball screw linear actuator. The ball screw has a threaded rod that is turned by a motor. A slide is a part of the mechanism that fits over the threads on the treaded rod, and when the rod rotates clockwise, the slide moves toward the motor; when the rod is rotated counterclockwise the slide moves away from the motor. The part of the solar panel axis controller system that needs to have linear action applied to it is mounted on the slide. As the ball screw rotates, the slide moves closer to or away from the motor and provides the linear motion that moves the face of the solar panel so it catches more of the sunlight.

Typically the motor that is connected to the ball screw mechanism is a stepper motor, and its shaft rotation can be controlled precisely, which means that the position of the slide on the ball screw can also be controlled precisely. If a

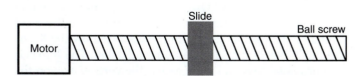

FIGURE 11–41 A ball screw that has a threaded rod that is turned by a motor. The part of the system that needs to have linear action applied to it is mounted on the slide. As the ball screw rotates, the slide moves closer to or away from the motor.

ball screw uses a stepper motor, a stepper motor controller must be used to drive the stepper motor.

This type of positioning system can be operated as an open-loop system by sending a predetermined number of pulses to the stepper motor at specific times to change the position of the solar panels, or it can be set up as a closed-loop system, and a sensor can be mounted on or near the panel to detect the amount of sunlight. The sensor can provide feedback to the stepper motor. If the sensor is used, it will determine the direction the panel should be positioned to receive the most sunlight. Typically a computer control is used as the closed-loop control. Some types of ball screw mechanisms also include a limit switch at each end of the screw to detect when the slide has moved completely to one end or the other. The limit switch is used to cause the stepper motor to reverse its direction of rotation, which will cause the slide to begin the move in the opposite direction.

In some applications, such as control of vertically mounted solar panels, all that is needed to move the panel is a motor. The motor is mounted with a gear on its shaft that meshes with the gear that causes the movement to the panel. The gear ratio is designed so that a small motor that has high torque can rotate the solar panels. In some applications a stepper motor can be used as the drive motor, because it has enough torque to rotate the shaft, and its position can be precisely controlled. The motion control for the vertical axis solar panels can be closed-loop or open-loop system, depending on the size of the system and the amount of money that is available for controls.

11.7.4 Multi (Two-Axis) Solar Panel Tracking Systems

In some larger solar farm applications, the position of solar panels is adjusted on a continual basis to ensure that they harvest the maximum amount of solar energy. Adjusting the panels continually makes sure the panels' return on investment remains high. If the panels were placed in fixed positions, they would produce less energy and bring less income over time. The additional money that is obtained by adjusting the panels more than offsets the cost of the solar panel tracking system. Typically large ground-mounted panels use the tracking technology, whereas the extra weight of the tracking hardware makes the system too heavy to use with solar panels that are roof mounted.

Some tracking systems adjust both the horizontal and vertical locations of the panel in order to receive the most sunlight that is available. Figure 11–42 shows how the X axis and the Y axis of the solar panel are adjusted. The X axis is the horizontal axis and the Y axis is the vertical axis on the panel. The *two-axis solar tracking system* is more complex and requires two motors that independently change the position of the X and Y axes. Typically panels that can have their

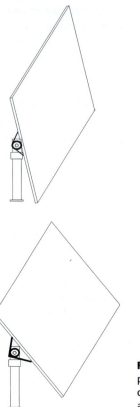

FIGURE 11–42 A two-axis solar panel tracking system. These two diagrams show the X and Y axes adjustments to the solar panel's face.

X and Y axes adjusted are mounted on a pedestal or pole. It is important to understand that even though the hardware is capable of moving the panel through a variety of positions, each point that it stops the panel is securely locked so that it will not be damaged or moved by high winds.

11.7.5 Advantages and Disadvantages of Solar Tracking Systems

One of the major advantages of solar tracking systems is getting a larger return for the investment in the panels. If a large number of panels are used, the amount of extra income from the solar panels will more than offset the cost of operating the automated tracking systems. If one or two sets of panels are used for residential use, a manual tracking system may improve the output of the panels so that additional panels are not needed for the system. In some cases the tracking system will improve the amount of energy harvested to the point where the return on investment is lowered by one or two years.

Some of the disadvantages of having tracking hardware are that it is an additional cost and an additional maintenance element that must be checked from time to time. If the tracking hardware wears out, it may cause the system output to become lower, and it becomes an additional cost to replace a worn or broken part. In some cases the tracking hardware is not strong enough to hold the panels in the correct position during high winds that might cause the panels to become damaged. One other disadvantage of using tracking equipment is that it adds an additional expense to purchasing the equipment and additional expense and time in installing and testing the system. In some cases additional panels can be purchased for the cost of the tracking hardware and software, and the output of the additional panels will be equal to the improved efficiency caused by the tracking system.

11.8 TROUBLESHOOTING SOLAR PHOTOVOLTAIC PANELS

Solar photovoltaic panels are very reliable, but there are times when problems occur and the panels must be troubleshooted and repaired. Some of the problems solar panel have are brought on by temperature extremes that may range from below freezing to above 90°F. Problems are also caused by moisture getting into panels or into the electrical connections at the panels and when wind causes stress on the panels. In some cases where panels are mounted on the rooftop and other locations, they may be struck by lightning or have problems with static electricity, which can cause damage to terminal connections and the panels themselves over time. The glass or outside covering of the panel can also be damaged by hail or other debris that can cause it to crack or fracture. In some cases the electrical output of the solar panel is diminished by dirt covering the outside glass. In this case if it does not rain often enough, the panels will have to be washed periodically to remove the excess dirt.

One of the typical problems that solar panels encounter is a broken wire connection where the wires are attached to the solar panel PV cells. Typically these connections are soldered in place and sealed to keep weather out, which makes it difficult to test and repair them.

The volt-ohm-milliamp (VOM) meter is the most-used meter for troubleshooting electrical problems with solar panels. Information about using the VOM to measure voltage, current, and resistance is provided in Chapter 12. The VOM can check for DC voltage at the panel and determine which wire is the positive terminal and which is the negative terminal. This test is called determining the polarity of the panel wires. The VOM can also check for AC voltage that comes from the inverter.

The VOM can also check the continuity of wires and their connections. A continuity test is actually a resistance test. When the resistance test is made on wires, the amount of resistance should be very low or near zero ohms. If the test indicates the wire has low resistance, the wire is said to have continuity. If the continuity test indicates the resistance is high, the wire will not conduct electricity easily, and it will cause higher than normal current and temperature and must be replaced. If the wire is broken or has an open circuit, its resistance will be infinite, and the wire will have to be replaced. The continuity test is also useful for testing fuses. A good fuse

should have low resistance and an open fuse will have infinite resistance.

If the external glass or the external covering for the solar panel becomes cracked or fractured, it will allow moisture to enter the panel. You will become aware of this condition when you see condensation on the inside of the external glass or covering. When moisture gets inside the panel, it will block some of the sunlight from coming through, but it will also cause water damage and corrosion in the solar cells and anywhere there are electrical connections. In some cases some small damage can be repaired, but in most cases the solar panel will need to be replaced because the seal cannot be restored and eventually the water or moisture will enter and damage the panel.

You will be able to perform electrical tests to determine if a solar panel has loose solder connections, because the voltage will be intermittent. You can attach a voltmeter to the panel and measure its output. While the meter is connected, you can bump the panel or apply pressure to it at various points to see if the problem can be duplicated. This would be similar to the stress the panel would receive when the wind is blowing on it. If the voltage changes when stress is applied, it may indicate the terminal connection is going bad. You can also put some stress on the electrical connection and see if the voltage output drops. If there is a problem, it may be possible to clean the connection and re-solder it and replace the silicon that is used to seal it. Some products are available that are similar to a glue or conducting epoxy that has solder material in it, and this glue is used to hold the wires together, and the solder provides the electrical connection so that electrical current can be conducted through the connection. For this material to work the temperature must be warm, the wire terminal ends must be clean, and the material must adhere tightly to the wire. Some technicians have successfully made electrical repairs with this material, whereas others claim that this type of repair does not work over the long period of time the panel must be in production. If the panel cannot be repaired, it will eventually need to be replaced.

If the solar panel you suspect of not operating correctly is connected in a string or in an array, you can test to see how much electrical current the string is producing and how much each panel is producing. You can connect a DC ammeter to one of the panel's wires. Some VOM meters have the ability to read milliamps (one millionth of an amp) up to 10 amps. Other DC ammeters can read current up to 100 amps. You will see how to place the meter to measure DC current in Chapter 12. When you have the meter in place, you should be able to measure the DC current and voltage the panel is producing. When you have an established measurement, you can put a piece of cardboard over the panel that you suspect is bad, and the amount of voltage and current that you are measuring should go down. If the voltage and current go down, it indicates that the panel that you have put the cardboard on is producing electrical power. If the measurements for the string do not change when you cover the panel, that panel is not producing any power and must be removed from the string and replaced. This test is called a shading test, and you can test the relative output of each panel by moving the piece of cardboard to cover each panel individually and watching the voltmeter or ammeter readings.

If the panel you suspect of malfunctioning is a single panel or if you can isolate it, you can do the same test by covering the panel with a piece of cardboard. If the panel is producing some small amount of electrical power when full sunlight is shining on it, you should suspect it has a problem and you will need to check the wire connections to make sure they are clean. If the panel is isolated and it does not produce any electrical power when its surface is exposed to light, you must suspect that it has a problem with the connection of the positive or negative wires. Since an individual panel only has two wires, it will not matter which one is bad, since both wires need to be good for current to flow and for voltage to be present at the end of the leads. You can visually inspect the connection for any damage, and you can check each individual wire for continuity with the ohmmeter. If the test indicates either wire is open, it must be replaced if it is possible to make a new connection at the panel. If both lead wires test OK, the problem is inside the panel and typically the panel will need to be replaced. Be sure you check the recycling procedure for each type of panel.

11.8.1 Failure of the Bypass Diode

In a solar photovoltaic panel a bypass diode is connected across the positive and negative terminals of the two wires from the solar panel. Typically the diode is mounted in the combiner box across the terminal of the positive and negative terminal wires. The diode is connected in reverse bias, which means the diode's negative terminal (cathode) is connected to the positive wire of the photovoltaic panel, and the diode's positive terminal (anode) is connected to the negative terminal of the panel. During normal operation when the solar photovoltaic panel is producing the proper amount of electrical power when the sun is shining, the diode does not do anything in the circuit. When one or more solar panels in a string become shaded and the amount of voltage it produces becomes less than that the other panels in the string are producing, that panel can clamp down the amount of current the string produces if it is connected in series with the other panels. If this happens, the output of the entire string becomes reduced to the point that the panels will be ineffective. If the solar panels have a bypass diode, when one of the panels becomes shaded, the bypass diode will allow current to pass around that panel (bypass it) until

the condition passes and the panel receives more sunlight and begins to produce the same amount of voltage as the other panels in the string.

A problem occurs when the bypass diode shorts out and causes the electrical power production of that solar panel to be nullified. When a diode is shorted, a voltage differential cannot be created across the positive and negative terminals of the photovoltaic panel terminals and that panel cannot create any power. If the bypass diode fails because it becomes open, the only problem that can occur is that the string that the panel is in will no longer be protected against shading problems. If the bypass diode becomes shorted, you will notice a drop in the power output of the string. This means that if a string is producing less power than it is rated for, you will need to test the bypass diode of each panel. An easy way to test the diode to determine if it is shorted is to put a lead from your voltmeter on each lead of the diode. If the diode is shorted, you will measure zero volts across the two terminals because the shorted diode will have no voltage differential across its leads. If you measure any voltage across the diode terminals, you will know that it is not shorted. It is important to understand that this test cannot detect an open diode, because both the open diode and a diode that is working correctly will have a voltage differential across their two terminals. If you suspect the diode is open, you can shade the panel it is connected to and the current from this string should be reduced tremendously if the diode is open because the shaded panel will draw down all the panels in the string that are connected in series with it. If the diode is working correctly, the current for the string will bypass the shaded panel. A further test can be made of the diode if you suspect it is open by removing it from the circuit and testing it for resistance in both directions. This is called testing the diode in forward bias and reverse bias. When a VOM is used as an ohmmeter, it has an internal battery. When you touch the leads of the meter to the leads of the diode, and then switch the lead, one way the battery positive terminal will be connected to the diode negative terminal and the diode will have high resistance. When the leads of the meter are reversed, the diode should have low resistance. This test is explained in detail in Chapter 12, and it is called testing the front to back ratio of resistance. It is also called testing the forward resistance and the reverse resistance since you switch the meter leads that have battery voltage applied to them, which means that positive voltage is placed on the positive anode terminal of the diode, and then the negative meter terminal is touched to the negative cathode terminal of the diode, which puts reverse polarity voltage on the terminals. In reality you will find that it is just as easy to change the diode out and replace it with a new one if you suspect it is bad, since the diodes are not expensive and they are easy to access at the terminals in the combiner box.

11.8.2 Problems with Corroded or Burnt Terminals

Another problem occurs with solar panels when their electrical terminals become loose or corroded. A loose or corroded terminal will cause a point of high resistance and excessive current will flow at that point, which will cause the terminal to overheat and burn. This problem becomes worse over time as the terminal gets hotter and begins to deteriorate to a point where it will no longer pass any current. This problem will come to your attention when the output from the string goes to zero. The output of an individual panel will also be zero, and you would be called to check out the panels. When you check out the panel and its connections in the combiner box, you should be able to find the burnt or corroded terminals with a visual inspection.

11.8.3 Problems with Voltage Fading in Higher Temperatures

Typically solar photovoltaic panels will lose some of their ability to produce voltage at very high temperatures. If the voltage loss becomes excessive, the panel must be replaced. Most data sheets for the panels will indicate the amount of voltage loss the panel has for each degree centigrade above the rated temperature.

These data are the current/voltage (IV curve) and indicate how much loss can be expected in the higher temperatures. This problem is called *voltage fading,* and it becomes more of a problem when the photovoltaic panels are used to charge batteries. When the voltage at the battery is different from the voltage that is fading at the solar panels, the current flow from the solar panel will become less. It is important to have as large a voltage differential as possible between the voltage produced by the panel and the voltage at the batteries. When the voltage difference is small, the amount of current will be lower.

Another way to determine the voltage loss of the solar panel is to measure the temperature on the panel and identify how much voltage is being produced. Next use some material to shade the panel for a short period of time or use a fan to cool off the panel. In some cases you can also cool the panel down by allowing cool water to flow over its face. Be aware that the glass on the front of the panel can be shocked if the water is too cold so you may need to use a small amount of water or allow the water to warm up slightly before allowing it to come into contact with the glass. If the amount of voltage output increases when the panel cools down, you have determined that the panel is suffering from temperature or heat fade. If the amount of loss is too great and the panel is in severe heat for extended days of the year, you may need to change the panel out and replace it to get the amount of power from the string back to its specified level.

11.8.4 Periodic Maintenance of Photovoltaic Panels

Photovoltaic panels need to be inspected periodically to ensure that they are operating properly. When the photovoltaic system is inspected, you must inspect the entire electrical system, including the terminal connections at the panels, the terminal connections in the combiner boxes and at the junction boxes, and then the connections at the inverter and all of the disconnect switches. During these tests you must check the tightness of all terminal lugs and terminal connections and inspect them for any damage that might have occurred. You may also need to make voltage and current measurements to ensure that the panels are operating correctly.

Photovoltaic panel manufacturers provide a complete list of checkpoints for their panels and the electrical equipment. This checklist will include the points that are to be checked and the time interval between the checks. The results of the test should be entered into the maintenance records so that any problems can be identified and tracked. The main concept of periodic maintenance is to find problems and fix them while they are still small.

Questions

1. Explain the steps involved in installing a residential solar PV system.
2. Explain the steps involved in installing commercial solar photovoltaic panels on a rooftop.
3. What is a bypass diode and what function does it provide?
4. Explain the steps involved in installing commercial solar photovoltaic panels that are mounted at ground level.
5. Identify the electrical components in a commercial photovoltaic panel installation.
6. Explain why you would perform a shading test and what problem should it show.
7. Explain the parts of a closed-loop control system.
8. What is a stepper motor and why is it used with solar panel tracking systems?
9. Explain the parts of the single-axis and double-axis solar panel tracking system.
10. Identify the tests to troubleshoot problems with solar photovoltaic panels.

Multiple Choice

1. A closed-loop system _____
   a. Has a feedback sensor that tells the control system how close it is getting to the set point.
   b. Does not have a feedback sensor that tells the control system how close it is getting to the set point.
   c. Changes its output based on a table rather than on a feedback sensor.

2. A stepper motor _____
   a. Allows its shaft to turn exactly 10 times when an electrical pulse is received.
   b. Uses a capacitor to increase its torque.
   c. Is designed to turn its shaft a part of a turn when it receives a pulse, which allows it to have precise positioning capability.

3. A solar photovoltaic tracking system _____
   a. Is designed to keep the face of the solar panel fixed in one position.
   b. Is designed to move the face of the solar panel to ensure that it receives the most amount of sunlight.
   c. Can only be changed manually.

4. The bypass diode can be tested _____
   a. With an ohmmeter by testing the forward resistance and reverse resistance.
   b. By checking for voltage across the terminals when the panel is producing power.
   c. By shading the panel that the diode is connected to and seeing if that panel goes to bypass mode for the string.
   d. All of the above.

5. A shading test is a test _____
   a. To determine whether a solar photovoltaic panel is operating correctly.
   b. To determine how well the inverter converts DC to AC voltage.
   c. That helps determine the size of solar panels that should be selected for a solar panel application.

Electricity and Electronics for Solar Energy Systems

OBJECTIVES FOR THIS CHAPTER

When you have completed this chapter, you will be able to

- Explain the operation of a three-phase transformer.
- Identify the basic parts of a DC motor and explain its operation.
- Explain how to change the speed of a DC motor
- Explain how to change the rotation of a DC motor.
- Explain the differences among series, shunt, and compound DC motors.
- Identify the basic parts of a three-phase AC motor and explain its operation.
- Explain how to change the speed of an AC motor.
- Explain how to change the rotation of an AC three-phase motor.
- Identify the basic parts of a single-phase AC motor and explain its operation
- Explain which two materials are combined to make P-type and N-type material.
- Explain the operation of a diode (PN) junction and show the input AC waveform and the output DC waveform.
- Identify a four-diode and six-diode full-wave bridge rectifier.
- Explain the operation of a power electronic frequency converter (inverter).

TERMS FOR THIS CHAPTER

AC motor
Armature
Auxilliary Contacts Brushes
Circuit Breaker
Commutator
Compound DC Motor
Conventional Current Flow
DC Motor
Delta-Connected Three-Phase Transformer
Diode
Electron Current Flow
Forward Bias
Free Electron
Fuse
Fuse Disconnect
High-Leg Delta Voltage
Hole
Induction Motor
Inverter

Ladder Diagram
Light Emitting Diode (LED)
Load Center
Motor Starter
N-Material
Overcurrent Protection
Overload
P-Material
Power Circuit
Rectifier
Reverse Bias
Rotor
Series DC Motor
Shunt DC Motor
Slip
Transformer
Wye-Connected Three-Phase
    Transformer
Wound Rotor

12.0 OVERVIEW FOR THIS CHAPTER

This chapter will continue the explanation of electrical fundamentals that was presented in Chapter 7. The electrical technology covered in this chapter continues the discussion of the operation of transformers that began in Chapter 7. This chapter will cover advanced transformer concepts including three-phase transformers and how these transformers provide delta and wye voltages.

This chapter will also provide an introduction to DC and AC motors and motor starters. You will learn how DC and AC motors are used in a variety of applications, including tracking motors for solar photovoltaic panels and for pumps and fans in solar heating systems. This section will also explain controls such as motor starters used to turn these motors on and off. Other components such as fuses and overloads will be explained.

The final part of this chapter covers all the electronic components that you will find in power supplies and inverters and other electronic controls for solar energy. This part of the chapter will explain diode rectifiers, silicon controlled rectifiers (SCRs), transistors, and other components used in electronic controls. You will learn the names of these components, the names of their terminals, and the theory of their operation and how to troubleshoot these electronic devices.

12.1 THREE-PHASE TRANSFORMERS

Figure 12-1 shows three-phase transformers which you will encounter on the job. The transformer may be mounted near the solar energy equipment, or it may be located in a transformer vault (a special room where transformers are mounted). In some cases the transformers are mounted on a utility pole just outside the commercial site. Figure 12-2 shows a three-phase transformer with its cover removed so you can see three separate transformer windings. The

FIGURE 12-2   A typical three-phase transformer with its cover removed, showing the three independent transformers and connection terminals.   (Courtesy Clay Carroll)

three-phase transformer operates exactly like a single-phase transformer in that AC voltage is applied to the primary side of the transformer, and induction causes voltage to be created in the secondary winding. In the case of the three-phase transformer the 120° phase shift of the three-phase voltage applied to the primary windings will be maintained in the secondary side of the transformer.

When you are working on a commercial system, you may need to make connections on the secondary side of the transformer to the disconnect box or between the secondary side of the transformer and a load center. You may also need to make voltage tests at the terminals of the transformer. You can use the diagram that is provided on the side of the transformer to identify the correct terminals to use. In most cases you will not be required to make any electrical connections at the primary side of the transformer, since the primary voltage may be rather high (12,470 VAC or higher). The electrical technicians for the electric utility company or technicians from a high-voltage service company will make the primary voltage connections for the transformer. You will be expected to make the connections for the solar energy equipment and at the various switch gears where the voltage is below 480 volts.

12.1.1 The Wye-Connected Three-Phase Transformer

Figure 12-3 is a diagram of a *wye-connected three-phase transformer*. Note that the physical shape of the transformer windings resembles the letter Y, which gives this type of configuration its name. You should also notice that this diagram shows only the secondary coils of the transformer. It is traditional to show only the primary-side connections or only the secondary-side connections when discussing a transformer power distribution system, since showing both the primary and secondary connections in the same diagram tends to become confusing.

FIGURE 12-1   Typical three-phase transformers with their covers in place. (Les Lougheed/Pearson Education/PH College)

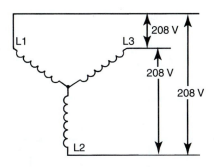

FIGURE 12-3 A diagram showing the secondary winding of the three-phase transformer connected in a Y configuration. The voltage available at L1–L2, L2–L3, and L3–L1 is 208 V for this transformer.

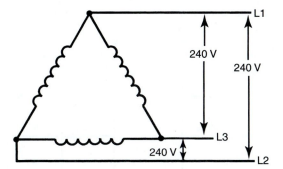

FIGURE 12-4 Electrical diagram of the secondary windings of a transformer connected in a delta configuration. Notice that the voltage at L1–L2, L2–L3, and L3–L1 is 240 V.

The amount of voltage measured at the L1–L2, the L2–L3, and the L3–L1 terminals on the secondary winding will be 208 V for the wye-connected transformer if its turns ratio is set for low voltage. If the turns ratio is set for high voltage, the amount of voltage between each terminal will be 480 V. The voltage indicated between the windings for the transformer shown in the diagram is 208 V, which means that the turns ratio for this transformer will provide the lower voltage. If 480 V were needed, a transformer with a higher turns ratio would be used. The primary and secondary voltages for each transformer are provided on their data plate and can be specified when the transformer is purchased and installed.

12.1.2 The Delta-Connected Three-Phase Transformer

Figure 12-4 shows a *delta-connected three-phase transformer*. Note that the shape of the transformer windings in this diagram looks like a triangle (Δ). This shape is the Greek letter D, which is named "delta." This diagram also shows only the secondary side of the three-phase transformer.

The amount of voltage measured at L1–L2, L2–L3, and L3–L1 is 240 V for the delta-connected transformer if it is wired for its lower voltage. If the delta-connected transformer is wired for its higher voltage, the voltage at L1–L2, L2–L3, and L3–L1 will be 480 V. If the delta-connected

transformer is wired for its lower voltage, it is easy to differentiate it from a wye-connected transformer. If the delta-connected transformer and wye-connected transformer are both connected for their higher voltages, you cannot tell them apart, since both will provide 480 V between their terminals.

If you need to know whether the transformer windings are connected as delta or wye, you can check the physical connections or you can make an additional voltage measurement between each line and the neutral terminal of the transformer if one is provided. The next section will explain these measurements.

12.1.3 Delta- and Wye-Connected Transformers with a Neutral Terminal

Figure 12-5 shows a wye-connected transformer and a delta-connected transformer, each with a neutral terminal. The neutral terminal on the wye-connected transformer is at the point where the three ends of the individual windings are connected. This point is called the "wye point."

Note that the neutral point for the delta-connected transformer is actually the midpoint of the secondary winding that is connected between L1 and L3. This point is essentially the center tap of one of the transformer windings. Traditionally it is the winding that is connected between L1 and L3 if a neutral connection is used with the three-phase transformer.

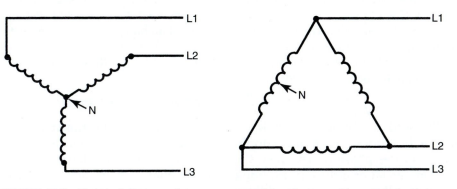

FIGURE 12-5 Electrical diagram of a wye-connected transformer with a neutral point and a delta-connected transformer with a neutral point.

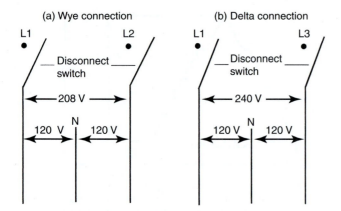

FIGURE 12-6 (a) Voltage from L1 to L2 for a wye- connected transformer is 208 V and from L1 to N or L2 to N is 1120 V. (b) Voltage from L1 to L3 for a delta-connected transformer is 240 V and from L1 to N or L1 to L3 is 120 V.

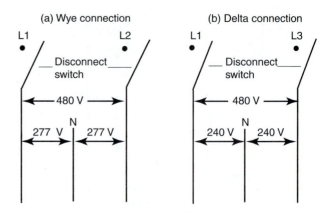

FIGURE 12-7 (a) Voltage from L1 to L2 for a wye-connected transformer is 480 V and from L1 to N is 277 V. (b) Voltage from L1 to L3 for a delta-connected transformer is 480 V and from L1 to N or L3 to N is 240 V.

Figure 12-6a shows the amount of voltage that you would measure between terminals L1 and L2 of a wye-connected transformer and between terminals L1 and N and L2 and N. Notice that the voltage between L1 and L2 is 208 V, so this transformer is wired for the lower voltage. The voltage between L1 and N and between L2 and N is shown as 120 V. Figure 12-6b shows the amount of voltage that you would measure between terminals L1 and L3 and between L1 and N and L3 and N for a delta-connected transformer. The voltage between L1 and L3 is 240 V, so this transformer is connected for its lower voltage. The voltage between L1 and N and between L3 and N is shown as 120 V. Notice that since the neutral point for the delta-connected transformer is exactly halfway on the transformer winding, the voltage L1–N and L3–N will always be exactly half the voltage between L1 and L3.

This is the main difference between a wye-connected transformer and a delta-connected transformer. The voltage at L1–N and at L3–N for any delta-connected transformer is always exactly half, and the L1–N or L2–N voltage for a wye-connected transformer will always be more than half the voltage at L1–L2. The exact amount of voltage at L1–N can be calculated by dividing the voltage between L1 and L2 by 1.73, which is the square root of $3 (\sqrt{3} = 1.73)$. The square root of 3 is used because of the relationship among the phase shifts of the three phases. Note that 208 divided by 1.73 is 120 V.

The same relationship of voltage between L1 and L2 and L1 and N exists when the transformers have turns ratios for their higher voltage. Figure 12-7a shows a wye-connected voltage between L1 and L2 at 480 V, and between L1 and N at 277 V. The 277 V can be calculated by dividing 480 by 1.73. The 277 V that come from L1–N or L2–N are generally used for fluorescent lighting systems in commercial buildings. Since the supply voltage originates from a three-phase transformer, the lighting system in a commercial or industrial building will also use L3–N, so that voltage is used from all three legs of the transformer. This voltage is referred to as

single-phase voltage, since only one line of the three-phase transformer is used at each circuit. For example, L1–N is a single-phase circuit.

It is also important to understand at this time that L1–L2, L2–L3, or L3–L1 could each be used to supply complete power for an electrical system. Even though this type of power supply uses two legs of the transformer, it is still called a single-phase power supply because only one phase of voltage is used at any instant in time. For example, if the system is powered with voltage from L1–L2, during any given half-cycle of AC voltage, the power source for the system will come from L1, and then during the next half-cycle it will come from L2. The source of power will continue to oscillate between L1 and L2, but only one phase is in use during any instant in time.

Figure 12-7b shows that the higher voltage for the delta system between L1 and L3 is also 480 V, but this time the voltage between L1 and N or L3 and N is 240 V. Again the L1–N or L3–N voltage is exactly half the supply voltage, since the neutral point on the transformer is the center tap of one of the transformers. It is important to understand that it is easy to distinguish between a wye-connected power source and a delta-connected power source by measuring the voltages at L1–L3 and L1–N. If the L1–N voltage is half the L1–L3 voltage, the system is a delta-connected system. If the L1–N voltage is more than half, it is a wye-connected system. You should remember that the L1–N wye voltage can always be calculated by dividing the L1–L2 voltage by 1.73.

12.1.4 The High-Leg Delta System

When a three-phase transformer system is used for the power source for an electrical system, it may have the neutral tap, or neutral. It is important to remember that a three-phase system does not need to have a neutral to operate correctly. The neutral is added only if the lower voltage (120 V) is needed for some part of the system.

Equipment manufacturers generally make all the components in a three-phase system a higher voltage, or they may supply a small transformer inside the equipment power panel to drop the higher voltage to the necessary voltage level. For example if the equipment needs 208 V three phase for the motors, a control transformer can be provided in the power panel of the equipment to drop the 208 V L1–L2 to 120 V or 24 V. The 120 V or 24 V is used to provide power for the control circuit to power relay coils. Since the primary side of the control transformer can be powered by L1–L2 (208 V), a neutral is not needed in this system. If the control transformer or motors require 120 V for power, a neutral tap or neutral will be needed, and this component will be connected between L1 and N, L2 and N, or L3 and N.

If the transformer is connected as a delta transformer, it is important to understand that a different voltage becomes available between L2 and N. Figure 12-8 diagrams this voltage. The secondary winding of this three-phase transformer is connected as a delta transformer. The voltage from L1–N and L3–N is 120 V, so the voltage between L1 and L2, L2 and L3, or L3 and L1 is 240 V.

The different voltage occurs between L2 and N. Since the winding between L1 and L3 has the center tap, it stands to reason that the voltage between L1 and N or between L3 and N will be exactly half the voltage at L1–L3. Since the center tap is not between L2 and L1, the voltage for L2–N must come from one complete phase (240 V) and half the next phase (120 V). Since this voltage uses two phases, the 240 V and the 120 V are out of phase, and the resultant voltage from them is 208 V.

Since the 208 V between L2 and N come from two phases, it will cause the transformer to overheat if it is used to power any components that require 208 V. For this reason the voltage is called the *high-leg delta voltage* to indicate that it is derived from L2–N of a delta-connected transformer and that it should not be used to power 208-V components. It is very important to understand that the L2 leg of the transformer is usable when it is used with L1–L2 or L2–L3 as part of the three-phase system or 240-V single-phase system. The only problem occurs

when the L2 terminal is used in conjunction with the neutral, which creates the L2–N voltage of 208 V.

The L2 terminal in any power distribution box for a delta-wired system should always be marked with an orange wire to identify it as the high-leg delta. In some areas of the country the high-leg delta is also called the wild leg, red leg, or stinger leg.

12.2 INTRODUCTION TO DC MOTORS

DC motors are commonly used to operate parts of tracking systems for solar panels and other applications such as pumps. It is important to remember that the earliest machines required speed control and DC motors could have their speed changed by varying the voltage sent to them. The earliest speed controls for DC motors were nothing more than large resistors. DC motors required DC voltage for operation. This means that a source for the DC voltage is needed or the DC voltage from the solar panels can be used.

As you read this section of the chapter, you will learn about controlling a DC motor's speed and its torque, and how to reverse the direction of its rotation. You will gain a good understanding of the basic parts of the DC motor so that when you must troubleshoot a DC motor circuit, you will be able to recognize a malfunctioning component and make repairs or replace parts as quickly as possible. A review of magnetic theory is provided to help you understand the operation of the DC motor.

12.2.1 Magnetic Theory

DC motors operate on the principles of basic magnetism. You should remember that a coil of wire can be magnetized when current is passed through it. When this principle was used in relay coils, the polarity of the current was not important. When the current is passed through a coil of wire to make a field coil for a motor, the polarity of the current will determine the direction of rotation for the motor.

The polarity of the current flowing through the coil of wire will determine the location of the north and south magnetic poles in the coil of wire. Another important principle involves the amount of current that is flowing through the coil. The amount of current was not important as long as enough current was present to move the armature of the relay or solenoid. In a DC motor, the amount of current in the windings will determine the speed (rpm) of the motor shaft and the amount of torque that it can produce.

The direction of current flow through a coil of wire will determine the magnetic polarity of the coil. The direction of current flow will determine which end of a coil of wire is negative or positive. This will determine which end of the coil will be the north pole of the magnet and which end will be the south pole. If you change the

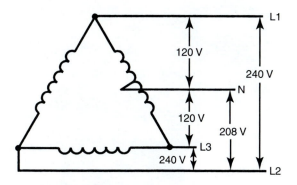

FIGURE 12-8 A high-leg delta voltage occurs between L2 and N because the center tap to produce the neutral is the center tap of the L1–L3 winding.

direction of the current flow in the coil of wire, the magnetic poles will be reversed in the coil. This is important to understand because the direction of the motor's rotation is determined by the changing magnetic field.

Another basic concept about magnets that you should remember is the relationship between two like poles and two unlike poles. When the north poles of two different magnets are placed close together, they will repel each other. When the north pole of one magnet is placed near the south pole of another magnet, the two poles will attract each other very strongly.

The third principle that is important to understand with the coil of wire is that the strength of the magnetic field can be varied by changing the amount of current flowing through the wire in the coil. If a small amount of current is flowing, a small number of flux lines will be created and the magnetic field will be relatively weak. If the amount of current is increased, the magnetic field will become stronger. The strength of the magnetic field can be increased to the point of saturation. A magnetic coil is said to be saturated when its magnetic strength cannot be increased by adding more current.

Saturation is similar to filling a drinking glass with water. You cannot get the level of the glass any higher than full. Any additional water that is put into the glass when it is full will not increase the amount of water in the glass. The additional water will run over the side of the glass and be wasted. The same principle can be applied to a magnetic coil. When the strength of the magnetic field is at its strongest point, additional electric current will not cause the field to become any stronger.

12.2.2 DC Motor Theory

The DC motor has two basic parts, which include the rotating part that is called the *armature,* and the stationary part that includes coils of wire that are called the *field coils.* The stationary part is also called the *stator.* Figure 12-9 shows a typical DC motor, Figure 12-10 shows a DC armature, and Figure 12-11 shows a typical stator. From Figure 12-10 you can see that the armature is made of coils of wire wrapped around a core, and the

FIGURE 12-10 An armature from a DC motor.

core has an extended shaft that rotates on bearings. You should also notice that the ends of each coil of wire on the armature terminate at one end of the armature. The termination points are called the *commutator,* and this is where the *brushes* make electrical contact to bring electrical current from the stationary part to the rotating part of the machine.

The picture of the stator in Figure 12-11 shows the location of the coils that are mounted inside it. These coils will be referred to as field coils in future discussions, and they may be connected in series or parallel with each other to create changes of torque in the motor. You will find the size of wire in these coils and the number of turns of wire in the coil will depend on the effect that is to be achieved.

It will be easier to understand the operation of the DC motor from a basic diagram that shows the magnetic interaction between the rotating armature and the stationary field coils. Figure 12-12 shows three diagrams that explain the DC motors' operation in terms of the magnetic interaction. In Figure 12-12a a bar magnet has been mounted on a shaft so that it can spin. The field winding is one long coil of wire that has been separated into two sections. The top section is connected to the positive pole of the battery, and the bottom section is connected to the negative pole of the battery. It is important to understand that the battery represents a source of voltage for this winding. In the actual industrial type

FIGURE 12-9 A DC motor.

FIGURE 12-11 DC motor stator.

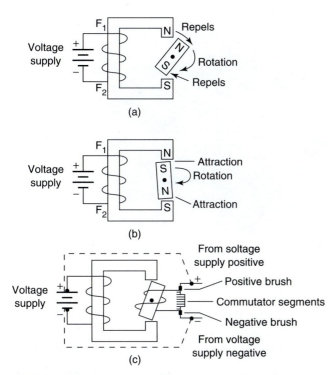

FIGURE 12-12 (a) Magnetic diagram that explains the operation of a DC motor. The rotating magnet moves clockwise because like poles repel. (b) The rotating magnet is being attracted because the poles are unlike. (c) The rotating magnet is now shown as the armature coil, and its polarity is determined by the brushes and commutator segments.

motor this voltage will come from the DC voltage source for the motor. The current flow in this direction makes the top coil the north pole of the magnet and the bottom coil the south pole of the magnet.

The bar magnet represents the armature and the coil of wire represents the field. The arrow shows the direction of the armature's rotation. Notice that the arrow shows the armature starting to rotate in the clockwise direction. The north pole of the field is repelling the north pole of the armature, and the south pole of the field coil is repelling the south pole of the armature.

As the armature begins to move, the north pole of the armature comes closer to the south pole of the field, and the south pole of the armature comes closer to the north pole of the field. As the two unlike poles come near each other, they begin to attract. This attraction becomes stronger until the armature's north pole moves directly in line with the field's south pole, and its south pole moves directly in line with the field's north pole (Figure 12-12b).

When the opposite poles are at their strongest attraction, the armature will be "locked up" and will resist further attempts to continue spinning. For the armature to continue its rotation, the armature's polarity must be switched. Since the armature in this diagram is a permanent magnet, you can see that it would lock up during the first rotation and not work. If the armature is an electromagnet, its polarity can be changed by changing the direction of current flow through it. For this reason the

armature must be changed to a coil (electromagnet), and a set of commutator segments must be added to provide a means of making contact between the rotating member and the stationary member. One commutator segment is provided for each terminal of the each magnetic coil. Since this armature has only one coil, it will have only two terminals, so the commutator has two segments.

Since the armature is now a coil of wire, it will need DC current flowing through it to become magnetized. This presents another problem; since the armature will be rotating, the DC voltage wires cannot be connected directly to the armature coil. A stationary set of carbon brushes is used to make contact to the rotating armature. The brushes ride on the commutator segments to make contact so that current will flow through the armature coil.

Figure 12-12c shows the DC voltage applied to the field and to the brushes. Since negative DC voltage is connected to one of the brushes, the commutator segment on which the negative brush rides will also be negative. The armature's magnetic field causes the armature to begin to rotate. This time when the armature gets to the point where it is locked up with the magnetic field, the negative brush begins to touch the end of the armature coil that was previously positive and the positive brush begins to touch the end of the armature coil that was negative. This action switches the direction of current flow through the armature, which also switches the polarity of the armature coil's magnetic field at just the right time so that the repelling and attracting continue. The armature continues to switch its magnetic polarity twice during each rotation, which causes it to continually be attracted and repelled with the field poles.

This is a simple two-pole motor that is used primarily for instructional purposes. Since the motor has only two poles, it will operate rather roughly and not provide too much torque. Additional field poles and armature poles must be added to the motor for it to become useful for industry.

12.2.3 DC Motor Components

The armature and field in a DC motor can be wired three different ways to provide different amounts of torque or different types of speed control. The armature and field windings are designed slightly different for different types of DC motors. The three basic types of DC motors are the *series motor*, the *shunt motor*, and the *compound motor*. The series motor is designed to move large loads with high starting torque, in applications such as a crane motor or lift hoist. The shunt motor is designed slightly different since it is made for applications such as pumping fluids, where constant-speed characteristics are important. The compound motor is designed with some of the series motor's characteristics and some of the shunt motor's characteristics. This allows the compound motor to be used in applications where high starting torque and controlled operating speed are both required.

FIGURE 12-13 Exploded view of DC motor.

It is important that you understand the function and operation of the basic components of the DC motor, since motor controls will take advantage of these design characteristics to provide speed, torque, and direction of rotation control. Figure 12-13 shows an exploded view of a DC motor. The basic components include the armature assembly, which includes all rotating parts; the frame assembly, which houses the stationary field coils; and the end plates, which provide bearings for the motor shaft and a mounting point for the brush rigging. Each of these assemblies is explained in depth so that you will understand the design concepts used for motor control.

12.2.4 Armature

The armature is the part of a DC motor that rotates and provides energy at the end of the shaft. It is basically an electromagnet since it is a coil of wire that has to be specially designed to fit around core material on the shaft. The core of the armature is made of laminated steel and provides slots for the coils of wire to be pressed onto. Figure 12-14a is a sketch of a typical DC motor armature. Figure 12-14b shows the laminated steel core of the armature without any coils of wire. This gives you a better look at the core. The armature core is made of laminated steel to prevent the circulation of magnetic currents. If the core were solid, magnetic currents would be produced that would circulate in the core material near the surface and cause the core metal to heat up.

These magnetic currents are called eddy currents. When laminated steel sections are pressed together to make the core, the eddy currents cannot flow from one laminated segment to another, so they are effectively canceled out. The laminated core also prevents other magnetic losses called flux losses. These losses tend to make the magnetic field weaker so that more core material is required to obtain the same magnetic field

strengths. The flux losses and eddy current losses are grouped together by designers and called core losses. The laminated core is designed to allow the armature's

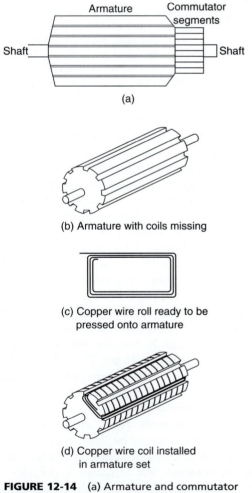

FIGURE 12-14 (a) Armature and commutator segments. (b) Armature prior to the coil's wire being installed. (c) Coil of wire prior to being pressed into the armature. (d) A coil pressed into the armature. The end of each coil is attached to a commutator segment.

magnetic field to be as strong as possible since the laminations prevent core losses. Notice that one end of the core has commutator segments. There is one commutator segment for each end of each coil. This means that an armature with four coils will have eight commutator segments. The commutator segments are used as a contact point between the stationary brushes and the rotating armature. When each coil of wire is pressed onto the armature, the end of the coil is soldered to a specific commutator segment. This makes an electrical terminal point for the current that will flow from the brushes onto the commutator segment and finally through the coil of wire. Figure 12-14c shows the coil of wire before it is mounted in the armature slot, and Figure 12–14d shows the coil mounted in the armature slot and soldered to the commutator segment.

The shaft is designed so that the laminated armature segments can be pressed onto it easily. It is also machined to provide a surface for a main bearing to be pressed on at each end. The bearing will ride in the end plates and support the armature when it begins to rotate. One end of the shaft is also longer than the other since it will provide the mounting shaft for the motor's load to be attached. Some shafts have a key way or flat spot machined into them so that the load that is mounted on it can be secured. You must be careful when handling a motor that you do not damage the shaft since it must be smooth to accept the coupling mechanism. It is also possible to bend the shaft or cause damage to the bearings so that the motor will vibrate when it is operating at high speed. The commutator is made of copper. A thin section of insulation is placed between each commutator segment. This effectively isolates each commutator segment from all others.

12.2.5 Motor Frame

The armature is placed inside the frame of the motor where the field coils are mounted. When the field coils and the armature coils become magnetized, the armature will begin to rotate. The field winding is made by coiling up a long piece of wire. The wire is mounted on laminated pole pieces called field poles. Similar to an armature these poles are made of laminated steel or cast iron to prevent eddy current and other flux losses. Figure 12-15a shows the location of the pole pieces inside a DC motor frame, and Figure 12-15b shows the laminated field core for the DC motor.

The amount of wire that is used to make the field winding will depend on the type of motor that is being manufactured. A series motor uses heavy-gage wire for its field winding so that it can handle the large field currents. Since the wire is a large gage, the number of turns of wire in the coil will be limited. If the field winding is designed for a shunt motor, it will be made of small-gage wire and many turns can be used.

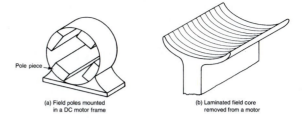

FIGURE 12-15 (a) The location of the pole pieces in the frame of a DC motor. (b) An individual pole piece. You can see that it is made of laminated sections. The field coils are wound around the pole pieces.

After the coils are wound, they are coated for protection against moisture and other environmental elements. After they have been pressed onto the field poles, they must be secured with shims or bolts so that they are held rigidly in place. Remember that when current is passed through the coil, it will become strongly magnetized and attract and repel the armature magnetic poles. If the field poles are not rigidly secured, they will be pulled loose when they are attracted to the armature's magnetic field and then pressed back into place when they become repelled. This action will cause the field to vibrate, damage the outer protective insulation, and cause a short circuit or a ground condition between the winding and the frame of the motor.

The ends of the frame are machined so that the end plates will mount firmly into place. An access hole is also provided in the side of the frame or in the end plates so that the field wires can be brought to the outside of the motor, where DC voltage can be connected.

The bottom of the frame has the mounting bracket attached. The bracket has a set of holes or slots provided so that the motor can be bolted down and securely mounted on the machine it is driving. The mounting holes will be designed to specifications by frame size. The dimensions for the frame sizes are provided in tables printed by motor manufacturers. Since these holes and slots are designed to a standard, you can pre-drill the mounting holes in the machinery before the motor is put in place. The slots are used to provide minor adjustments to the mounting alignment when the motor is used in belt-driven or chain-driven applications. It is also important to have a small amount of mounting adjustment when the motor is used in direct-drive applications. It is very important that the motor be mounted so that the armature shaft can turn freely and not bind with the load.

12.2.6 End Plates

The end plates of the motor are mounted on the ends of the motor frame. The end plates are held in place by four bolts that pass through the motor frame. The bolts can be removed from the frame completely so that the end plates can easily be removed for maintenance. The end plates also house the bearings for the armature shaft. These bearings can be either sleeve or ball type. If the bearing is a ball-bearing type, it is normally permanently lubricated; if it is a sleeve bearing, it will require a light film of oil to operate properly. The end plates that house a sleeve bearing will have a lubrication tube and wicking material. Several drops of lubricating oil are poured down the lubrication tube, where they will saturate the wicking material. The wicking is located in the bearing sleeve so that it can make contact with the armature shaft and transfer a light film of oil to it. Other types of sleeve bearings are made of porous metal so that they can absorb oil to be used to create a film between the bearing and the shaft. It is important that the end plate for a sleeve bearing be mounted on the motor frame so that the lubricating tube is pointing up. This position will ensure that gravity will pull the oil to the wicking material. If the end plates are mounted so that the lubricating tube is pointing down, the oil will flow away from the wicking and it will become dry. When the wicking dries out, the armature shaft will rub directly on the metal in the sleeve bearing, which will cause it to quickly heat up, and the shaft will seize to the bearing. For this reason it is also important to follow lubrication instructions and oil the motor on a regular basis.

12.2.7 Brushes and Brush Rigging

The brush rigging is an assembly that securely holds the brushes in place so that they will be able to ride on the commutator. It is mounted on the rear end plate so that the brushes will be accessible by removing the end plate. An access hole is also provided in the motor frame so that the brushes can be adjusted slightly when the motor is initially set up. The brush rigging uses a spring to provide the proper amount of tension on the brushes so that they make proper contact with the commutator. If the tension is too light, the brushes will bounce and arc; if the tension is too heavy, the brushes will wear down prematurely.

Figure 12-16 shows the location of a carbon brush as it touches the commutator segments, and brush rigging that holds the brushes in place on the DC motor is shown in Figure 12-17. Notice that it is mounted on the rear end plate. Since the rigging is made of metal, it must be insulated electrically when it is mounted on the end plate. The DC voltage that is used to energize the armature will pass through the brushes to the commutator segments and into the armature coils. Each brush has a wire connected to it. The wires will be connected to either the positive or negative terminal of the DC power supply. The motor will always have an even number of brushes. Half the brushes will be connected to positive voltage and half will be connected to negative voltage. In most motors the number of brush

FIGURE 12-16 The position where a carbon brush rides on the commutator segments.

sets will be equal to the number of field poles. It is important to remember that the voltage polarity will remain constant on each brush. This means that for each pair, one of the brushes will be connected to the positive power terminal, and the other will be connected permanently to the negative terminal.

The brushes will cause the polarity of each armature segment to alternate from positive to negative. When the armature is spinning, each commutator segment will

FIGURE 12-17 Brush rigging holds carbon brushes in place so they ride on the commutator. The spring on the top ensures that the brushes are held against the commutator with the correct amount of pressure.

come in contact with a positive brush for an instant and will be positive during that time. As the armature rotates slightly, that commutator segment will come in contact with a brush that is connected to the negative voltage supply and it will become negative during that time. As the armature continues to spin, each commutator segment will be alternately powered by positive and then negative voltage.

The brushes are made of carbon-composite material. Usually they have copper added to aid in conduction. Other material is also added to make them wear longer. The end of the brush that rides on the commutator is contoured to fit the commutator exactly so that current will transfer easily. The process of contouring the brush to the commutator is called seating. Whenever a set of new brushes is installed, it should be seated to fit the commutator. The brushes are the main part of the DC motor that will wear out. It is important that their wear be monitored closely so that they do not damage the commutator segments. Most brushes have a small mark on them called a wear mark or wear bar. When a brush wears down to the mark, it should be replaced. If the brushes begin to wear excessively or do not fit properly on the commutator, they will heat up and damage the brush rigging and spring mechanism. If the brushes have been overheated, they can cause burn marks or pitting on the commutator segments and also warp the spring mechanism so that it will no longer hold the brushes with the proper amount of tension.

If the spring mechanism has been overheated, it should be replaced and the brushes should be checked for proper operation. If the commutator is pitted, it can be turned down on a lathe. After the commutator has been turned down, the brushes will need to be reseated.

After you have an understanding of the function of each of the parts or assemblies of the motor, you will be able to better understand the operation of a basic DC motor. Operation of the motor involves the interaction of all the motor parts. Some of the parts will be altered slightly for specific motor applications. These changes will become evident when the motor's basic operation is explained.

12.2.8 DC Motor Operation

The DC motor you will find in modern industrial applications operates similarly to the simple DC motor described earlier in this chapter. Figure 12-18 shows an electrical diagram of a simple DC motor. Notice that the DC voltage is applied directly to the field winding and the brushes. The armature and the field are both shown as a coil of wire. In later diagrams a field resistor will be added in series with the field to control the motor speed.

When voltage is applied to the motor, current begins to flow through the field coil from the negative terminal to the positive terminal. This sets up a strong magnetic field in the field winding. Current also begins

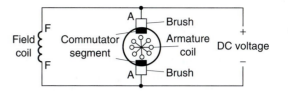

FIGURE 12-18 Simple electrical diagram of a DC shunt motor. This diagram shows the electrical relationship between the field coil and armature.

to flow through the brushes into a commutator segment and then through an armature coil. The current continues to flow through the coil back to the brush that is attached to other end of the coil and returns to the DC power source. The current flowing in the armature coil sets up a strong magnetic field in the armature.

The magnetic field in the armature and field coil causes the armature to begin to rotate. This occurs from the unlike magnetic poles attracting each other and the like magnetic poles repelling each other. As the armature begins to rotate, the commutator segments will also begin to move under the brushes. As an individual commutator segment moves under the brush connected to positive voltage, it will become positive; when it moves under a brush connected to negative voltage it will become negative. In this way the commutator segments continually change polarity from positive to negative. Since the commutator segments are connected to the ends of the wires that make up the field winding in the armature, it causes the magnetic field in the armature to change polarity continually from the north pole to the south pole. The commutator segments and brushes are aligned in such a way that the switch in polarity of the armature coincides with the location of the armature's magnetic field and the field winding's magnetic field. The switching action is timed so that the armature will not lock up magnetically with the field. Instead the magnetic fields tend to build on each other and provide additional torque to keep the motor shaft rotating.

When the voltage is de-energized to the motor, the magnetic fields in the armature and the field winding will quickly diminish and the armature shaft's speed will begin to drop to zero. If voltage is applied to the motor again, the magnetic fields will strengthen and the armature will begin to rotate again.

12.2.9 Types of DC Motors

Three basic types of DC motors are used in solar energy systems today: the series motor, the shunt motor, and the compound motor. The series motor is capable of starting with a large load attached, such as solar panel tracking systems. The shunt motor is able to operate with rpm control while it is at high speed.

The compound motor, a combination of the series motor and the shunt motor, is able to start with fairly large loads and have some rpm control at higher speeds.

In the remaining parts of this section we show a diagram for each of these motors and discuss their operational characteristics. As a technician you should understand methods of controlling their speed and ways to change the direction of rotation, because these are the two parameters of a DC motor you will be asked to change for drive motors that control solar panel positioning or for some pump motors. It is also important to understand the basic theory of operation of these motors because you will be controlling them with solid state electronic circuits and you will need to know if problems that arise are the fault of the motor or the solid state circuit.

12.2.10 DC Series Motors

The series motor provides high starting torque and is able to move large shaft loads when it is first energized. The DC series motor may be used in solar energy system applications where very large torque loads are required. For example the axis drives for large solar tracking systems may use a DC series motor. Figure 12-19 shows the wiring diagram of a series motor. The field winding in this motor is wired in series with the armature winding. This is the attribute that gives the series motor its name.

Since the series field winding is connected in series with the armature, it will carry the same amount of current that passes through the armature. For this reason the field is made from heavy-gage wire that is large enough to carry the load. Since the wire gage is so large, the winding will have only a few turns of wire. In some larger DC motors, the field winding is made from copper bar stock rather than the conventional round wire that you use for power distribution. The square or rectangular shape of the copper bar stock makes it fit more easily around the field pole pieces. It can also radiate more easily the heat that has built up in the winding due to the large amount of current being carried.

The amount of current that passes through the winding determines the amount of torque the motor shaft can produce. Since the series field is made of large conductors, it can carry large amounts of current and produce large torques. For example the starter motor that is used to start an automobile's engine is a series motor that can draw over 100 A when it is turning the engine's crankshaft on a cold morning. Series motors used to

power hoists or cranes can draw currents of thousands of amperes during operation.

The series motor can safely handle large currents since the motor does not operate for an extended period. In most applications the motor will operate for only a few seconds while this large current is present. Think about how long the starter motor on the automobile must operate to get the engine to start. This period is similar to that of industrial series motors.

12.2.11 Series Motor Operation

Operation of the series motor is rather easy to understand. The field winding is connected in series with the armature winding. This means that power will be applied to one end of the series field winding and to one end of the armature winding (connected at the brush).

When voltage is applied, current begins to flow from the negative power supply terminals through the series winding and armature winding and back to the positive terminal. The armature is not rotating when voltage is first applied, so the only resistance in this circuit will be provided by the large conductors used in the armature and field windings. Since these conductors are so large, they will have a small amount of resistance. This causes the motor to draw a large amount of current from the power supply. When the large current begins to flow through the field and armature windings, it causes a strong magnetic field to be built. Since the current is so large, it will cause the coils to reach saturation, which will produce the strongest magnetic field possible.

12.2.12 Reversing the Rotation of the Motor

The direction of rotation of a series motor can be changed by changing the polarity of either the armature or the field winding. It is important to remember that if you simply changed the polarity of the applied voltage, you would be changing the polarity of both field and armature windings and the motor's rotation would remain the same.

Since only one of the windings needs to be reversed, the armature winding is typically used because its terminals are readily accessible at the brush rigging. Remember that the armature receives its current through the brushes, so that if their polarity is changed, the armature's polarity will also be changed. A reversing motor starter is used to change wiring to cause the direction of the motor's rotation to change by changing the polarity of the armature windings. Figure 12-20 shows a DC series motor that is connected to a reversing motor starter. The armature's terminals are marked A1 and A2 and the field terminals are marked S1 and S2.

When the forward motor starter is energized, the top contact identified as F closes so the A1 terminal is connected to the positive terminal of the power supply, and the bottom F contact closes and connects terminal A2

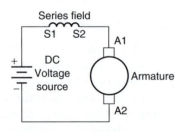

FIGURE 12-19 Electrical diagram of series motor. Notice that the series field is identified as S1 and S2.

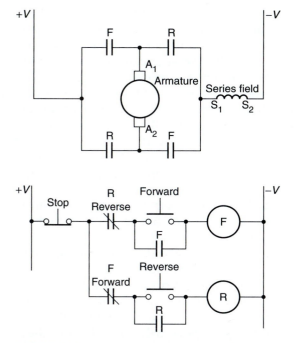

FIGURE 12-20 DC series motor connected to forward and reverse motor starter.

pushbutton is depressed. You will see a number of other ways to control the forward and reverse motor starter in later discussions and in the chapter on motor controls.

12.2.13 DC Series Motor Used as a Universal Motor

The series motor is used in a wide variety of power tools, such as electric hand drill, saws, and power screwdrivers. It is also used in solar energy applications for small pumps and fans. In most of these cases the power source for the motor is AC voltage. The DC series motor will operate on AC voltage. If the motor is used in a hand drill that needs variable speed control, a field rheostat or other type of current control is used to control the speed of the motor. In some newer tools the current control uses solid-state components to control the speed of the motor. The motors used for these types of power tools have brushes and a commutator, and these are the main parts of the motor that will wear out. You can use the same theory of operation provided for the DC motor to troubleshoot these types of motors.

12.2.14 DC Shunt Motors

The shunt motor is different from the series motor in that the field winding is connected in parallel with the armature instead of in series. A parallel circuit is often referred to as a shunt. Since the field winding is placed in parallel with the armature, it is called a shunt winding and the motor is a shunt motor. Figure 12-21 is a diagram of a shunt motor. Notice that the field terminals are marked F1 and F2, and the armature terminals are marked A1 and A2. The shunt field is represented with multiple turns using a thin line.

The shunt winding is made of small gage-wire with many turns on the coil. Since the wire is so small, the coil can have thousands of turns and still fit in the slots. The small-gage wire cannot handle as much current as the heavy-gage wire in the series field, but since this coil has many more turns of wire, it can still produce a strong magnetic field.

12.2.15 Shunt Motor Operation

The shunt motor can be used in pumps for solar energy heating systems and for drives for solar panel axis drives.

to S1. Terminal S2 is connected to the negative terminal of the power supply. When the reverse motor starter is energized, terminals A1 and A2 are reversed. A2 is now connected to the positive terminal. Notice that S2 remains connected to the negative terminal of the power supply terminal. This ensures that only the armature's polarity has been changed and the motor will begin to rotate in the opposite direction.

You will also notice the normally closed set of "R" contacts that are connected in series with the forward pushbutton, and the normally closed set of "F" contacts connected in series with the reverse pushbutton. These contacts provide an interlock that prevents the motor from being changed from forward to the reverse direction without stopping the motor. The circuit can be explained as follows: when the forward pushbutton is depressed, current will flow from the stop pushbutton, through the normally closed "R" interlock contacts, and through the forward pushbutton to the forward motor starter coil. When the forward motor starter coil is energized, it will open its normally closed contacts that are connected in series with the reverse pushbutton. This means that if someone depresses the reverse pushbutton, current could not flow to the reverse motor starter coil. If the person depressing the pushbuttons wants to reverse the direction of rotation of the motor, he or she will need to depress the stop pushbutton first to de-energize the forward motor starter coil, which will allow the normally closed "F" contacts to return to their normally closed position. When the reverse motor starter coil is energized, its normally closed "R" contacts that are connected in series with the forward pushbutton will open and prevent the current flow to the forward motor starter coil if the forward

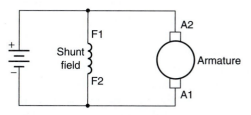

FIGURE 12-21 A DC shunt motor. The shunt coil is identified as a coil of fine wire with many turns that is connected in parallel (shunt) with the armature.

The shunt motor has slightly different operating characteristics than the series motor. Since the shunt field coil is made of fine wire, it cannot produce the large current for starting as does the series field. This means that the shunt motor has low starting torque, which requires that the shaft load be rather small.

When voltage is applied to the motor, the high resistance of the shunt coil keeps the overall current flow low. The armature for the shunt motor is similar to that of the series motor, and it will draw enough current to produce a magnetic field strong enough to cause the armature shaft and load to start turning.

Like the series motor, when the armature begins to turn, it will produce back electromotive force (EMF). The back EMF will cause the current in the armature to begin to diminish to a very small level. The amount of current the armature will draw is directly related to the size of the load when the motor reaches full speed. Since the load is generally small, the armature current will be small. When the motor reaches full rpm, its speed will remain rather constant.

12.2.16 Controlling the Speed of the Motor

When the shunt motor reaches full rpm, its speed will remain fairly constant. The speed remains constant because of the load characteristics of the armature and shunt coil. You should remember that the speed of a series motor could not be controlled since it was totally dependent on the size of the load in comparison to the size of the motor. If the load were large for the motor size, the speed of the armature would be very slow. If the load were light compared to the motor, the armature shaft speed would be much faster; if no load were present on the shaft, the motor could run away.

The shunt motor's speed can be controlled. The ability of the motor to maintain a set rpm at high speed when the load changes is due to the characteristic of the shunt field and armature. Since the armature begins to produce back EMF as soon as it starts to rotate, it will use the back EMF to maintain its rpm at high speed. If the load increases slightly and causes the armature shaft to slow down, less back EMF will be produced. This will allow the difference between the back EMF and applied voltage to become larger, which will cause more current to flow. The extra current provided to the motor with the extra torque is required to regain the armature's rpm when this load increases slightly.

The shunt motor's speed can be varied in two ways. These include varying the amount of current supplied to the shunt field and controlling the amount of current supplied to the armature. Controlling the current to the shunt field allows the rpm to be changed 10% to 20% when the motor is at full rpm.

This type of speed control regulation is accomplished by slightly increasing or decreasing the voltage applied to the field. The armature continues to have full voltage applied to it, while the current to the shunt field is regulated by a rheostat that is connected in series with the shunt field. When the shunt field's current is decreased, the motor's rpm will increase slightly. When the shunt field's current is reduced, the armature must rotate faster to produce the same amount of back EMF to keep the load turning. If the shunt field current is increased slightly, the armature can rotate at a slower rpm and maintain the amount of back EMF to produce the armature current to drive the load. The field current can be adjusted with a field rheostat or a silicon controlled rectifier (SCR) current control.

The shunt motor's rpm can also be controlled by regulating the voltage that is applied to the motor armature. This means that if the motor is operated on less voltage than is shown on its data plate rating, it will run at less than full rpm. You must remember that the shunt motor's efficiency will drop off drastically when it is operated below its rated voltage. The motor will tend to overheat when it is operated below full voltage, so motor ventilation must be provided. Remember that the motor's torque is reduced when it is operated below the full voltage level.

Since the armature draws more current than the shunt field, the control resistors would be much larger than those used for the field rheostat. SCRs are used for this type of current control. The SCR is able to control the armature current since it was capable of controlling several hundred amperes.

12.2.17 Reversing the Rotation of the Motor

The direction of rotation of a DC shunt motor can be reversed by changing the polarity of either the armature coil or the field coil. In this application the armature coil is usually changed, as was the case with the series motor. Figure 12-22 shows the electrical diagram of a DC shunt motor connected to a forward and reversing motor starter. You should notice that the F1 and F2 terminals of the shunt field are connected directly to the power supply, and the A1 and A2 terminals of the armature winding are connected to the reversing starter.

When the forward starter is energized, its contacts connect the A1 lead to the positive power supply terminal and the A2 lead to the negative power supply terminal. The F1 motor lead is connected directly to the positive terminal of the power supply, and the F2 lead is connected to the negative terminal. When the motor is wired in this configuration, it will begin to run in the forward direction. The control part of the diagram shows that when the forward motor starter coil is energized, the reverse motor starter coil is locked out.

When the reversing starter is energized, its contacts reverse the armature wires so that the A1 lead is connected to the negative power supply terminal and the A2 lead is connected to the positive power supply terminal. The field leads are connected directly to the power supply, so

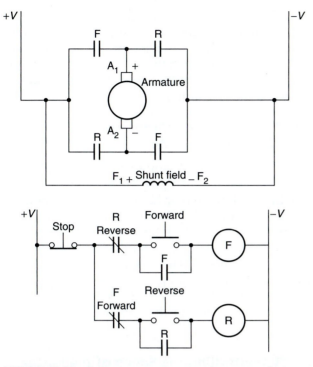

FIGURE 12-22 A shunt motor connected to a reversing motor starter. Notice that the shunt field is connected across the armature, and it is not reversed when the armature is reversed.

their polarity is not changed. Since the field's polarity has remained the same and the armature's polarity has reversed, the motor will begin to rotate in the reverse direction. The control part of the diagram shows that when the forward motor starter coil is energized, the reverse motor starter coil is locked out.

12.2.18 DC Compound Motors

The DC compound motor is a combination of the series motor and the shunt motor. It has a series field winding that is connected in series with the armature and a shunt field that is in parallel with the armature. The combination of series and shunt windings allows the motor to have torque characteristics of the series motor and regulated speed characteristics of the shunt motor. Figure 12-23 is a diagram of the compound motor. Several versions of the compound motor are also shown in this diagram.

12.2.19 Cumulative Compound Motors

Figure 12-23a shows a diagram of the cumulative compound motor. It is called cumulative because the shunt field is connected so that its coils are aiding the magnetic fields of the series field and armature. The shunt winding can be wired as a long shunt or as a short shunt. Figure 12-23a and Figure 12-23b show where the shunt field is connected in parallel with only the armature; this means that the motor is connected as a short shunt DC motor. Figure 12-23c shows the motor connected as a long shunt where the shunt field is connected in parallel with the series field, interpoles, and the armature.

Figure 12-23a also shows the short shunt motor as a cumulative compound motor, which means that the polarity of the shunt field matches the polarity of the armature. You can see in this figure that the top of the shunt field is positive polarity and it is connected to the positive terminal of the armature. In Figure 12-23b you can see that the shunt field has been reversed so that the negative terminal of the shunt field is now connected to the positive terminal of the armature. This type of motor is a differential compound motor because the polarities of the shunt field and the armature are opposite.

The cumulative compound motor is one of the most common DC motors because it provides high starting torque and good speed regulation at high speeds. It is called cumulative because the shunt field is wired with similar polarity in parallel with the magnetic field aiding the series field and armature field. When the motor is

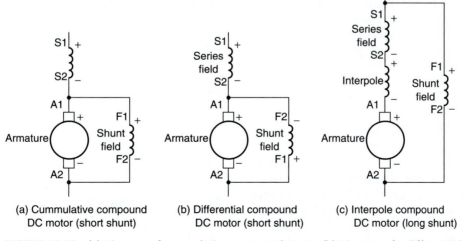

(a) Cummulative compound DC motor (short shunt)

(b) Differential compound DC motor (short shunt)

(c) Interpole compound DC motor (long shunt)

FIGURE 12-23 (a) Diagram of a cumulative compound motor. (b) Diagram of a differential compound motor. (c) Diagram of an interpole compound motor.

connected this way, it can start even with a large load and then operate smoothly when the load varies slightly.

You should recall that the shunt motor can provide smooth operation at full speed, but it cannot start with a large load attached, and that the series motor can start with a heavy load, but its speed cannot be controlled. The cumulative compound motor takes the best characteristics of both the series motor and shunt motor, which makes it acceptable to most applications.

12.2.20 Differential Compound Motors

Differential compound motors use the same motor and windings as the cumulative compound motor, but they are connected to them in a slightly different manner to provide slightly different operating speed and torque characteristics. Figure 12-23b shows a differential compound motor with the shunt field connected so its polarity is reversed to the polarity of the armature. Since the shunt field is still connected in parallel with only the armature, it is considered a short shunt.

In this diagram you should notice the F1 and F2 are connected in reverse polarity to the armature. In the differential compound motor the shunt field is connected so that its magnetic field opposes the magnetic fields in the armature and series field. When the shunt field's polarity is reversed like this, its field will oppose the other fields and the characteristics of the shunt motor are not as pronounced in this motor. This means that the motor will tend to over speed when the load is reduced just like a series motor. Its speed will also drop more than that of the cumulative compound motor when the load increases at full rpm. These two characteristics make the differential motor less desirable than the cumulative motor for most applications.

12.2.21 Interpole Compound Motors

The interpole compound motor is build slightly differently from the cumulative and differential compound motors. This motor has interpoles added to the series field (Figure 12-23c). The interpoles are connected in series between the armature and series winding. It is physically located behind the series coil in the stator. It is made of wire that is the same gage as the series winding, and it is connected so that its polarity is the same as the series winding pole it is mounted behind. Remember that these motors may have any number of poles to make the field stronger.

The interpole prevents the armature and brushes from arcing due to the buildup of magnetic forces. These forces are created from counter EMF and are called armature reaction. They are so effective that normally all DC compound motors that are larger than 1/2 hp will utilize them. Since the brushes do not arc, they will last longer and the armature will not need to be cut down as often. The interpoles also allow the armature to draw heavier currents and carry larger shaft loads.

When the interpoles are connected, they must be tested carefully to determine their polarity so that it can be matched with the series winding. If the polarity of the interpoles does not match the series winding it is mounted behind, it will cause the motor to overheat and may damage the series winding.

12.2.22 Reversing the Rotation of the DC Compound Motor

Each of the compound motors shown in Figure 12-23 can be reversed by changing the polarity of the armature winding. If the motor has interpoles, the polarity of the interpole must be changed when the armature's polarity is changed. Since the interpole is connected in series with the armature, it will be reversed when the armature is reversed. The interpoles are not shown in the diagram to keep it simplified. The armature winding is always marked as A1 and A2, and these terminals should be connected to the contacts of the reversing motor starter.

12.2.23 Controlling the Speed of the Motor

The speed of a compound motor can be changed easily by adjusting the amount of voltage applied to it. In fact it can be generalized that prior to the late 1970s, any industrial application that required a motor to have a constant speed would be handled by an AC motor, and any application that required the load to be driven at variable speeds would automatically be handled by a DC motor. This statement was true because the speed of a DC motor was easier to change than that of an AC motor. Since the advent of solid-state components and microprocessor controls, this condition is no longer true. Today a solid-state AC variable frequency motor drive can vary the speed of an AC motor as easily as that of a DC motor. Now you must understand methods of controlling the speed of both AC and DC motors.

12.3 AC MOTORS

AC motors are similar to DC motors except that they generally do not use brushes or commutator segments; instead the current that makes the magnetic fields in the rotating member is created by induction, similar to a transformer. This section will help you understand the theory of operation and the basic parts of AC induction motors. The only AC motor that uses brushes is the wound rotor motor that uses slip rings instead of commutator segments. You will learn more about all of these motors in this section.

12.3.1 Characteristics of Three-Phase AC Voltage

When you are trying to understand the basic operation of a three-phase AC induction motor, it is important to understand the operation of AC voltage. The AC induction

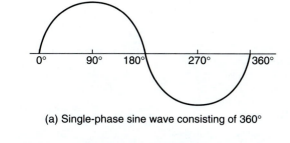

(a) Single-phase sine wave consisting of 360°

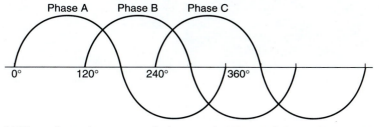

(b) Three-phase sine wave; each sine wave is 120° out of phase with the next

FIGURE 12-24   (a) Diagram of single-phase voltage. (b) Diagram of three-phase voltage.

motor is designed specifically to take advantage of the characteristics of the three-phase voltage that it uses for power. Figure 12-24a shows a diagram of single-phase voltage and Figure 12-24b shows a diagram of three-phase voltage. Note in Figure 12-24b that each of the a diagram of three phases represents a separately generated voltage like the single-phase voltage. The three separate voltages are produced out of phase with one another by 120°. The units of measure for this voltage are electrical degrees. The sine wave has 360°. The sine wave may be produced once during each rotation of the generator's shaft rotating in 360°. If the sine wave is produced by one rotation of the generator's shaft, 360 electrical degrees are equal to 360 mechanical degrees. All of the discussions in this section will be based on 360 electrical degrees being equal to 360 mechanical degrees.

The first voltage shown in the diagram is called A phase; it is shown starting at 0° and peaking positively at the 90° mark. It passes through 0 V again at the 180° mark and peaks negatively at the 270° mark. After it peaks negatively, it returns to 0 V at the 360° mark, which is also the 0° point. The second voltage is called B phase and starts its 0-voltage point 120° later than the A phase. The B phase peaks positively, passes through 0 voltage and negative peak voltage as the A phase does, except that it is always 120° later than the A phase. This means that the B phase is increasing in the positive direction when the A phase is passing through its 0 voltage at the 180° mark.

The third voltage shown on this diagram is the C phase. It starts at its 0-voltage point 240° after the A phase starts at its 0-voltage point. This puts the B phase 120° out of phase with the A phase and the C phase 120° out of phase with the B phase.

The AC motor takes advantage of this characteristic to provide a rotating magnetic field in its stator and rotor

that is very strong because three separate fields rotate 120° out of phase with one another. Since the magnetic fields are induced from the applied voltage, they will always be 120° out of phase with one another. The induced magnetic field is 180° out of phase with the voltage that induced it, which occurs naturally with induced voltages. The 180° phase difference is not as important as the 120° phase difference between the rotating magnetic fields, which helps create torque for the motor shaft.

Since the magnetic fields are 120° out of phase with one another and are rotating, one will always be increasing its strength when one of the other phases is losing its strength by passing through the 0-voltage point on its sine wave. This means that the magnetic field produced by all three phases never fully collapses, and its average is much stronger than that of a field produced by single-phase voltage. A detailed discussion of three-phase voltage and transformers is provided at the end of this chapter.

12.3.2 Basic Parts of a Three-Phase Motor

AC motors are available to operate on single-phase or three-phase supply voltage systems. Most single-phase motors are less than 3 hp; although some larger ones are available, they are not as common. Three-phase motors are available up to several thousand horsepower, although motors that are less than 50 hp are most common.

The AC induction motor has three basic parts: the stator, which is the stationary part of the motor; the rotor, which is the rotating part of the motor; and the end plates, which house the bearings that allow the rotor to rotate freely. This section will provide information about each of these parts of an AC motor. Figure 12-25 shows an exploded view of a three-phase motor that includes the location of its basic parts.

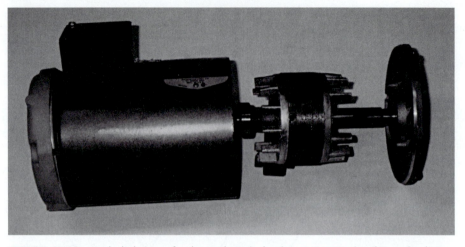

FIGURE 12-25 Exploded view of a three-phase induction motor. Notice the squirrel cage rotor with its shaft resting in the bearing in the end plate.

12.3.3 Stator

The stator is the stationary part of the motor and is made of several parts. Figure 12-26 shows the stator of a typical induction motor. The stator is the frame for the motor housing the stationary winding with mounting holes for installation. The mounting holes for the motor are sized according to National Electrical Manufacturers Association (NEMA) standards for the motor's frame type. Some motors will also have a lifting ring in the stator to provide a means for handling larger motors. The lifting ring and mounting holes are actually built into the frame or housing part of the stator.

An insert is set inside the stator that provides slots for the stator coils to be inserted. This insert is made of laminated steel to prevent eddy current and flux losses in the coils. The stator windings are made by wrapping a predetermined length of wire on preformed brackets in the shape of the coil. These windings are then wrapped with insulation and installed in the stator slots. A typical

FIGURE 12-26 The stator of an AC motor.

four-pole, three-phase motor will have three coils mounted consecutively in the slots to form a group. The three coils will be wired so that each receives power from a separate phase of three-phase power supply. Three groups are connected to form one of the four poles of the motor. This grouping is repeated for each of the other three poles so that the motor has a total of 36 coils to form the complete four-pole stator. It is not essential that you understand how to wind the coils or put them into the stator slots; rather you should understand that these coils are connected in the stator, and 3, 6, 9, 12, or 15 wires from the coil connections will be brought out of the frame as external connections. The external connection wires can be connected to allow the motor to be powered by 208/230 or 480 V, or they allow the motor to be connected to provide the correct torque response for the load. Other changes can also be made to these connections to allow the motor to start so it uses less locked-rotor current.

After the coils are placed in the stator, their ends (leads) will be identified by a number that is used to make connections during the installation procedure. The coils are locked into the stator with wedges that keep the coils securely mounted in the slots and allow them to be removed and replaced easily if the coils are damaged or become defective due to overheating.

12.3.4 Wound Rotor

The rotor in an AC motor can be constructed from coils of wire wound on laminated steel, or it can be made entirely from laminated steel without any wire coils. A rotor with wire coils is called a *wound rotor* and it is used in a wound-rotor motor. The wound-rotor motor can produce more torque than a similar-size induction motor because it uses brushes and slip rings to transfer current to the rotor. Since the wound-rotor motor has brushes and slip rings, it requires more maintenance than an induction motor. Figure 12-27 shows the two slip rings on the armature of a wound-rotor motor.

FIGURE 12-27   A wound-rotor armature with two slip rings.

FIGURE 12-29   Squirrel cage rotor for an induction motor.

12.3.5 Laminated Steel Rotor

The AC induction motors that use a laminated steel rotor are called *induction motors* or squirrel-cage induction motors. The core of the rotor is made of die-cast aluminum or copper in the shape of a cage. A diagram of a squirrel-caged rotor is shown in Figure 12-28. The rotor is called a caged rotor because the aluminum or copper bars are held in place by an end ring. The caged rotor is sometimes called a squirrel-caged rotor because the overall shape of the rotor is the same shape as an exercise wheel used for hamsters or squirrels.

Laminated sections are pressed onto this core or the core is molded into laminated sections when the squirrel cage rotor is manufactured. Figure 12-29 is a picture of a squirrel cage rotor and you can see that the laminated steel sections are pressed onto the skeleton of the cage core. Laminated steel is a thin piece of steel that has an insulation coating to prevent it from having an electrical connection with the piece of steel next to it. The reason the thin pieces of steel are pressed onto the rotor bars,

rather than the rotor bar being a solid piece of iron, is that the individual pieces of steel are easily magnetized to one polarity, demagnetized, and then re-magnetized to the opposite polarity. The changing magnetic field is very important when AC current is used to create the magnetic field in the rotor. If the rotor were made of solid iron or steel, it would magnetize easily, but it would not give up the magnetic field easily, which would become a problem when AC current is applied to the rotor.

The fins or blades are built into the rotor for cooling the motor and it is important that these fan blades are not damaged or broken, since they provide all of the cooling air for the motor and they are balanced so that the rotor will spin evenly without vibrations.

12.3.6 Motor End Plates

The bearings for the motor shaft are located in the end plates. If the motor is a fractional-horsepower motor, it will generally use sleeve-type bearings; if the motor is one of the larger types, it will use ball bearings. Some ball bearings on smaller motors will be permanently lubricated; the larger motor bearings will require periodic lubrication. All sleeve bearings will require a few drops of lubricating oil periodically.

The end plates are mounted on the ends of the motor and held in place by long bolts inserted through the stator frame. When nuts are placed on the bolts and tightened, the end plates will be secured in place. If the motor is an open type, the end plates will have louvers to allow cooling air to circulate through the motor. An access plate may also be provided in the rear end plate

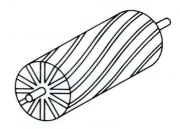

FIGURE 12-28   Diagram of a squirrel cage rotor.

to allow field wiring if one is not provided in the stator frame.

If the motor is not permanently lubricated, the end plate will provide an oiler tube or grease fitting for lubrication. It is important that the end plates are mounted on the motor so that the oiler tube or grease fitting is above the shaft so that gravity will allow lubrication to reach the shaft. If the end plate is rotated so that the lubrication point is mounted below the shaft, gravity will pull all the lubrication away from the shaft and the bearings will wear out prematurely. If you need to remove the end plates for any reason, they should be marked so that they will be replaced in the exact position from which they were removed. This also helps align the holes in the end plate with the holes in the stator so that the end plates can be reassembled easily.

12.3.7 Operation of a Wound-Rotor AC Motor

The wound-rotor motor has a set of coils that are mounted in the stationary part of the motor called the stator. The rotor for the wound-rotor motor is made with coils of wire pressed onto the rotor section. The terminal ends of each of these coils are connected to a slip ring, and a carbon brush rides on the slip ring. The reason brushes and slip rings are used is to get electrical current from the stationary part of the motor through the brushes to the slip ring, which is connected to the coil on the rotating part. If the wound-rotor motor is single phase, it will have two slip rings; if it is a three-phase motor, it will have three slip rings.

Since electrical current is provided to the rotor directly from the brushes to the slip rings, the magnetic field in the rotor is very strong. When power is applied to the motor, electrical current will flow through the stationary windings and create a strong magnetic field. Electrical current will flow from the brushes to the slip rings through the rotor coil, which will also become an electromagnet. Since the current supplied to the stationary windings is AC, it will cause a magnetic field to move around the sets of coils in a rotating fashion. The polarity of the coils in the rotor is opposite that of the coils in the stator, so the rotor's magnetic field will be attracted to the magnetic field that is rotating through the stationary coils. This will cause the rotor to begin to spin, or chase the rotating magnetic field in the stator. The speed of the rotor will be set by the frequency of the AC voltage and the number of field coils in the stator. Table 12-1 shows the speed of a 2-pole through a 12-pole motor when 60 Hz AC voltage is applied to them.

The main drawback of a wound-rotor motor is that the brushes will wear down, and slip rings will become rough due to electrical contact and electrical arcing. This means that the brushes will need to be changed and the slip rings will need to be resurfaced periodically.

12.3.8 Operation of an AC Induction Motor

The basic principle of operation of an inductive motor is based on the fact that the rotor receives its current by induction rather than with brushes and slip rings like the wound-rotor motor. The rotor in the induction motor is made of laminated steel, so it can easily be magnetized by an induced current, since it rotates in close proximity to the stator. When current is flowing through the stator coils, a strong magnetic field is created. When the laminated steel sections of the rotor pass through this magnetic field, current is induced into the laminated steel rotor, causing it to become magnetized. Just as in the wound-rotor motor, the magnetic field in the laminated steel rotor is the opposite polarity of the magnetic field in the stator. Since the magnetic field in the stator is caused by AC current flowing through it, it causes the magnetic field to rotate from coil to coil around the stator. The magnetic current that created the magnetic field in the laminated steel rotor will be 180° out of phase with the current that produced it. This will cause the magnetic field in the rotor to be the opposite polarity of the magnetic field rotating through the stator. Since the magnetic field in the rotor is the opposite polarity to the magnetic field rotating through the stator, it will cause the rotor magnetic field to be attracted to the stator magnetic field, which causes the rotor to begin to spin.

Since the induction motor uses induced current to create the magnetic field in the rotor, there are some extra losses that are not found in the wound-rotor motor, so the induction motor is slightly less efficient than the wound-rotor motor. Since the induction motor does not need brushes or slip rings, it will not need as much maintenance.

12.3.9 Using Slip to Create Torque in the Induction Motor

When the rotor in the induction motor begins to turn, it will begin to create a voltage that is out of phase with the voltage that is applied to the stator. This voltage is called an electromotive force (EMF), and since it is out of phase with the applied voltage it is called counter EMF or CEMF. The amount of electrical potential difference

TABLE 12-1 Rated motor speeds at 60 Hz by the number of poles.						
Motor Speeds at 60 Hz						
Number of poles	2 poles	4 poles	6 poles	8 poles	10 poles	12 poles
RPM	3600	1800	1200	900	720	600

between the applied voltage and the CEMF that the rotor creates will determine the amount of current that the motor draws. The amount of current that the motor draws will determine the amount of torque the motor shaft (rotor) will provide. The amount of CEMF is determined by the amount of slip the rotor has. *Slip* is the difference between the rated speed and the actual speed of a motor, and the more slip the motor has the more torque it can create at its shaft so it can move larger loads.

Theoretically the rotor and the induction motor should spin at the same rate as a wound-rotor motor, and the speed would be determined by the number of poles and the frequency of the applied voltage. In reality if the motor rotor turned at its rated speed, the rotor would produce a counter EMF that was equal to the applied voltage, so the amount of current the motor would draw would be zero. If the amount of current the motor is drawing were zero, the motor shaft would not have any torque and would not be useful. In order for the motor to have torque at its shaft, there must be a difference between the applied voltage and the counter EMF created by the rotor. This difference can only occur when the speed of the rotor is slower, then its rated speed. The larger the difference in speed, the more slip occurs, and the more torque the AC induction motor will have. Since the speed of the AC motor will be slightly less than its rated speed, it will be less efficient than a wound-rotor motor.

12.3.10 Synchronous AC Motor

Another type of AC motor is the synchronous motor, and it combines the strengths of the wound-rotor AC motor with the strengths of the induction AC motor. The synchronous motor has a wound rotor, which means it has coils of wire in the rotor and it has slip rings and brushes. It also has a small squirrel-cage winding embedded in the wound rotor. When AC voltage is first applied to the synchronous motor, it starts as an induction motor because the small squirrel-cage winding receives an induced current that causes the rotor to begin to turn and come up to speed. When the speed of its rotor is at about 95%, a small amount of DC current is applied to coils in the wound rotor. This small DC current flows through the brushes and slip rings, and they cause a strong magnetic field to be created in the rotor coils. Since the magnetic field in the rotor coils is strong, the rotor speed synchronizes with the rotating field that the AC voltage causes in the stator winding. This synchronization causes the rotor to spin at its rated speed based on the number of poles and the frequency of the AC voltage without any slip. This causes the synchronous motor shaft to provide maximum torque and makes it more efficient. Since the synchronous motor has brushes and slip rings, it will require some periodic maintenance like the wound-rotor motor.

12.4 WHY MOTOR STARTERS ARE USED

Motor starters are magnetically controlled devices that are usually used in larger commercial applications. The motor starter is a larger version of a relay or contactor and is used to control larger motors. The motor starter also has *overcurrent protection* for motors built into it, and relays and contactors do not. This overcurrent protection is called an *overload* and is sized to trip if the amount of current drawn by the motor exceeds the designated limit.

Earlier you learned about relays and contactors that are used to start motors in electrical systems. These relays and contactors are designed to close their contacts and provide current to the motors in the system. Their main function is to turn on and off to provide current to these motors. The motors in these applications are protected against overcurrent by fuses or internal overloads that are built into the winding.

These motors are protected by fuses or circuit breakers in the disconnect for short-circuit protection, and by the overloads in motor starters to protect against slow overcurrents. If an open motor has a problem such as loss of lubrication or overload, it will draw extra current, which will damage the motor if the overload is allowed to continue for any length of time. The overloads in the motor starter sense the excess current and trip the motor starter so that its contacts open and stop all current flow to the motor until a technician manually resets the overloads.

12.4.1 The Basic Parts of a Motor Starter

Figure 12-30 shows a typical motor starter with all its parts identified. This is a three-pole starter that is used in three-phase circuits. The incoming voltage is connected at the top at the terminals identified as L1, L2, and L3, and the motor leads are connected to the bottom terminals, which are identified as T1, T2, and T3. The three major sets of contacts are located in the top part of the motor starter, and the overload assembly is mounted in the lower part of the motor starter. The coil is located in the middle of the motor starter, and it has an indicator that shows the word *ON* when the coil is energized, and *OFF* when the coil is de-energized.

The motor starter has three sets of contacts that are in series with a heater. This ensures that all the current that flows to the motor must pass through a heater. If the current flowing to the motor is normal, the heater does not provide sufficient heat to cause the overload to trip. If the motor draws excess current, that current flowing through the heater will cause it to create excess heat that will trip the normally closed overload contacts. Since the normally closed overload contacts are connected in series with the motor starter coil, current to the motor starter coil will be interrupted when the overload contacts open. A reset button on the motor starter must be manually

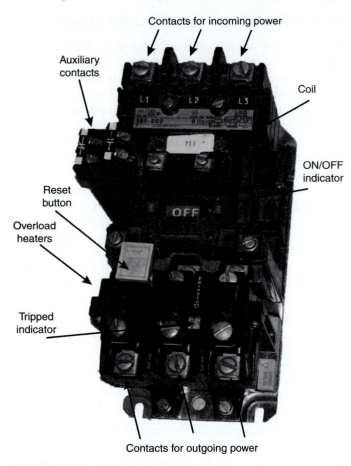

FIGURE 12-30 A typical motor starter.

pushed to set the overload contacts back to their normally closed position.

The control circuit is shown as a *ladder diagram* at the bottom of this figure. In this diagram you can see that the motor starter coil is energized when the start push button is depressed. The manual push button is used for this example because it is easier to understand, but you should be aware that most motor starters in electrical applications are controlled by manual switches or other types of controls.

The motor starter also has one or more additional sets of normally open contacts called *auxiliary contacts.* The auxiliary contacts are connected in parallel with the start push button. These contacts serve as a seal-in circuit after the motor starter coil is energized. The start push button is a momentary switch, which means that it is spring loaded in the normally open position. When the start push button is depressed, current flows from L1 through the normally closed stop push button contacts and through the start push button contacts to the motor starter coil. This current causes the coil to become magnetized so that it pulls in the three major sets of load contacts and the auxiliary set of contacts. When the auxiliary contacts close, they create a parallel path around the start push button contacts so that current still flows around the start push button contacts to the coil when the push

button is released. Since the stop push button is connected in series in this circuit, the current to the coil is de-energized and all of the contacts drop out when the stop push button is depressed.

12.4.2 The Operation of the Overload

The overload for a motor starter consists of two parts. The heater is the element that is connected in series with the motor, and all the motor current passes through it. The heater is actually a heating element that converts electrical current to heat. The second part of the overload device is the trip mechanism and overload contacts. The trip mechanism is sensitive to the heat. If it detects excess heat from the heater, it trips and causes the normally closed overload contacts to open. Since the motor starter coil is connected in series with the normally closed overload contacts, all current to the coil is interrupted and the coil becomes de-energized when the overload contacts are tripped to their open position. When the coil becomes de-energized, the motor starter contacts return to their open position, and all current to the motor is interrupted. When the overload contacts open, they remain open until the overload is reset manually. This ensures that the overloaded motor stops running and cools down until someone investigates the problem and resets the overloads.

Figure 12-31 shows a typical heating element for a motor starter overload device, which is called the heater. In the diagram for this figure you can see that the heater is actually a heating element that converts electrical current into heat as it passes through the heating element. You should also notice the knob protruding from the bottom of the heater. This knob has a shaft that is held in position inside the mechanism.

Figure 12-32 shows all of the parts of the overload assembly, including the reset button, the shaft the heating element mounts on, the ratchet and pawl mechanism, the overload contacts, and the terminals that connect to the contacts. The trip mechanism consists of the ratchet from the heater and a pawl. The ratchet is actually the toothed knob that protrudes from the bottom of the heater. The pawl has spring pressure that tries to rotate the ratchet. Because the heater holds the shaft of the ratchet tightly

FIGURE 12-31 A typical heater assembly. Notice the ratchet that protrudes from the bottom of the heater.

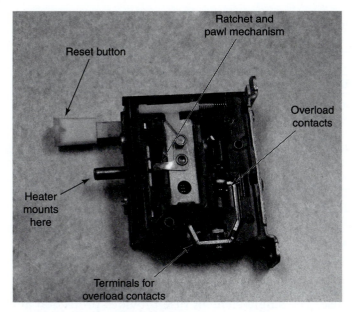

FIGURE 12-32 Overload with all its parts identified.

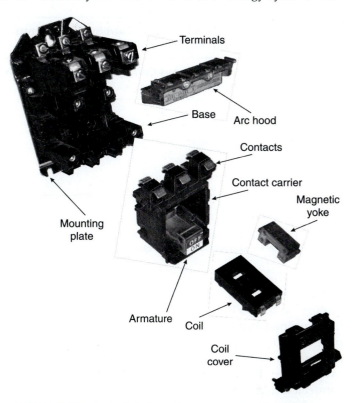

FIGURE 12-33 Exploded view of a motor starter. The stationary contacts are shown at the far left, and the movable contacts are shown to the right of the stationary contacts. The coil and magnetic yoke are shown in the middle of the picture. The overload mechanism is shown at the far right side of the picture.

with solder, the pawl cannot move. When the heater overheats, it melts the solder that holds the ratchet in place and allows it to spin freely. When the ratchet spins, it allows the pawl to move past it, which in turn allows the normally closed contacts to move to their open position.

After the overload condition has occurred and the overload contacts have opened, the motor starter is de-energized and the motor stops running. When the motor stops running, the heating element is allowed to cool down. After the heating element cools down for several seconds, the reset button can be depressed, which moves the pawl back to its original position, and the normally closed overload contacts move back to their closed position. If the motor continues to draw excess current when it is restarted, the excess current will cause the heater to trip the overload mechanism again. If the motor current is within specification, the heaters do not produce enough heat to cause the overload mechanism to trip.

Since the over current condition must last for several minutes to cause the overload mechanism to trip, the overloads allow the motor to draw high locked-rotor amperage (LRA) during the few seconds the motor is trying to start, without tripping. If the motor has an over current condition while it is running, the overloads allow the condition to last several minutes before the motor is de-energized. If the problem continues, the overloads will sense the over current and trip to protect the motor.

12.4.3 Exploded View of a Motor Starter

It is easier to see all the parts of the motor starter in an exploded view picture, as in Figure 12-33. At the far left you can see that the main contacts of the motor starter are much larger than those in a traditional relay. The coil is shown in the middle of the picture. The coil has two

square holes in it that allow the magnetic yoke to be mounted through it. The magnetic yoke and coil are mounted in the movable contact carrier. When the coil is energized, it pulls the magnetic yoke upward, which causes the movable contacts to move upward until they make contact with the stationary contacts that are shown at the far left. The overload mechanism is shown at the far right in this figure. It is mounted at the lower part of the motor starter, and all current that flows through the contacts must also flow through the overload mechanism.

12.4.4 Sizing Motor Starters

At times you will need to select the proper size motor starter for an application. The size of motor starters is determined by the National Electrical Manufacturers Association (NEMA). The ratings refer to the amount of current the motor starter contact can safely handle. The sizes are shown in the table provided in Figure 12-34. You can see that the smallest size starter is a size 00, which is rated for 9 A and is sufficient for a 2-hp motor connected to 480 V three phase or for a 1-hp single-phase motor connected to 240 V. You can see that the size 1 motor starter is rated for 27 A, which is sufficient for a 10-hp three-phase motor connected to 480 V or a $7\frac{1}{2}$-hp single-phase motor connected to 240 V. A size 00 motor starter is about 4 in. high, and a size 2 motor starter is

Horsepower (Hp) and Locked-Rotor Current (LRA) ratings for single-phase full-voltage magnetic controllers for limited plugging and jogging duty, 50 or 60 Hz

Size of Controller	Continuous Current rating (A)	115 V		230 V		Service Limit Current Rating* (A)
		Hp	LRA	Hp	LRA	
00	9	1/3	50	1	45	11
0	18	1	80	2	65	21
1	27	2	130	3	90	32
1P	36	3	140	5	135	42
2	45	3	250	7.5	250	52
3	90	7.5	500	15	500	104

*See clause 4.1.2.

FIGURE 12-34   NEMA sizes for single-phase full-voltage magnetic controllers. Notice that the smallest motor starter is a size 00, and the largest is a size 3.   (Courtesy National Electrical Manufacture Association NEMA.)

approximately 8 in. high. The size 4 starter will protect motors up to 100 hp. It is important to understand that the overload heaters for the motor starter can also be purchased for a specific current rating. This means that each motor starter can have a heater that is rated specifically to the amount of current that the motor that is connected to it draws.

12.5 FUSES

Fuses perform a function similar to that of an overload, except a fuse uses an element that is destroyed when the over current occurs. A fuse provides short-circuit protection, and the overload is not designed to protect the motor when short-circuit current occurs. The fuse provides a thermal sensing element that is capable of carrying current. When the amount of current becomes excessive, the heat that is generated is sensed by the fuse element, which melts when the temperature is high enough.

Fuses are available in a variety of sizes and shapes for different applications. Figure 12-35 shows various types of cartridge-type fuses. Each fuse is sized for the amount of current it will limit. When the amount of

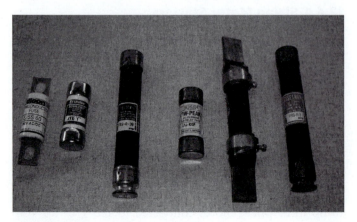

FIGURE 12-35   Various types of fuses.

current is exceeded, the fuse link melts and opens the fuse. The single-element fuse provides protection at a single level. This type of fuse is generally used for non-inductive loads such as heating elements or lighting applications. The dual-element fuse provides two levels of protection. The first level is called slow over current (overload) protection, and it consists of a fusible link that is soldered to a contact point and attached to a spring. When a motor is started, it draws locked-rotor amperage (LRA) for several seconds. This excess current causes heat to build in the fuse, and this heat is absorbed in the slow over current link. If the motor starts and the current drops to the full load amperage (FLA) level, the link will cool off, and the fuse will not open. If the LRA current continues for 30 seconds, the amount of heat generated will cause the solder that holds the slow over current link to melt. When the solder melts, the spring pulls the link open and interrupts all current flowing through the fuse. The slow over current link allows the fuse to sustain over current conditions for short periods of time, and if the condition clears, the fuse will not open. If the condition continues to exist, the fuse will open.

The second type of element is called the short-circuit element. This element opens immediately when the amount of current exceeds the level of current the link is designed to handle. In the dual-element fuse, the short-circuit link is sized to be approximately five times the rating of the fuse. A short circuit is by definition any current that exceeds the full-load current rating by five times. (Some manufactures use the rating of 10 times.)

The single-element fuse has only a short-circuit element in it. This type of fuse is generally not used in circuits to start motors, since the motor draws locked-rotor current. If a single-element fuse is used to protect a motor circuit, it must be sized large enough to allow the motor to start, and then it is generally too large to protect the motor against an over current condition of 20%, which will eventually damage the motor if it is allowed to occur for several hours.

FIGURE 12-36   A fused disconnect has the fuses in series with the remainder of the circuit.

12.5.1 Fused Disconnect Panels

The cartridge-type and screw-base fuses are generally mounted in a panel called a *fused disconnect*. The fused disconnect is normally mounted near the equipment it is protecting, and it serves two purposes. First, it provides a location to mount the fuses; second, it provides a means of disconnecting the electrical supply voltage to a circuit. A three-phase disconnect is shown in Figure 12-36. The fused disconnect has a switch handle that is used to disconnect main power from the circuit and the fuses so that you can safely remove and replace or test the fuses.

SAFETY NOTICE!

It is recommended that you always use plastic fuse pullers to remove fuses from and replace them in a disconnect to protect you from electrical shock even when the fuse disconnect switch is in the off position. It is important to remember that even though the switch handle is open, line voltage is still present at the top terminal lugs in the disconnect. You should never use metal pliers or screwdrivers to remove fuses.

12.5.2 Circuit Breakers and Load Centers

The *circuit breaker* is similar to the thermal overload in that it is an electromechanical device that senses both over current and excess heat. Some circuit breakers also sense magnetic forces. Circuit breakers are mounted in electrical panels called *load centers*. The load center can be designed for three-phase circuits or for single-phase circuits. The single-phase panel provides 240-V AC and 120-V AC circuits. Figure 12-37 shows a typical load center without any circuit breakers mounted in it.

Circuit breakers are manufactured in three basic configurations for single-phase 120-V AC applications that require one supply wire, 240-V AC single-phase applications that require two supply wires, and three-phase applications that require three wires. The three-phase circuit breakers can be mounted only in a load

FIGURE 12-37   A load center that is used to mount circuit breakers. The load center is sometimes called a circuit breaker panel.

center that is specifically manufactured for three-phase circuits. The two-pole and single-pole breakers can be mounted in a single-phase or three-phase load center.

The operation of the circuit breaker is similar to that of the thermal overload in that it senses excessive current that will trip its circuit after a specific amount of time. The main difference between the circuit breaker and the thermal overload is that the circuit breaker is mounted in the load center to protect both the circuit wires as well as the load. The circuit breakers are sized for the total current rating of the wire and all the loads that are connected to the wire. In some cases this means that the circuit breaker is sized for the current flowing to the motors. In these circuits it may be necessary to use a circuit breaker to protect the entire circuit, with overloads at each motor to protect them individually from overheating. This means that the main job of the circuit breaker is to protect the circuit against short-circuit conditions and to protect the entire circuit against over current conditions rather than to protect individual motors against over current. This is why many circuits have a combination of circuit breakers and overload devices. Figure 12-38 shows examples of single-phase and three-phase circuit breakers.

FIGURE 12-38   Single-pole, two-pole, and three-pole circuit breakers for load centers.

12.6 ELECTRONIC COMPONENTS USED IN INVERTERS AND CIRCUITS

Electronics are used in all areas of solar energy control systems, such as power electronic frequency converter (*inverter*) that converts DC voltage to AC voltage. Other electronic components are used in boards to control variable-frequency drives that control motor and pump speeds. Electronics make systems more efficient and more reliable. In this section you will find that electronic components provide functions that are similar to those electrical components and systems you already understand. The information in this section will provide names for all components and explain the operation of electronic devices such as diodes, which are the basic component for power supplies and inverters, transistors, and SCRs that have become commonplace in electrical systems because they provide better control than electromechanical devices and are less expensive to manufacture. Electronic components are also called solid-state components. You will also gain an understanding of P-material and N-material, which are the building blocks of all electronic components. After you get a basic understanding of P-material and N-material, you will be introduced to diodes, transistors, SCRs, and triacs, and you will see application circuits of each of these types of components. The theory of operation and troubleshooting techniques for each type of device will also be presented so that you will have a good understanding of how to determine whether a device or circuit board is working correctly. At first, most solar technicians think that it is difficult to troubleshoot electronic devices because you cannot see their operation. After you fully understand the theory of operation of each component and basic circuit, you will find it easy to determine if it is working correctly or if it is faulty and must be replaced. You will also find in this section that if you determine that a component or circuit is faulty, you will change out the complete device or board, rather than try to repair an individual component. You will also begin to understand that every component has a specific test that can prove that an electronic device is good or faulty.

12.6.1 Conductors, Insulators, and Semiconductors

Atoms have protons and neutrons in their nucleus and electrons that move around the nucleus in regions of space called shells. The number of electrons in the atom is different for each element. For example, you learned that copper has 29 electrons and three of them are located in the outermost shell. The outermost shell is called the valence shell and the electrons in that shell are called valence electrons. The atoms of the most stable material have eight valence electrons, which are found as four pairs. This means that an atom may have five, six, or seven valence electrons and it will take less energy to add

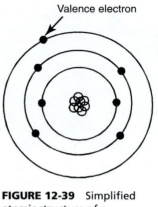

FIGURE 12-39 Simplified atomic structure of a conductor.

electrons to get a full shell (eight), or it may have one, two, or three electrons, and it will take less energy to give up these electrons to achieve the stable configuration of a full shell with eight electrons.

A conductor is a material that allows electrons (electrical current) to flow easily through it, and an insulator does not allow current to flow through it. An example of a conductor is copper that is used for electrical wiring. An example of an insulator is rubber or plastic. The atomic structure of a conductor makes it easier for electrons to flow through it, and the atomic structure of an insulator makes it nearly impossible for any electrons to flow through it.

Figure 12-39 shows a simplified atomic structure of a conductor. Atoms of conductors can have one, two, or three valence electrons. The atom in this example has one valence electron. Since all atoms will gain or lose sufficient electrons to achieve a configuration of eight electrons (four pairs) in their valence shell, it takes less energy for conductors to give up these electrons (one, two, or three) so that the valence shell will become empty. At this point the atom becomes stable because the previous shell becomes the new valence shell and it has eight electrons. The electrons that are given up are free to move as current flow.

Figure 12-40 shows the simplified atomic structure of an insulator. Insulators will have five, six, or seven

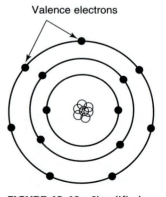

FIGURE 12-40 Simplified structure of an insulator.

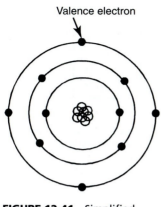

Valence electron

FIGURE 12-41 Simplified atomic structure of a semiconductor material.

valence electrons. In this example you can see that the atom has seven valence electrons. This structure makes it easy for insulators to take on extra electrons to get eight valence electrons. The electrons that are captured to fill the valence shell are electrons that would normally be free to flow as current.

Semiconductors are materials whose atoms have exactly four valence electrons. They can take on four new valence electrons like an insulator to get a full valence shell, or they can give up four valence electrons like a conductor to get an empty outer shell. Then the previous shell that has eight electrons becomes the valence shell. Figure 12-41 shows the simplified atomic structure of a semiconductor material.

12.6.2 Combining Atoms

When solid-state material or other material is manufactured, large numbers of atoms are placed together. The structure that becomes most stable at this point is called a lattice structure. Figure 12-42 shows the lattice structure that occurs when atoms are combined. In this diagram atoms of silicon (Si), which is a semiconductor material with four valence electrons, combine so that one valence electron from each of the neighbor atoms is shared, so that all atoms look and act as if they have eight valence electrons each.

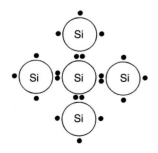

FIGURE 12-42 Atoms of silicon semiconductor material combine to create a lattice structure.

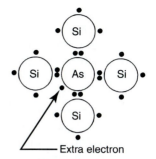

Extra electron

FIGURE 12-43 N-type material formed by combining four silicon atoms with a single arsenic atom.

12.6.3 Combining Arsenic and Silicon to Make N-Type Material

Other types of atoms can be combined with semiconductor atoms to create the special material that is used in solid-state transistors and diodes. Figure 12-43 shows four silicon atoms combined with one atom of arsenic. Arsenic has five valence electrons; when the silicon atoms are combined with it, they create a strong lattice structure. You can see that each silicon atom donates one of its valence electrons to pair up with each of the valence electrons of the arsenic atom. Since the arsenic atom has five valence electrons, one of the electrons will not be paired up and will become displaced from the atom. This electron is called a *free electron* and it can go into conduction with very little energy. Since this new material has a free electron, it is called *N-type material.*

12.6.4 Combining Aluminum and Silicon to Make P-Type Material

An atom of aluminum can also be combined with semiconductor atoms to create the special material called *P-type material.* Figure 12-44 shows four silicon atoms combined with one atom of aluminum. Aluminum has three valence electrons. When the silicon atoms combine with it, they create a strong lattice structure. You can see

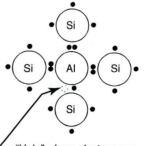

"Hole" where electron was

FIGURE 12-44 P-type material formed by combining four silicon atoms with a single aluminum atom.

that each silicon atom donates one of its valence electrons to pair up with each of the valence electrons of the aluminum atom. Since the aluminum atom has three valence electrons, one of the four aluminum electrons will not be paired up and it will have a space where any free electron can move into it to combine with the single electron. This free space is called a *hole,* and it is considered to have a positive charge since it is not occupied by a negatively charged electron. Since this new material has an excess hole that has a positive charge, it is called P-type material.

12.6.5 The PN Junction

One piece of P-type material can be combined with one piece of N-type material to make a PN junction. Figure 12-45 shows a typical PN junction. The PN junction creates an electronic component called a *diode.* The diode is the simplest electronic device. When DC voltage is applied to the PN junction with the proper polarity, it causes the junction to become a good conductor. Conversely, if the polarity of the voltage is reversed, the PN junction becomes a good insulator.

12.6.6 Forward Biasing the PN Junction

When DC battery voltage is applied to the PN junction so that positive voltage is connected to the P-type material and negative voltage is connected to the N-type material, the junction is forward biased. Figure 12-46a shows a battery connected to the PN junction so that it is forward biased. The positive battery voltage causes the majority of holes in the P-type material to be repelled, so that the free holes move toward the junction, where they will come into contact with the N-type material. At the same time the negative battery voltage also repels the free electrons in the N-type material toward the junction. Since the holes and free electrons come into contact at the junction, the electrons recombine with the holes to cause a low-resistance junction, which allows current to flow freely through it. When a PN junction has low resistance, it will allow current to pass just as if the junction were a closed switch.

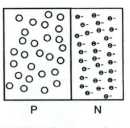

FIGURE 12-45 A piece of P-type material joined to a piece of N-type material.

It is important to understand that up to this point in this text, all current flow has been described in terms of *conventional current flow,* which is based on a theory that electrical current flows from a positive source to a negative return terminal. At this point you can see that this theory will not support current flow through electronic devices. For this reason *electron current flow theory* must be used when discussing electronic devices. In electron current flow theory, current is the flow of electrons and it flows from the negative terminal to the positive terminal in any electronic circuit.

12.6.7 Reverse Biasing the PN Junction

Figure 12-46b shows a battery connected to the PN junction so that it is *reverse biased.* The positive battery voltage is connected to the N-type material, and the negative battery voltage is connected to the P-type material. The negative voltage on the P-type material attracts the majority of holes in the P-type material, so they move away from the junction. Thus, they cannot come into contact with the N-type material. At the same time the positive battery voltage that is connected to the N-type material attracts the free electrons away from the junction. Since the holes and free electrons are both attracted away from the junction, no electrons can recombine with any holes. Thus a high-resistance junction is formed that will not allow any current flow. When the PN junction has high resistance, it will not allow any current to pass, just as if the junction were an open switch.

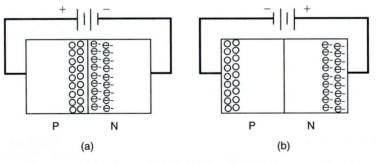

(a) (b)

FIGURE 12-46 (a) A forward-biased PN junction. (b) A reverse-biased PN junction.

12.6.8 Using a Diode for Rectification

It is difficult to manipulate AC voltage in electronic circuits. With the advent of electronics, DC voltage and current can be manipulated (changed) to control a variety of components in an electrical system such as inverters, proportional hydraulic valves, and other frequency control systems. Since most equipment is supplied with AC voltage, electronic components are required to change the voltage into DC voltage one or more diodes are used to change AC voltage to DC voltage. An electronic diode is a simple PN junction. Figure 12-47a shows the PN junction for the diode, and Figure 12-47b shows the symbol for a diode. The symbol for the diode looks like an arrowhead that is pointing against a line. The arrowhead is called the anode, and it is also the P-type material of the PN junction. The other terminal of the diode is the cathode, and it is the N--type material. Since the anode is made of positive P-type material, it is identified with a (+) sign. The cathode is identified with a (−) sign, since it is made of N-type material. The AC power source produces a sine wave that has a positive half-cycle followed by a negative half-cycle. The diode converts the AC sine wave to half-wave DC voltage by allowing current to pass when the AC voltage

provides a forward bias to the PN junction, and it blocks current when the AC voltage provides a reverse bias to the PN junction. The forward-bias condition occurs when the AC voltage sine wave provides a positive voltage to the anode and a negative voltage to the cathode. During this part of the AC cycle, the diode is forward biased and has very low resistance, so current can flow. When the other half of the AC sine wave occurs, the diode becomes reverse biased, with negative voltage applied to the anode and positive voltage applied to the cathode. During the time the diode is reverse biased, a high-resistance junction is created, and no current will flow through it.

Rectification is the process of changing AC voltage to DC voltage. One of the main jobs of the diode is to convert AC voltage to DC voltage. Most electronic circuits used in equipment need some DC voltage to operate. Since the equipment is connected to AC voltage, a power supply is required to provide regulated DC voltage for the solid-state circuits, and the diode is part of the power supply that rectifies the AC voltage to DC.

12.6.9 Half-Wave and Full-Wave Rectifiers

When one diode is used in a circuit to convert AC voltage to DC voltage, it is called a half-wave *rectifier,* since only the positive half of the AC voltage is allowed to pass through the diode, and the negative half is blocked. The rectifier shown in Figure 12-48 is a half-wave rectifier. The half-wave rectifier is not very efficient, since half of the AC sine wave is wasted.

If four diodes are used in the circuit, they can convert both the positive half-wave and the negative half-wave of the AC sine wave. Figure 12-49 shows a circuit

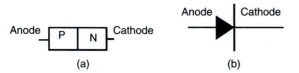

FIGURE 12-47 (a) PN junction for a diode. (b) Electronic symbol of a diode. The anode is the arrowhead part of the symbol, and the cathode is the other terminal.

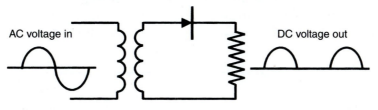

FIGURE 12-48 A single diode used in a circuit to convert AC voltage to DC voltage.

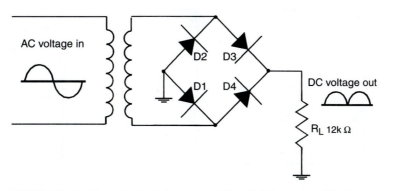

FIGURE 12-49 Four-diode, full-wave rectifier. This type of rectifier is often called a full-wave bridge rectifier, since the diodes are connected in a bridge circuit.

with four diodes used to rectify AC voltage to DC voltage. Since the four diodes can convert both the positive half and the negative half of the AC sine wave, this type of rectifier is called a full-wave rectifier and it is sometimes called a full-wave bridge. The full-wave bridge is a circuit that consists of four diodes. On some circuit boards they are four individual diodes on the board, whereas other boards package the four diodes into an integrated circuit (IC) that has four leads. Two of the leads are marked AC and this is where the AC voltage is supplied to the bridge. The other two leads are identified as DC+ and DC−, and these leads will provide DC voltage out of the bridge. If only one of the diodes in a bridge rectifier goes bad, the bridge will provide approximately half its rated voltage. If two or more diodes go bad in the bridge rectifier, the DC output voltage is usually zero.

12.6.10 Three-Phase Rectifiers

Three-phase AC motors can have their speed changed to run more efficiently by changing the frequency of the voltage supplied to them. In these applications, six diodes are used to convert three-phase AC voltage to DC voltage, and then a microprocessor-controlled circuit converts the DC voltage back to three-phase AC voltage. The frequency of this voltage can be adjusted to change the speed of the motors. Figure 12-50 shows a six-diode three-phase rectifier. Notice that the supply voltage to the diodes is three-phase AC voltage, and the output voltage from the rectifier consists of six positive half-waves.

Most power supplies for battery charging applications and motor drives use three-phase AC voltage instead of single-phase voltage. This means that the rectifier for these circuits must use a three-phase bridge, which has six diodes to provide full-wave rectification (two diodes for each line of the three phases). Figure 12-51a

shows the electrical diagram for a three-phase bridge rectifier. The secondary winding of the three-phase transformer is shown connected to the diode rectifier. Phase A of the three-phase voltage from the transformer is connected where the cathode of diode 1D is connected to the anode of diode 2D. Phase B is connected where the cathode of diode 3D is connected to the anode of diode 4D, and phase C is connected where the cathode of diode 5D is connected to the anode of diode 6D. The anodes of diodes 1D, 3D, and 5D are connected to provide a common point for the DC negative terminal of the output power. The cathodes of diodes 2D, 4D, and 6D are connected to provide a common point for the DC positive terminal of the output power.

A good rule of thumb for determining the connections on diode rectifiers is that the AC input voltage will be connected to the bridge where the anode and cathode of any two diodes are joined. Since this occurs at two points in the bridge, in a four-diode bridge, the two AC lines will be connected there without respect to polarity since the incoming AC voltage does not have a specific polarity. The positive terminal for the power supply will be connected to the bridge where the two cathodes of the diodes are joined, and the negative terminal will be connected to the bridge where the two anodes of the diodes are joined.

The diagram in Figure 12-51b shows the waveforms for the three-phase sine waves that supply AC power to the bridge, and Figure 12-51c shows the six half-waves of the output pulsing DC voltage that comes out of the bridge rectifier to the load. You should notice that since the six half-waves overlap, the DC voltage does not have a chance to get to the zero voltage point; thus the average DC output voltage is higher than the two-diode single-phase rectifier and the four-diode single-phase rectifier.

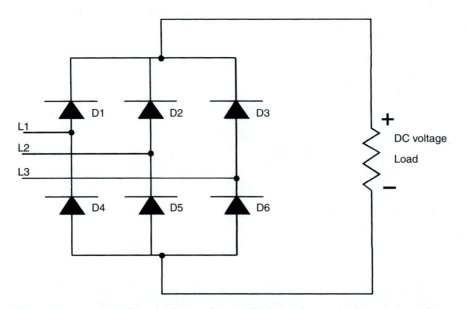

FIGURE 12-50   A six-diode bridge used to rectify three-phase AC voltage to DC voltage.

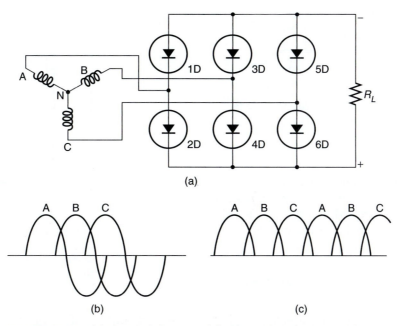

FIGURE 12-51   (a) Electrical diagram of the three-phase bridge rectifier that is connected to the secondary winding of a three-phase transformer. (b) Three-phase input sine waves. (c) Six half-waves for the DC output.

The three-phase full-wave bridge rectifier is used where the required amount of DC power is high and the transformer efficiency must be high. Since the output waveforms of the half-waves overlap, they provide a low ripple percentage. In this circuit, the output ripple is six times the input frequency. Since the ripple percentage is low, the output DC voltage is usable without much filtering. This type of rectifier is compatible with transformers that are wye or delta connected.

12.6.11 Testing Diodes

One of the tasks that you must perform as a technician is testing diodes to see if they are operating correctly. One way to do this is to apply AC voltage to the input of the diode circuit and test for DC voltage at the output of the diode circuit. If the amount of DC voltage at the power supply is half what it is rated for in a four-diode bridge rectifier circuit, you can suspect that one of the diode pairs has one or both diodes opened. If this occurs, you can turn off all voltage to the diodes and use an ohmmeter to test each diode to determine which one is faulty.

When you are testing the diodes with an ohmmeter, it is important that all power to the diode circuit be turned off. You should remember from earlier chapters that the ohmmeter uses an internal battery as a DC voltage source. Since you know the diode can be tested for forward bias and reverse bias with a DC voltage source, you can use the ohmmeter as the voltage source and the meter to test for high resistance and low resistance through the diode junction. Figure 12-52a shows an example of putting the positive ohmmeter terminal on the anode of the diode and the negative

ohmmeter terminal on the cathode of the diode to cause the diode to go into forward bias. During this test the diode is forward biased and the ohmmeter should measure low resistance. When the ohmmeter leads are reversed as in Figure 12-52b so that the negative meter lead is connected to the anode of the diode and the positive meter lead is connected to the anode of the diode, the diode is reverse biased. When the diode is reverse biased, the ohmmeter should measure infinite (∞) resistance. If the diode indicates high resistance

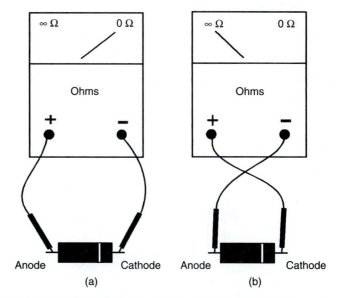

FIGURE 12-52   (a) Using the battery in an ohmmeter to forward bias a diode. The diode should have low resistance during this test. (b) Using the battery in the ohmmeter to reverse bias a diode. The diode should have high resistance during this test.

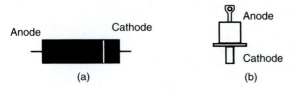

FIGURE 12-53 (a) Typical diode with anode and cathode identified. (b) Power diode with anode and cathode identified.

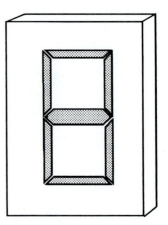

FIGURE 12-55 LEDs used in seven-segment displays. The seven-segment display can display numbers 0–9.

when it is reversed biased and low resistance when it is forward biased, it is good. If the diode indicates low resistance during both the forward and the reverse bias tests, the diode is shorted. If the diode shows high resistance during both tests, it is open.

12.6.12 Identifying Diode Terminals with an Ohmmeter

Since the ohmmeter can be used to determine if a diode is good or faulty, the same test can be used to determine which lead of a diode is the anode and which lead is the cathode. When you use the ohmmeter to test the diode for forward and reverse bias, you should notice that the ohmmeter indicates high resistance when the diode is reverse biased and low resistance when the diode is forward biased. When the meter indicates low resistance, you know the diode is forward biased, so the positive lead is touching the anode and the negative lead is touching the cathode. This method will work when you are testing any diode. If the diode has markings, you can identify the cathode end of the diode because it has a strip around it. Figure 12-53 shows two types of diodes with the anode and cathode identified in each.

12.6.13 Light-Emitting Diodes

A *light-emitting diode (LED)* is a special diode that is used as an indicator because it gives off light when current flows through it. Figure 12-54a shows a typical LED, and Figure 12-54b shows its symbol. You can see that the LED looks like a small indicator lamp. You will

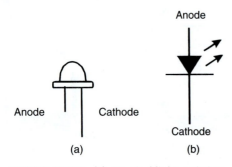

FIGURE 12-54 (a) A typical light-emitting diode (LED). Notice that the LED looks like a small indicator lamp. (b) The symbol for an LED.

likely encounter LEDs on various controls such as thermostats. The major difference between an LED and an incandescent lamp is that the LED does not have a filament, so it can provide thousands of hours of operation without failure.

The LED must be connected in a circuit in forward bias. Since the typical LED requires approximately 20 mA to illuminate, it is usually connected in series with a 600-Ω to 800-Ω resistor that limits the current. If the voltage is higher, the size of the resistance will be larger. Figure 12-55 shows a set of seven LEDs connected to provide a seven-segment display. The seven-segment display has the capability to display all numbers 0–9. Seven-segment displays are used to display numbers on electronic displays.

LEDs are also used in opto isolation circuits where larger field voltages are isolated from smaller computer signals. The LED is encapsulated with a phototransistor. When the input signal is generated, it causes current to flow through the LED, and light from the LED shines on the phototransistor, which goes into conduction and passes the signal on to the computer.

12.6.14 PNP and NPN Transistors

Two pieces of N-type material can be joined with a single piece of P-type material to form an NPN transistor. A PNP transistor can be formed by joining two pieces of P-type material with a single piece of N-type material. Figure 12-56 shows the electronic symbol and the material for both the PNP and the NPN transistors. The terminals of the transistor are identified as the emitter, collector, and base. The base is the middle terminal, and the emitter is the terminal identified by the arrowhead.

12.6.15 Operation of a Transistor

A transistor can be connected in a circuit to perform a wide variety of functions. The simplest function for a

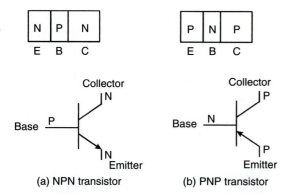

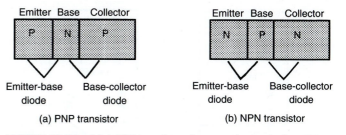

FIGURE 12-56 (a) Electronic symbol and diagram of an NPN transistor. (b) Electronic symbol and diagram of a PNP transistor.

FIGURE 12-58 (a) A PNP transistor shown as its equivalent PN junctions. Each PN junction can be tested for forward bias and reverse bias. (b) An NPN transistor shown as its equivalent PN junctions.

transistor to provide is that of an electronic switch. Figure 12-57 shows a transistor as an electronic switch. In this type of application, the base terminal of the transistor acts like the coil of a relay, and the emitter–collector circuit acts like the contacts of a relay. When the proper amount and polarity of DC voltage are applied to the base of the transistor, the resistance between the collector and emitter is relatively low, which allows the maximum amount of circuit current to flow through the emitter–collector circuit. The transistor at this time acts like a relay with its coil energized.

When the polarity of the voltage on the base of the transistor is reversed, the emitter–collector circuit is changed to a high-resistance circuit, which acts like the relay when the coil is de-energized. The major advantage of the transistor is that a small amount of voltage or current on the base can switch the transistor from high resistance to low resistance. Since the base current is small and the current flowing through the collector is large, the transistor is called an amplifier. Transistors are used in a variety of applications including motor protection circuits, inverters, and variable frequency motor drives.

Figure 12-58 shows a PNP transistor and an NPN transistor as two PN diode circuits. The equivalent diode circuits are shown with each transistor to give you an idea of how the two junctions work together inside each transistor. Since each transistor is made from two PN junctions, each junction can be tested just like a single-junction diode for forward bias (low resistance) and reverse bias (high resistance). If you work on a number of systems that have electronic circuits, you can purchase a commercial-type transistor tester that allows you to test the transistor either in the circuit or out of the circuit.

12.6.16 Typical Transistors

You will be able to identify transistors by their shape. Small transistors are used for switching control circuits, and larger transistors are mounted to heat sinks so that they can easily transfer heat. Figure 12-59 shows types of transistors.

12.6.17 Troubleshooting Transistors

Transistors can be tested by checking each P and N junction for front-to-back resistance. Figure 12-60 shows these tests. You can see that each time the battery in the ohmmeter forward biases a PN junction, the resistance is

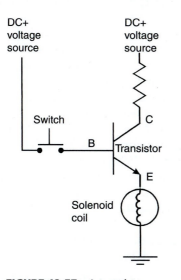

FIGURE 12-57 A transistor used as an electronic switch.

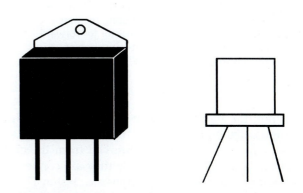

FIGURE 12-59 Typical transistors that are used for power control and switching.

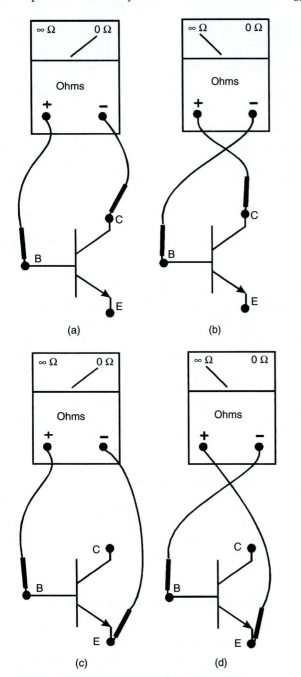

FIGURE 12-60 (a) Testing the base-collector junction of an NPN transistor for forward bias. (b) Testing the base-collector junction of an NPN transistor for reverse bias. (c) Testing the base-emitter junction of an NPN transistor for forward bias. (d) Testing the base-emitter junction of an NPN transistor for reverse bias.

low; when the battery reverse biases the junction, the meter indicates high resistance. You can test a transistor in this manner if it has been removed from the circuit. You can also test transistors while they are connected in circuit with a commercial-type transistor tester. The transistor tester performs similar front-to-back resistance tests across each junction.

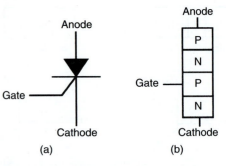

FIGURE 12-61 (a) Electronic symbol for the silicon controlled rectifier. The terminals of the SCR are the anode, cathode, and gate. (b) P-type material and N-type material in an SCR. The P-type material and N-type material are combined to make a PNPN junction.

12.6.18 The Silicon Controlled Rectifier (SCR)

The silicon controlled rectifier (SCR) is made by combining four PN sections of material. Figure 12-61a shows the PN material for the SCR, and Figure 12-61b shows its electronic symbol. The terminals on the SCR are identified as anode, cathode, and gate. Since the SCR is basically a diode that is controlled by a gate, its symbol uses the arrow from the rectifier diode that you studied at the beginning of this chapter. When the SCR is turned on, it can conduct large amounts of DC voltage and current (more than 1000 V and 1000 A) through its anode and cathode. The major difference between the SCR and the junction diode is that the junction diode is always able to pass current in one direction when the diode is forward biased. The SCR is forward biased by applying positive voltage to its anode and negative voltage to its cathode. At this point the SCR still has high resistance at its anode–cathode junction. If a positive voltage pulse is applied to the SCR gate, the SCR's anode–cathode junction will have low resistance, and the SCR will be in conduction. When the pulse is removed from the gate of the SCR, it will remain in conduction because positive current that comes through the anode will replace the voltage the gate provided. The only way to turn the SCR off is to provide reverse bias voltage to the anode–cathode or to reduce the current flowing through the anode–cathode to zero. You should remember that the AC sine wave has zero voltage right before it provides the negative half of its waveform. This means that if the SCR is powered with AC voltage, the SCR will be turned off when the AC waveform goes through 0 V and then to its negative half-cycle. When the AC voltage waveform goes positive again, a gate pulse can be provided and the SCR can go into conduction again. The gate provides a pulse that is used to cause the SCR to go into conduction.

12.6.19 Operation of the Silicon Controlled Rectifier

Figure 12-62 shows an SCR connected in a circuit to control voltage to a DC load. The source voltage for this

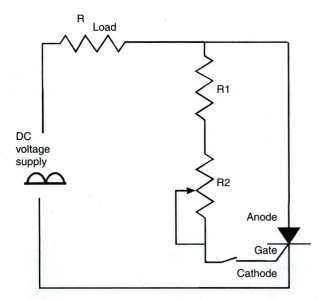

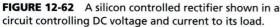

FIGURE 12-62 A silicon controlled rectifier shown in a circuit controlling DC voltage and current to its load.

circuit is AC voltage. The main advantage of the SCR is that it will not go into conduction until it receives a pulse of voltage to its gate. The timing of the pulse can be controlled so that it can be delivered anytime during the half-cycle, which controls the amount of time the SCR will be in conduction. The amount of time the SCR is in conduction will control the amount of current that flows through the SCR to its load. If the SCR is turned on immediately during each half-cycle, it will conduct all the half-wave DC voltage just like a normal diode rectifier. If the gate delays the point at which the SCR turns on and goes into conduction at the 45° point of the half-wave, the amount of voltage and current the SCR conducts will be 50% of the fully applied voltage.

The other important feature of the SCR is that it will go into conduction only when its anode and cathode are forward biased. This means that if the supply voltage is AC, the SCR can go into conduction only during the positive half-cycle of the AC voltage. When the negative half-cycle occurs, the anode and cathode will be reverse biased and no current will flow. This means that the SCR will automatically be turned off when the negative half of the AC sine wave occurs. Since the positive half of the AC sine wave occurs for 180°, the SCR can provide control of only 0°–180° of the total 360° AC sine wave.

It is also important to understand that since the turnoff point of the SCR is fixed to the point where the sine wave begins to go negative, the SCR can be controlled only by adjusting the point where it is turned on. The point where the SCR is turned on and goes into conduction is called the *firing angle.* If the SCR is turned on at the 10° point in the AC sine wave, its firing angle is 10°. If the SCR is turned on at the 45° point, its firing angle is 45°. The number of degrees the SCR remains in

conduction is called the *conduction angle.* If the SCR is turned on at the 10° point, its conduction angle will be 190°, which is the remainder of the 180° of the positive half of the AC sine wave.

12.6.20 Controlling the SCR

Figure 12-63 shows an SCR in a circuit with a unijunction transistor (UJT) connected to its gate. The load in this circuit is a DC motor. This type of DC motor is often used as a small control motor like a pump motor. The circuit is powered by AC voltage, and the variable resistor in the oscillator (capacitor–resistor) circuit sets the timing for the pulse that is used to energize the gate of the SCR. You should notice that a diode rectifier provides pulsing DC voltage for the capacitor, which charges to set the timing for the pulse that comes from the UJT. Since this DC voltage comes from the original AC supply voltage, it will have the same timing relationship of the original sine wave. This means adjusting the pulse from the UJT to turn on the SCR gate at just the right time to control the firing angle of the SCR from 0°–180°. In reality the firing angle is usually controlled from 0°–90°, which gives sufficient range of control to adjust the output DC voltage that is sent to the DC motor. You should remember that the speed of the DC motor can be controlled by adjusting the voltage sent to the armature and field. This diagram shows the waveform for the voltage at each point in this circuit. The load in this circuit could also be any other DC-powered load.

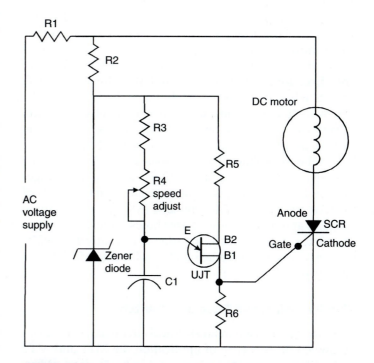

FIGURE 12-63 An SCR used to control a DC motor. A UJT is connected to the gate of the SCR to control its firing angle.

12.6.21 Testing the SCR

You will need to test the SCR to determine if it is faulty. Since the SCR is made of PN junctions, you can use forward-bias and reverse-bias tests to determine if it is faulty. In this test you should put the positive probe on the anode and the negative probe on the cathode. At this point the ohmmeter will still indicate that the SCR has high resistance. If you use a jumper wire and connect positive voltage from the anode to the gate, you should notice that the SCR will go into conduction and have low resistance. The SCR will remain in conduction until the voltage applied to its anode and cathode is reverse biased, or until the voltage applied to the anode and cathode is reduced to zero. This means that you can turn the ohmmeter polarity switch to the opposite setting, or you can remove one of the probes and the SCR will stop conducting. It is important to understand that the amount of current to keep the SCR in conduction is approximately 4 to 6 mA. This means that some high-impedance digital volt/ohmmeters will not have enough current when set to the ohms range to keep the SCR in conduction. If this is the case, you may need to test the SCR with an analog ohmmeter. The analog ohmmeter is a type of ohmmeter with a needle and scale. You should also test the SCR for reverse bias to ensure that it has high resistance. Sometimes an SCR will not go into conduction because it has developed an open in its anode–cathode circuit. Other SCRs may stay in conduction at all times, which means the SCR is shorted.

12.6.22 The Triac

The *triac* is basically two SCRs that have been connected back to back in parallel so that one of the SCRs will conduct the positive part of an AC signal and the other will conduct the negative part of an AC signal. As you know, the SCR can control voltage and current in only one direction, which means that it is limited to DC circuits when it is used by itself. Since the triac acts like two SCRs that are connected inverse parallel, one section of the triac can control the positive half of the AC voltage, and the other section of the triac can control the negative half of the AC voltage. Figure 12-64a shows the electronic symbol of the triac, and Figure 12-64b shows the arrangement of its P-type and N-type materials. The terminals of the triac are called main terminal 1 (MT1), main terminal 2 (MT2), and gate. Figure 12-64c shows typical triac semiconductor devices. Since the triac is basically two SCRs that are connected inverse parallel, MT1 and MT2 do not have any particular polarity.

12.6.23 Using the Triac as a Switch

The triac can be used in an electrical circuit as a simple on-off switch. In this type of application the MT1 and MT2 terminals are connected in series with the AC load. When the gate gets a positive pulse signal, the triac turns on for

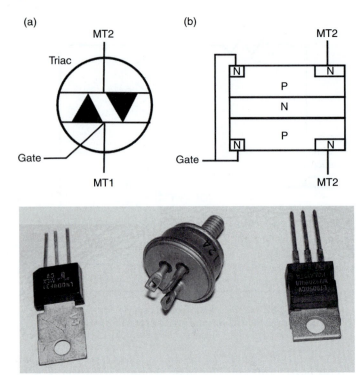

FIGURE 12-64 (a) Electronic symbol for the triac. (b) P-type and N-type materials in the triac. (c) Typical triac semiconductor devices.

the positive half of the AC cycle. When the AC voltage waveform returns from its positive peak to 0 V, the triac turns off. Next, the negative half-cycle of the AC voltage waveform reaches the triac, and it receives a negative pulse on its gate and goes into conduction again.

This means that the triac looks as though it turns on and stays on when AC voltage is applied. The load connected to the triac will receive the full AC sine wave just as if it were connected to a simple single-pole switch. The major difference is that the triac switch can be used for millions of on-off switching cycles. The other advantage of the triac switch is that the gate pulse can be a very small amount of voltage and current. This allows the triac to be used in temperature control circuit for a solar energy system where the temperature-sensing part of a thermostat can be a small solid-state sensing element called a thermistor. The sensing element can also be a narrow strip of mercury in a glass bulb that is very accurate but can carry only a small amount of voltage or current.

Another useful switching application for a triac is the solid-state relay. Figure 12-65 shows a thermostat used to control a triac that energizes the oil cooler solenoid coil to ensure that the hydraulic oil for the hydraulic tracking system for the large solar panels stays at the correct temperature. The temperature-sensing element is connected to the gate terminal of the triac. When the temperature increases, the thermostat sends a signal to the triac gate. The small amount of voltage flowing through the gate is sufficient to cause the triac to go into conduction and provide voltage to the solenoid valve.

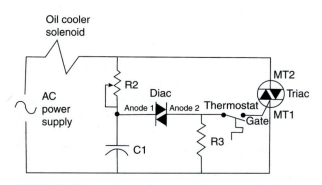

FIGURE 12-65 A triac used as a switch to turn on voltage to the oil cooler solenoid to ensure that the hydraulic oil for the hydraulic tracking system for the solar panels stays cool. The triac receives its gate signal from a small amount of voltage that moves through the temperature-sensing element.

12.6.24 Using the Triac for Variable Voltage Control

The triac can also be used in variable voltage control circuits, since it can be turned on any time during the positive or negative half-cycle, in much the same way that the SCR is controlled for DC circuit applications. In this type of application a diac is an electronic component that is used to provide a positive and a negative pulse that can be delayed from 0° to 180° to control the amount of current flowing through the triac. This type of circuit can be used to control the amount of current and voltage supplied to electric heating elements that are used as back up heat in a solar heating system. This allows the amount of current and voltage to be controlled from zero to maximum by adjusting resistor R2, which in turn allows the temperature to be accurately controlled. A resistive type temperature sensor can be used instead of the fixed R2 resistor, and another application for this type of system will to hold the temperature for the solar energy system or possibly the oil temperature control for the parabolic trough type solar energy system. Figure 12-66 shows a triac used to control an electric heating element powered by an AC voltage source. Notice that a variable resistor is connected with a capacitor to provide an oscillator pulse to the diac. Since the resistor and capacitor are

connected to an AC voltage source, the pulse will be both positive and negative as the AC sine wave changes polarity. The triac is connected to the same AC voltage source so that the timing of the pulse from the diac to its gate will always be synchronized with the polarity of the voltage arriving at the main terminals of the triac.

12.6.25 Testing the Triac

Since the triac is made from P-type material and N-type material, it can be tested like other junction devices. The only point to remember is that since the triac is essentially two SCRs mounted inverse parallel to each other, some of the ohmmeter tests will not be affected by the polarity of the ohmmeter leads. In the first test of a triac an ohmmeter should be used to test the continuity between MT1 and MT2. When no gate pulse is present, the resistance between these terminals should be infinite regardless of which ohmmeter probe is placed on each terminal. Figure 12-67 shows how the ohmmeter should be connected to the triac. In Figure 12-67a you can see that the positive ohmmeter probe is connected to MT1, and the negative probe is connected to MT2. Since no voltage is applied to the gate, the resistance should be infinite. When voltage from MT2 is jumped to the gate, the triac will go into conduction and the ohmmeter will indicate low resistance.

In Figure 12-67b you can see that the positive ohmmeter probe is connected to MT1, and the negative probe is connected to MT2. When voltage from MT1 is jumped to the gate, the triac will go into conduction and the ohmmeter will indicate that the resistance is low. It is important to remember that the triac will stay in conduction only while the gate signal is applied from the same voltage source as MT1. As soon as the voltage source is removed, the triac will turn off. Figure 12-67c shows the gate receiving voltage from MT2, and the triac is in conduction.

12.6.26 Inverters: Changing DC Voltage to AC Voltage

Inverters are circuits specifically designed to change DC voltage to AC voltage. As you know, systems such as solar photovoltaic panels produce DC voltage that must be converted to AC power so that it is usable in AC electrical systems. In solar energy systems inverters convert DC voltage into usable AC voltage at exactly 60 Hz. In some solar energy systems the power is stored in a battery so it can be used later. Since the voltage is stored in a battery as DC voltage, it must go through an inverter to be changed to AC voltage so it is usable in a home where the frequency of the voltage needs to be a constant 60 Hz.

12.6.27 Single-Phase Inverters

The simplest inverter to understand is the single-phase inverter, which takes a DC input voltage and converts it to single-phase AC voltage. Single-phase inverters are used

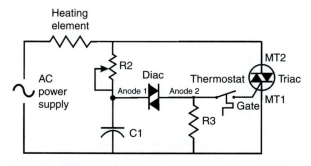

FIGURE 12-66 A triac used to control variable voltage to an AC electric heating element. Notice that a diac is used to provide a positive and a negative pulse to the triac gate.

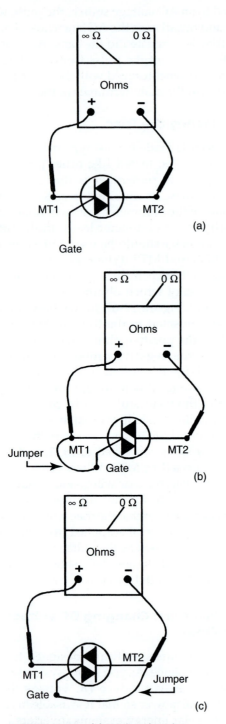

FIGURE 12-67 (a) Testing the triac by placing the ohmmeter positive probe on MT2 and the negative probe on MT1. When gate voltage is applied from the MT2 probe, the triac goes into conduction. (b) Testing the triac by placing the ohmmeter positive probe on MT1 and the negative probe on MT2. When gate voltage is applied from the MT2 probe, the triac goes into conduction. (c) The jumper is placed between MT2 and the gate.

in many solar panel systems, since the solar panels produces DC voltage instead of AC voltage. These smaller inverters are used in small residential systems where the DC voltage may be stored in batteries and used at times when the sunlight is not available, such as at night. The inverter simply converts DC voltage to AC voltage and ensures that the AC output voltage is exactly at 60 Hz.

12.6.28 Using Transistors for a Six-Step Inverter

Figure 12-68 shows the electrical diagram of a single-phase inverter that uses four transistors. Since the transistors can be biased to any voltage between saturation and zero, the waveform of this type of inverter can be more complex to look more like the traditional AC sine wave. The waveform shown in this figure is a six-step AC sine wave. Two of the transistors will be used to produce the top (positive) part of the sine wave, and the remaining two transistors will be used to produce the bottom (negative) part of the sine wave.

When the positive part of the sine wave is being produced, the transistors connected to the positive DC bus voltage are biased in three distinct steps. During the first step, the transistors are biased to approximately half

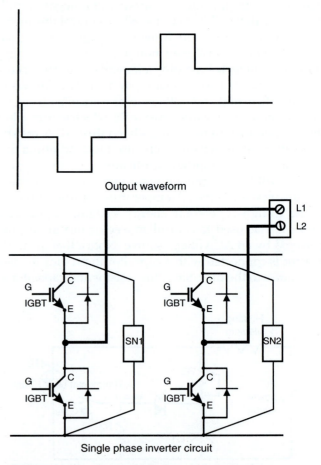

FIGURE 12-68 Electronic diagram of a single-phase insulated gate bipolar transistor (IGBT) inverter with the output waveforms for the AC voltage.

voltage for one-third of the period of the positive half-cycle. Then these transistors are biased to full voltage for the second third of the period of the positive half-cycle. The transistors are again biased at half voltage for the remaining third of the period. This sequence is repeated for the negative half-cycle. This means that the transistors that are connected to the negative DC bus are energized in three steps that are identical to the steps used to make the positive half-cycle.

Since six steps are required to make the positive and negative half-cycles of the AC sine wave, this type of inverter is called a six-step inverter. The AC voltage for this inverter will be available at the terminals marked M1 and M2. Even though the AC sine wave from this inverter is developed from six steps, the motor or other loads see this voltage and react to it as though it were a traditional smooth AC sine wave. The timing for each sine wave is set so that the period of each is 16 ms, which means it will have a frequency of 60 Hz. The frequency can be adjusted by adjusting the period for each group of six steps.

12.6.29 Three-Phase Inverters

Three-phase inverters are much more efficient for industrial applications where large amounts of voltage and current are required. The basic circuits and theory of operation are similar to those for the single-phase transistor inverter. Figure 12-69 shows a three-phase inverter

with three pairs of insulated gate bipolar transistors (IGBT). Each pair of transistors operates like the pairs in the single-phase six-step inverter. This means that the transistor of each pair that is connected to the positive DC bus voltage will conduct to produce the positive half-cycle, and the transistor that is connected to the negative DC bus voltage will conduct to produce the negative half-cycle.

The timing for these transistors is much more critical since they must be biased at just the right time to produce the six steps of each sine wave, and they must be synchronized with the biasing of the pairs for the other two phases so that all three phases will be produced in the correct sequence with exactly 120° between each phase.

12.6.30 Variable-Voltage Inverters (VVIs)

A variable-voltage inverter (VVI) is basically a six-step, single-phase or three-phase inverter. The need to vary the amount of voltage to the load became necessary when these inverter circuits were used with larger solar farms. Originally these circuits provided a limited voltage and limited variable-frequency adjustments because oscillators were used to control the biasing circuits. Also many of the early VVI inverters used thyristor technology, which meant that groups of SCRs were used with chopper circuits to create the six-step waveform. After microprocessors became inexpensive and widely

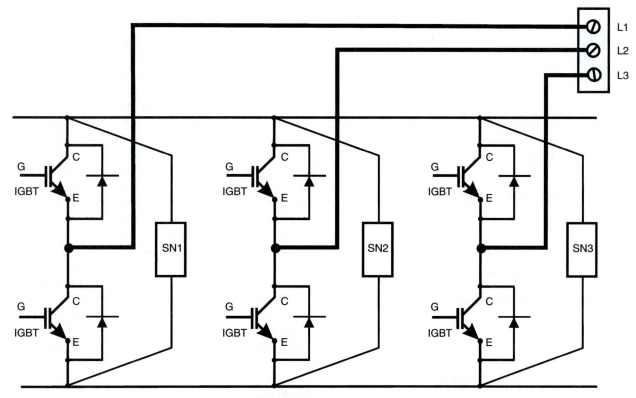

Three-phase inverter circuit

FIGURE 12-69 Electrical diagram of a three-phase inverter that uses six IGBTs.

used, they were used to control the biasing circuits for transistor-type inverters to give these six-step inverter circuits the ability to adjust the amount of voltage and the frequency through a much wider range. In the inverter for the solar energy system, the frequency needs to be exactly 60 Hz.

Figure 12-70 shows the voltage and current waveform for the VVI inverter. The voltage is developed in six steps and the resulting current looks like an AC sine wave. These are the waveforms that you would see if you placed an oscilloscope across any two terminals of this type of inverter.

12.6.31 Pulse-Width Modulation Inverters

Another method of providing variable-voltage and variable-frequency control for inverters is to use pulse-width modulation (PWM) control. This type of control uses transistors that are turned on and off at a variety of frequencies. This provides a unique waveform that makes multiple square-wave cycles that are turned on and off at specific times to give the overall appearance of a sine wave. The outline of the waveform actually looks similar to the six-step inverter signal. An example of this type of waveform is provided in Figure 12-71. The overall appearance of the waveform is an AC sine wave. Each sine wave is actually made up of multiple square-wave pulses that are caused by transistors being turned on and off very rapidly. Since the bias of these transistors can be controlled, the amount of voltage for each square-wave pulse can be adjusted so that the entire group of

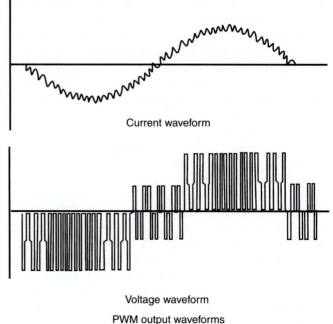

Current waveform

Voltage waveform

PWM output waveforms

FIGURE 12-71   Voltage and current waveforms for the pulse-width modulation (PWM) inverter. Notice that the overall appearance of each waveform is an AC six-step sine wave and that it is actually made of a number of square-wave pulses.

square waves has the overall appearance of the sine wave. If you look at the voltage waveform for the PWM inverter, you will notice that the outline of the AC sine wave still looks like the six-step sine wave originally used in the VVI inverters. The height of the steps of the AC sine wave is also increased when the voltage of the individual pulses is increased. This increases the total voltage of the sine wave that the PWM inverter supplies.

The width (timing) of each square-wave pulse can also be adjusted to change the period of the group of pulses that makes up each individual AC sine wave. When the width of the sine wave changes, it also changes the period for the sine wave. This means that the frequency is also changed and is controlled for the PWM inverter by adjusting the timing of each individual pulse. Since adjusting the voltage and frequency is fairly complex, the PWM inverter uses a microprocessor to control the biasing of each transistor. If thyristors are used as in SCRs, the microprocessor will control the phase angle for the firing circuit.

Early PWM circuits used thyristors such as SCRs to produce the square-wave pulses. The control circuit included triangular carrier waves to keep the circuit synchronized. This sawtooth waveform was sent to the oscillator circuit that controlled the firing angle for each thyristor. Today the PWM inverters use transistors mainly because of their ability to be biased from zero to saturation and back to zero at much higher frequencies. Modern circuits will more than likely use transistors for these circuits because they are now manufactured to handle larger currents that are well in excess of 1,500 A.

Variable voltage input (VVI) current

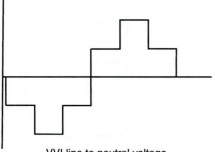

VVI line to neutral voltage

FIGURE 12-70   Voltage and current waveforms for the variable-voltage input (VVI) inverter.

12.6.32 Current-Source Input (CSI) Inverters

The current-source input (CSI) inverter produces a voltage waveform that looks more like an AC sine wave. The current waveform for this inverter looks similar to the original on/off square wave of the earliest inverters that cycled SCRs on and off in sequence. This type of inverter uses transistors to control the output voltage and current. The on time and off time of the transistor are adjusted to create a change in frequency for the inverter. The amplitude of each wave can also be adjusted to change the amount of voltage at the output. This means that the CSI inverter like the previous inverters can adjust voltage and frequency into fixed frequency applications or other applications that require variable voltage. Figure 12-72 shows the voltage and current waveform for the CSI inverter.

2.6.33 Applications for Inverters

Inverters are used today in solar energy systems to convert the DC voltage that is generated from the photovoltaic panels to AC voltage where it can be used in a home or commercial establishment, or it can be used to convert DC voltage from the photovoltaic solar panels

to AC voltage that is supplied to the electrical grid. When AC voltage is supplied to the grid, the conversion from DC voltage to AC voltage must control the frequency and the amount of voltage that is converted to ensure that it matches the voltage on the grid exactly.

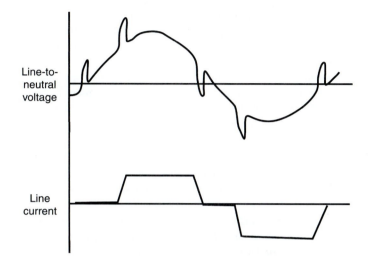

FIGURE 12-72   Voltage and current waveform for the current-source input (CSI) inverter.

Questions

1. Identify the basic parts of a DC motor.
2. Identify the basic parts of an AC motor.
3. Explain the basic operation of a DC motor.
4. Explain the basic parts of an AC motor.
5. Explain how three transformers can be connected as a wye or delta system to provide AC voltage.
6. Explain why fuses or a motor starter is needed in a circuit to protect a motor.
7. Explain which two materials are combined to make P-type and N-type materials.
8. Explain the operation of a diode (PN) junction and show the input AC waveform and the output DC waveform.
9. Identify the terminals of a transistor and explain its operation.
10. Explain the operation of a power electronic frequency converter (inverter).

Multiple Choice

1. An inverter _____
   a. converts DC voltage to AC voltage and back to DC voltage.
   b. converts AC voltage to DC voltage and back to AC voltage.
   c. converts DC frequency to a new DC frequency.
2. In a DC motor, the _____ rides on the commutator to make an electrical connection between the rotating member and the stationary part of the motor.
   a. armature
   b. carbon brush
   c. stator
3. An AC induction motor gets current into the rotor by _____

   a. induction.
   b. brushes riding on commutator segments.
   c. a wire that is soldered between the rotating segment and the stationary part of the motor.
4. The rectifier diode_____
   a. converts DC voltage to AC voltage.
   b. converts AC voltage to DC voltage.
   c. can be used as a seven-segment display for numbers.
5. An inverter _____
   a. changes AC voltage to DC voltage.
   b. uses two coils to step up or step down voltage.
   c. changes DC voltage to AC voltage.

GLOSSARY

Absorption stage battery charging A point during battery charging when the battery voltage is nearly at the full charge point. When the battery reaches the absorption stage, the controller then operates in constant voltage mode, holding the battery voltage at the absorption voltage setting for a preset time limit. During this time current falls gradually as the battery capacity is reached.

AC motor An electric motor that uses AC power to create a magnetic field that causes its rotor to turn. The number of poles in the motor and the frequency of the AC power will determine the speed the motor shaft turns.

AC voltage Voltages that are alternating in polarity, reversing positive and negative over time.

Active solar heating Solar energy systems that use electrical or mechanical energy such as fans and pumps to convert the sun's energy to heated water or heated air or to create electrical power.

Alternating current (AC) A circuit in which the electrons travel in one direction and then change direction and travel in the other direction. This movement by the electrons can best be shown by its characteristic waveform.

Amp-hour (ampere hour) A current of one ampere flowing for one hour.

Amps (amperes) Unit for electrical current.

Analog control A signal that can be varied between 0% and 100% and causes the system to respond between 0% and 100%; if the signal is from a sensor, it can measure any variable value from 0% to 100%.

Anti-islanding Control circuit that prevents a condition in which parts of the grid become disconnected and leave other parts still connected. Sometimes this occurs by accident, and other times the sections of the grid are isolated so that one or more sections can operate and supply power, while other parts are disconnected.

Apparent power Power caused by multiplying voltage times current when the current is caused by inductive reactance, capacitive reactance, and resistance. The amount of apparent power will appear to be larger than the true power because the true power has a phase shift caused by capacitance or inductance. True power, by comparison, calculates only current caused by pure resistance. *Note:* Apparent power has the same value as true power when a circuit contains only resistance.

Armature The moving part of a magnetic component such as a relay, solenoid, motor, or generator.

Atom The smallest part of an element that contains a nucleus with a proton and neutron in it and electrons that orbit around the outside of the nucleus.

Atomic number The number of protons in the nucleus of an atom.

Auxilliary contacts The extra sets of normally open contacts on a motor starter that are typically connected in parallel with the start pushbutton. These contacts serve as a seal-in circuit after the motor starter coil is energized.

Back film material A material that is placed on the back of solar panels and is dark colored. It is made of polyethylene terephthalate (PET) film. The back film provides support for the material resting on it and protects the back side of the solar panel from anything that may scratch or tear it during the installation process.

Backup generator A generator that is connected in parallel with the system power supply to a building. A master switch automatically starts the generator, which is sized to provide a percentage of the building's power that is typically used to provide backup to lighting circuits, computers, and other safety systems in the building until building power is returned.

Ball screw mechanism A mechanism that changes rotational motion to linear motion. This system allows a motor to turn the ball screw shaft, and the threads in the shaft make a positioning block move along the length of the screw. Typically used in positioning solar panels.

Brownout A reduction or cutback in electrical power, especially as a result of a shortage, a mechanical failure, or overuse by consumers. When the demand for electricity begins to outpace the supply on any given day, the voltage on the entire system will begin to lower or droop.

Brushes Carbon conductors that maintain an electrical connection between stationary and moving parts of a motor, generator, or other type of rotating machine; also known as slip rings.

Buck converter A step down DC to DC converter that is part of a switch mode power supply that uses a transistor, a capacitor, and an inductor with a diode to regulate the voltage level at levels below the supply voltage level. The advantage of the buck converter is that it provides DC power more efficiently than a power supply that uses a resistance as a voltage drop to regulate voltage at a reduced level.

Buck-boost regulator A circuit that utilizes the strong points of both the buck converter and the boost converter. In this circuit the transistor is connected in series like the buck converter, and the inductor has been moved to a position where it is connected in parallel with the output terminals. The freewheeling diode is connected as it was in the boost regulator.

Bulk stage charging The first stage of charging when the maximum amount of electrical power is sent to the batteries for charging. If the batteries are discharged, the controller operates in constant current mode, delivering its maximum current to the batteries until they are nearly fully charged during bulk stage charging.

Bypass diode A diode that is installed in a series string of solar panels. The diode protects a bad solar panel that is not producing power or one that is shaded and not producing as much as other panels. If a panel quits producing or is not producing sufficient power due to shading, the diode bypasses that solar panel so it cannot consume power and reduce the efficiency of the string. When the panel begins producing power again, it will produce voltage normally and the bypass diode does not affect it.

Cable connector A connector on the end of a wire that makes connections to other wires easily. Typically used on the wires for a solar panel that allow quick and easy connections from panel to panel.

Capacitive load A type of load that includes current flowing through capacitors that are used to correct power factor.

Capacitive loads are also used in some switch-mode computer power supplies.

Capacity factor The ratio of the actual output energy produced as compared to the rated energy output of a power plant or any alternative energy producing system.

Charge controller A controller that ensures that the batteries are charged to their peak potential and not allowed to overcharge. The charge controller also protects the batteries against being completely discharged.

Chopper An electronic control circuit that converts DC voltage to a different value of DC voltage. The chopper was originally specifically designed to convert a fixed DC voltage into variable DC voltages primarily used to control the speed of DC motors. In solar energy control circuits it is primarily used to convert one level of DC voltage to another.

Circuit breaker An electrical device that is designed to sense the amount of current flowing through it and break the electrical circuit if the amount of current exceeds the setting value of the circuit breaker. Excessive current can cause components to overheat and become permanently damaged.

Closed loop control A control system that uses a feedback signal and compares it to the set point. Any error is automatically adjusted for.

Coil A part of a relay or solenoid that is made by tightly winding a long piece of wire into loops. When current flows through the wire, it creates a strong magnetic field in the coil that can be used to open and close contacts in a relay or change the position of a solenoid valve from open to closed or vice versa. Coils of wire are also used in motor windings and transformers.

Cold cranking amps (CCA) A term that will specify the battery's ability to start an automobile or truck engine at a very cold temperature. This rating is based on a temperature of 0°F or 18°C, and it refers to the number of amperes a new and fully charged battery is able to deliver constantly for 30 seconds.

Combiner box An electrical connection box that allows multiple solar PV panels to be connected and fused as a single voltage source before the power is sent to the charge controller. The enclosure is rated NEMA 3R, which means that it is rainproof and sleet resistant for outdoor use or it is available in a NEMA 4, which is rated watertight for indoor uses. The combiner box can be installed directly on the rooftop where it is exposed to all types of outdoor weather, or it can be installed inside an enclosure on the roof or immediately inside the building where the power is brought in so it is not exposed to weather elements.

Commercial solar energy installation A solar energy installation specifically designed for commercial applications such as small stores, small restaurants, grocery stores, or other commercial locations.

Commutator A cylindrically shaped assembly that is fastened to the shaft of a motor or generator and is considered part of the armature assembly. It consists of segments of "bars" that are electrically connected to the two ends of one or more armature coils.

Commutator segments Bars of a commutator that are connected electrically to the two ends of one or more armature windings.

Compound DC motor A DC motor that has windings like a series motor and a shunt motor.

Conductance band A place inside the PN junction where electrons move when they are freed up from their shells. Once an electron makes it to the conduction band from its shells, it is free to move as electrical current flow.

Conductor A material that allows electrical current to flow through it. An example of a conductor is copper that is used for electrical wiring.

Contactor One or more sets of contacts that are opened or closed by a magnetic coil. The contactor is similar to a relay but normally is larger. By definition, contactor contacts are rated for more than 15 A.

Continuity test A test that measures the amount of resistance. The test is for two states, low resistance that indicates a fuse is good or a switch is closed, or high resistance that indicates a fuse or switch is open.

Conventional current flow A theory that explains that electrical current flows from the positive terminal to the negative terminal of a battery. This theory was proven to be incorrect when the electron microscope was able to determine that electrons had a negative charge and current flow actually flows from the negative terminal to the positive terminal of a battery. Conventional current flow theory is still used to explain many electrical diagrams and systems.

Converter A device or circuit for changing AC to DC. This is accomplished through a diode rectifier or thryristor rectifier circuit.

Current The flow of electrons that is measured in amperes.

Current-source input (CSI) This type of inverter produces a voltage waveform that looks more like an AC sine wave and current waveform that looks similar to the original on/off square wave of the earliest inverters that cycled SCRs on and off in sequence. This type of inverter uses transistors to control the output voltage and current.

Cycle life The total number of times a rechargeable battery can be charged and discharged over its lifetime.

DC interface enclosure An electrical enclosure that houses the disconnect switch and the contactor that is controlled by a magnetic coil. The DC interface provides a means of disconnecting all DC power from the solar array from the main controller. Any time major repair work must be accomplished, the disconnect will allow that part of the system to be disconnected and locked out.

DC motor A motor that is specifically designed to operate on DC power. The motor armature will turn faster as voltage level to the motor is increased.

DC voltage The electrical force (potential) that has polarity and causes the flow of electrons in one direction.

Deep cycle battery A battery that can be used for solar applications that can be charge for long periods of time during the day and discharged for long periods of time overnight.

Delta-connected three-phase transformer A transformer that has three windings connected in such a way that all three of the windings are connected in series with each other in the shape of a triangle. The name "delta" comes from the Greek letter "D," which has the shape of a triangle.

Department of Energy (DOE) Cabinet-level department of the U.S. government concerned with federal policies regarding energy. The DOE has a number of standards in place that cover some parts of the solar energy equipment, such as the electrical system and towers.

Depth of discharge The amount of energy that has been removed from the battery at any given time. It is usually measured as a percentage of the total capacity of the battery.

Diffuse light Light that does not strike the solar photocell directly.

Diode A two-terminal, solid-state semiconductor that allows current to flow in one direction.

Direct current (DC) The flow of electricity in a circuit in only one direction.

Direct sunlight Sunlight that is not filtered, reflected, or diffused.

Discharge a battery The condition when electricity is withdrawn from the battery.

Disconnect enclosure An enclosure that houses the electrical disconnect switch.

Doping Mixing elements with different numbers of valence electrons to create semiconductor P-type material or N-type material. An example is mixing silicon that has four valence electrons with elements that contain more or fewer valence electrons than silicon does.

Dry cell battery A battery that uses a paste or dry electrolyte.

Electrical contacts Contacts in a switch or relay that may be normally open or normally closed.

Electrical demand The maximum amount of electrical energy that is being consumed at a given time.

Electrical grid An interconnected network for delivering electricity from suppliers to consumers.

Electricity The flow of electrons.

Electrolyte A chemical in liquid or gel form that is used inside a battery to conduct electricity.

Electromagnet A magnet that is produced by passing current through a coil of wire.

Electron The negative part of an atom that flows to create electrical current.

Electron current flow A theory that explains that current is the flow of electrons and since an electron is negative, the flow of current starts at the negative terminal and returns to the positive terminal of a battery.

Electronic inverter An electronic circuit that converts AC electricity output from a wind turbine or other source that is at any uncontrolled frequency and converts it to DC voltage and then back to AC voltage at exactly 60 Hz. Some versions of the inverter are used to convert DC voltage output from a solar panel to AC voltage at exactly 60 Hz. *See also* Power electronic frequency converter.

Element The smallest part any material can be broken down to where the material has the same number of protons, neutrons, and electrons.

EMI (electromagnetic interference) Any unwanted signals or interference that gets into the AC voltage or current. An EMI filter helps keep the signal clean so that it can be used for computer loads or other sensitive loads.

Energy The capacity to do work.

Error The difference between the set point and the feedback signal when compared in the summing junction in a closed-loop control system.

EVA material A sheet of clear vinyl acetate that is laminated to the solar cell strings on each side to protect them from moisture or other harmful elements. The EVA material seals the solar strings between the two sheets. The solar cell strings are covered with a sheet from below and one sheet above, which are laminated during the manufacturing process.

Fail-safe system A secondary system that ensures continued operation even if the primary system fails.

Feedback (feedback signal) A signal from a sensor that measures a process variable in a closed-loop system that is sent to the summing junction. An example is a voltage or temperature sensor.

Flicker power A short-lived voltage variation in the electrical grid that might cause a load such as an incandescent light to flicker.

Float stage charging The third stage of charging a battery. During the float stage the battery is nearly fully charged, the voltage is controlled, and the current is limited. The battery controller lowers the amount of current flowing to the battery but keeps the voltage just below maximum level so it does not overcharge the battery.

Flooded cell battery A battery that has a liquid electrolyte and is also called a wet cell battery.

Forward bias A condition in which a positive electrical charge is applied to the positive terminal of a diode or any other PN material and causes a low-resistance condition at its PN junction, which allows current to flow through it.

Frame Aluminum that encases the entire group of components that make up the solar panel. The frame has all the mounting holes to attach brackets and other hardware to keep the solar panel mounted in place. Aluminum is used because it provides strength yet is lightweight.

Free electron An electron in the outermost (valence) shell of an atom that has received enough energy to break it loose from its orbit around its nucleus.

Frequency The number of periodic cycles per unit of time.

Full-bridge converter This converter adds two additional transistors to the half-bridge converter, which allows four transistors to be available to provide power to the output section, so this type of converter is used in power supplies in excess of 1,000 W.

Fuse A device designed as a one-time protection against over current or short-circuit current. The fusible link melts and causes an open circuit when its current rating is exceeded.

Fuse disconnect A special switch designed to provide a disconnect for electrical equipment as well as to provide a mounting for fuses. When the disconnect is in the open position, the fuses will be isolated from the source of power, so they can be removed and replaced or tested.

Galvanic Cell The simplest form of a battery that consists of two types of metal plates that allow ions from one metal to move to the other side of the battery when an electrolytic liquid is added to cover the metal plates. When electrons are released, they become electrical current.

Gel cell battery A valve-regulated lead-acid (VRLA) battery that has its electrolyte gelified. The gel consists of sulfuric acid mixed with silica gel. The vents on these batteries are typically designed so they cannot be removed. A gel battery may also be known as a *gel cell*.

Generator A device for converting energy into electrical power.

Gigawatt A unit of power equivalent to 1,000 megawatts or 1 million kilowatts.

Green technology A wide variety of technologies that involve developing energy that produces the minimal amount of pollution. Green technology today includes energy developed from wind, solar, bio fuels, geothermal sources, and other evolving technologies.

Grid An integrated network of electricity distribution and power lines, usually covering a large area. The grid allows power to be produced in one area and shipped to another area where it is needed.

Grid companies (Gridco's) Companies that manage the grid function, which is interconnection and routing of electricity through hardware cables so that no area has a brownout or a blackout due to insufficient power.

Grid-tied system Photovoltaic cells that are grouped together in large-scale power production and the electrical power is connected directly to the electrical grid.

Ground The negative side of a circuit. Ground also has the potential of the earth, as it is usually created by driving a copper rod into the ground approximately 8 feet. The ground circuit protects the circuit against electrical shock hazards, as it will lead any stray voltage to the ground and cause a circuit breaker or fuse to blow, creating an open in the circuit and removing any electrical shock hazard.

Ground-fault protection A system that continually checks the current that a generator produces and ensures that it is all going into the grid system. This type of system will also detect irregular voltage or currents that may be fed back into the system.

Grounding A method of connecting metal in a solar panel system to a ground (rod) stake that is pushed into the earth to a depth of 8 feet. Grounding makes everything connected to the ground rod have the electrical potential of the earth.

Heat Energy that occurs because of the motion of atoms or molecules. Heat can be transmitted or transferred by conduction through solids or liquids and by convection through liquids or gases and through empty space by radiation.

Heat transfer The movement of heat from a warmer area or object to a cooler area or object. Heat is transferred by conduction, convection, or radiation.

Heater An element in a motor starter that senses the amount of current flowing through it and converts it to heat. If the amount of current becomes too large, the heater produces enough heat to trip the motor starter and stop all power going to the motor.

High-leg delta voltage A voltage that is produced when a three-phase transformer is connected in a delta connection and the coil between L1 and L3 has a neutral point created. The voltage between the neutral point and L2 will produce 208 V that is called a high-leg delta voltage, which should not be used because it will cause an imbalance in the current for the transformer.

Hole A place in an atom from which an electron has moved. Since the electron has a negative charge, the place where electron has moved from will have a positive charge.

Hybrid lighting systems A lighting system that consists of solar lighting applications and conventional electrical lighting.

Hydrometer An instrument that can measure the specific gravity of an electrolyte, which is the exact percentages of sulfuric acid and water. The hydrometer has a rubber squeeze bulb mounted on the end of a glass tube. The glass tip of the tube is placed under the liquid level of the solution in the battery and when the bulb is collapsed and released a sample of the fluid is sucked up into the tube. When the tube is nearly full, a float in the glass tube will begin to float at a level that indicates the specific gravity of the fluid.

Induced current Current created by electromagnetic induction. When current flows through a coil of wire and builds a magnetic field and then the current stops (alternating current passing through its zero point), the magnetic field collapses and the flux lines cut across a second coil and induce a current in the second coil.

Induction Creating current in a coil of wire by passing it through a magnetic field.

Induction motor A motor that uses magnetic inductance to get current to flow through the rotating part of the motor.

Institute of Electrical and Electronics Engineers (IEEE) An engineering organization (pronounced I triple E) in the United States that develops, promotes, and reviews standards for the electronics, computer, and electric power industries.

Insulator A material that has high resistance and prevents the flow of electrons.

Inverter An electronic circuit that changes DC voltage to AC voltage and provides a set frequency such as 60 Hz.

Junction box An electrical box that it specifically designed to provide a space for multiple wires that are connected.

Kilovolt-ampere (kVA) 1,000 volt-amperes; a rating standard for transformers and other electrical components.

Kilowatt A unit of power, equal to 1,000 watts.

Kilowatt peak (KWp) power The value specifies the output power achieved by a solar module under full solar radiation (under set Standard Test Conditions).

Ladder diagram An industry standard for representing relay control logic. The ladder diagram shows the sequence of operation of the electrical circuit. The name comes from the fact that the overall form of the diagram looks like a ladder. Also called a *schematic diagram*.

Ladder logic An industry standard for representing relay control logic. The name comes from the fact that the overall form of the diagram looks like a wooden ladder. It is also referred to as a *ladder diagram*.

Lead-acid battery A battery that has lead electrodes and uses diluted sulfuric acid as the electrolyte. Each cell in a lead-acid battery generates about 2 V.

Light-emitting diode (LED) A two-lead solid-state PN junction device that produces a small amount of light when it is forward biased with electrical voltage.

Linear actuator   A pneumatic, hydraulic, or electric actuator that moves in a linear motion.

Linear power supply   This type of power supply uses a transformer and a four-diode full-wave bridge rectifier to convert AC voltage and produce a pulsing DC output waveform. A capacitor is used as a filter to smooth out the pulsing DC to pure DC. The remainder of the circuit contains a regulator and the load.

Lithium ion battery   A battery made from lithium, which is a lightweight metal that easily forms ions. The newer lithium-ion batteries can store almost twice as much electrical energy as comparable to NiCad rechargeable batteries.

Load center   An electrical panel that has the main disconnect switch for the panel and provides a means to mount individual circuit breakers.

Low voltage ride through (LVRT)   The safety system in place for conditions when low voltage occurs in the grid system where the solar panels are connected. The solar panels must have the capability to ride through this condition. It does not matter if the solar panel is causing the low voltage or if the low voltage exists in the grid where the solar panel is connected.

Magnet   Material that has an attraction to iron or steel.

Magnetic coil   A coil of wire that has electrical current passed through it and becomes a strong electromagnet. Magnetic coils are used in solenoids and relays.

Maximum power point tracking (MPPT)   A charge controller that is designed to change the position of the photovoltaic panels so that they harvest the maximum amount of electrical power at all times. This type of charge controller tracks the electrical maximum power point of a PV array and adjusts the position of the panels so they deliver the maximum available current for charging batteries or providing power to the grid.

Megawatt   A unit of power, equal to 1 million watts.

Milliampere   One thousandth (1/1000) of an ampere.

Motor starter   A large relay that has a single coil, multiple sets of contacts, and a set of overload heaters and contacts that are used to protect the motor against over current. The current that is supplied to the motor flows through the contacts of the motor starter and also flows through the heater element of the overload. When the current becomes too high for the motor to handle safely, the heat that is produced by the heating elements is sufficient to cause the overload contacts to "trip" and open. The contacts of the overload are connected in series with the motor starter coil, and when the overloads trip and open the contacts, the current supplied to the coil is interrupted so that the motor starter drops out its main contacts. When the main contacts open, the motor stops running.

Nanotechnology   A branch of engineering and science that deals with things that are smaller than 1,000 nanometers. A nanometer is 0.000,000,001 meter, which can also be defined as 1 times 10^{-9} meters. Nanotechnology is also called *nanotech* and it deals with the control of matter or devices at an atomic or molecular level.

National Electrical Code (NEC)   An electrical code written by the National Fire Protection Association (NFPA). Solar energy electrical systems and wiring in the United States must conform to the NEC. This includes all wiring in panels, switches, and controls. The NEC must also be followed for correct grounding of all electrical components and circuits to ensure personnel safety and prevent electrical shock hazards.

Net meter   A meter that runs in one direction when the solar panel is providing power to the grid and in the opposite direction when the application uses power from the grid rather than from the solar panel.

Neutron   The neutral part of an atom that resides in its nucleus.

Nickel-cadmium (NiCad) battery   A type of rechargeable battery that uses nickel oxide hydroxide and metallic cadmium as electrodes. The nickel-cadmium battery is commonly abbreviated as NiCad.

Nickel metal hydride battery   A battery made of nickel and metal hydride. This type of battery operates similarly to the NiCad battery in that it can be charged and discharged many times, but it is not as vulnerable to the so-called "memory effect." This means that nickel metal hydride batteries can be partially discharged and recharged, and during the next discharge, they are able to discharge completely and deeply without a problem.

Normally closed   A condition of switch contacts or relay contacts in which they have low resistance. A normal condition is considered to exist when no power is applied to the circuit.

Normally open   A condition of switch contacts or relay contacts in which they have high resistance. A normal condition is considered to exist when no power is applied to the circuit.

N-type material   Semiconductor material that has a negative charge and the majority of carriers are electrons. N-type silicon material is created by doping (mixing) the silicon (four valence electrons) with elements that contain one more valence electron than silicon does.

Nucleus   The center of an atom that contains the proton and neutron and has a neutral charge.

Open circuit   A circuit that has a point in it with extremely high resistance, which stops the flow of electrical current. The open can be created intentionally with a switch or contacts, or it can be created for safety reasons by a fuse or circuit breaker. Other fault conditions such as broken wires or broken components can also cause the high-resistance open.

Open fuse   A fuse that is bad because it has extremely high resistance and will not pass current. The open fuse is caused by a fault that causes high current that creates excessive heat through the fuse and causes its link to open.

Open loop   A type of control loop in which the output signal is controlled directly by the operator; this type of loop does not use a feedback signal.

Open switch   A switch that has high resistance between its contacts. When a switch is in the open condition, no current flows through it.

Over current protection   A fuse or circuit breaker that provides protection against too much current.

Overload   A part of a motor starter that consists of a heater element and a set of normally closed contacts.

Parallel circuit   A circuit that has more than one path for current to flow and two or more loads connected so that each load receives the same voltage. If two or more switches are connected in parallel, one switch can be opened and current can continue to flow through the other switches if they are closed.

Passive solar energy system   A system that collects heat and light by natural means with minimal use of mechanical and electrical devices, such as electrical controls, fans, or pumps to move the solar heat.

Passive solar heating A solar heating system that uses collectors on the roof and windows to collect heat energy. The heat energy is used directly or is moved to a storage area, which is usually made of rocks or dense material that can take on and release large amounts of heat energy quickly.

Peak electrical demand The largest amount of electrical energy that is needed during any hour at any time over a 24-hour period on any day of the week or month.

Peak sun hours The number of hours per day when solar energy produces more than 1kW per square meter of panel. The amount of sunlight does not have to be constant; rather it is the total hours that the sun energy exceeds that amount. The number of hours would be the same as if the sun shined continuously for that amount of time.

Periodic table of elements A chart or table that shows all of the elements with their abbreviations and the number of electrons and atomic weight of each element.

Photovoltaic array Three modules connected in series as a string and two strings connected in parallel to make a photovoltaic array.

Photovoltaic cell Cell that converts solar energy into DC electricity.

Photovoltaic module A grouping of photovoltaic cells that is assembled into a finished component (solar panel) that can be installed as a unit.

Photovoltaic panel A grouping of photovoltaic cells that is assembled into a finished component.

PN material Semiconductor material that is made from P-type material and N-type material.

Power circuit The part of an electrical circuit that has motors or other power loads.

Power electronic frequency converter An electronic circuit that changes DC voltage to AC voltage and provides a set frequency such as 60 Hz.

Power factor (PF) The ratio of true power (TP) to apparent power (AP). The formula for power factor is TP/AP; since true power is always less than apparent power, the power factor is always less than 1.00. The closer the PF is to 1.00, the more efficient the circuit is and the less the losses are.

Power quality The condition of electrical power provided to a residence, industrial application, or the grid with regard to frequency and voltage level. Power quality standards require the voltage to remain at the proper frequency and voltage level to within a small percentage. Events such as voltage sags, impulses, harmonics, and phase imbalance are not tolerated, and the system causing these problems must be either corrected or disconnected.

Pressure control A hydraulic control that adjusts the pressure of a system by allowing a portion of the hydraulic fluid to return to the tank and relieve the pressure of the system.

Primary winding One of two transformer windings by which voltage is taken into the transformer. Supply voltage is connected to the primary transformer winding.

Process variable (PV) The variable that is being sensed by the process sensors, such as temperature, pressure, or flow.

Program mode A mode for a programmable logic controller (PLC) that does not execute the scan cycle.

Programmable logic controller (PLC) A solid-state control system that continually scans its user program. The controller has a user-programmable memory for storage of instructions to implement specific functions.

Project development plan A plan that includes selecting the proper site for the solar energy system or solar energy farm; identifying available solar energy resources; predicting energy output for various-sized solar energy systems; identifying land owners and developing landowner agreements; identifying grid interconnection if necessary and utility companies that control the grid; and identifying government agencies that control the country, state, county, or city regulations that will affect your selection.

Proton The positive part of an atom that resides in the nucleus of the atom.

P-type material Semiconductor material in which a majority of carriers are holes that have a positive charge.

Public Utility Regulatory Policies Act (PURPA) A federal law passed in 1978 as part of the National Energy Act. This law requires utility companies to purchase power from independent providers, but the law does not determine the rate that should be paid for the power.

Pulse-width modulation (PWM) This type of control uses transistors to convert DC power to AC power by turning them on and off to provide a variety of AC frequencies. This circuit provides a unique waveform that makes multiple square-wave cycles that are turned on and off at specific times to give the overall appearance of an AC sine wave. The outline of the waveform actually looks very similar to the six-step inverter signal.

Push-pull converter A switch-mode power supply that gets more power from smaller components and has a higher efficiency. This power supply uses a center-tapped transformer so that both the top and bottom half-cycles are utilized. The primary winding of the transformer in this circuit is controlled by two transistors, which allows one of them to conduct during each half-cycle, so the output is receiving voltage directly through one of them at all times. The efficiency of this converter is approximately 90%.

Reactive power Electrical power produced by current flow through a capacitor or inductor.

Rectifier A device (usually a diode) that conducts current in only one direction, thereby transforming alternating current (AC) to direct current (DC).

Relay An electrical device that consists of a single coil and one or more sets of normally open or normally closed contacts.

Reserve capacity An important rating to help you identify how much amperage the battery can deliver on a continual basis. The reserve capacity is the number of minutes a fully charged battery will discharge 25 A on a continual basis at 80°F until its charge drops below a specific level. For a 12-V battery, voltage level is below 10.5 V.

Residential solar energy system A solar energy system specifically designed for residential applications.

Resistance The opposition of a body or substance to current passing through it, resulting in a change of electrical energy into heat or another form of energy.

Resistive load An electrical load that has electrical resistance. When current passes through the resistance, it creates heat.

Return on investment (ROI) The amount of money or other items of value that become available after money is invested into a project. For example, if a solar project costs $100,000 to install and start up, the money that the investors receive from the energy the solar project produces over a specific period of time is the ROI.

Reverse Bias A condition in which a negative electrical charge is applied to the positive terminal of a diode or any other PN material and causes a high-resistance condition at its PN junction, which stops current from flowing through it.

Rotary actuator A hydraulic, pneumatic, or electric actuator that has a shaft that is turned to provide rotary power.

Rotor The rotating part of a motor.

SCADA (Supervisory Control and Data Acquisition) A data collection system that provides information from operating solar energy farms such as sun direction, time of day, date, and the amount of electrical power that is being produced. When these data are gathered over a long period of time and stored in a computer data bank, they can be analyzed and very accurate projections can be developed.

Secondary winding A transformer has two windings, a primary winding and a secondary winding. Voltage is applied to the primary winding, and voltage comes out the secondary winding. The amount of voltage at the secondary winding is proportional to the number of turns in the primary and secondary windings.

Semiconductor Elements that have exactly four valence electrons and can be altered by combining other material with them in a process called doping, and the material that is added to the semiconductor material can make the new material either an insulator or a conductor.

Sensor A device that detects or measures something and generates a corresponding electrical circuit to an input circuit or controller.

Series circuit A circuit that has only one path. Whenever there is an open anywhere in a series circuit, all current flow stops.

Series DC Motor A DC motor that has its field winding and series winding connected in series.

Series-parallel circuit A circuit that has components connected both in series and in parallel.

Service drop The wires that connect from a building to the main power lines.

Servo system An automated control system that has a set point, a feedback signal called the process variable, a summing junction, and output signal. The output signal changes the output in relationship to the error, which is the difference between the set point and the feedback signal. Servo systems are used to control solar tracking systems.

Set point The setting for the maximum speed at which the blades can rotate.

Shell Another term for the orbit that an electron travels around an atom.

Shunt DC motor A DC motor that has its field winding connected in parallel with its series winding.

Single axis solar tracking system An electromechanical system that moves the face of a solar panel on one axis so that it tracks the sun and the solar panel receives the maximum amount of sunlight.

Six-step inverter An inverter that use six transistors to provide three-phase AC sine waves. One transistor conducts during the positive half-cycle and the other transistor conducts during the negative half-cycle. A pair of the transistors is used to produce each of the three AC phases.

SLI (starting, lighting, and ignition) battery A battery that is essentially an automotive battery that is designed to give a lot of current during starting, but then can be recharged immediately. The SLI battery can withstand several starts per hour if necessary and yet receive the recharge current from the alternator. This battery works well with solar energy systems.

Slip The difference between the rated speed and the actual speed of the armature (rotor) of an AC motor.

Smart grid Update of the present-day grid in which information will flow in both directions on the grid. These data will be able to be recorded and analyzed so that predictions can be made about the times energy will be consumed in the future.

Solar charge controller An electronic control system that controls the charge level of batteries. The charge controller for a solar photovoltaic system will control the rate of charge on the batteries and ensure that they do not completely discharge.

Solar energy Energy derived from the sun in the form of solar radiation.

Solar energy farm A location where a large number of solar panels are mounted.

Solar PV panel A complete assembly made of solar photovoltaic cells that are laminated by glass to protect them and encased in an aluminum frame that makes them ready for installation.

Solar tube lighting A solar lighting fixture that consists of a bubble lens that is mounted on the roof or exterior of a building and allows light to travel through the tube to an opening at the other end that is placed in the ceiling of a room. This allows solar light to be brought into a room that does not have any exterior walls for window locations.

Solar site assessment A preliminary report that consists of an evaluation of a potential solar energy site that includes the amount of solar energy available and the number of hours per day solar panels can produce electricity, proximity to existing electrical power lines for connection to the grid, and other information about the land and any other factors that would be positive or negative in regard to the acceptability of the site.

Specific gravity The density of a substance relative to the density of water. The specific gravity of water is rated as 1.000 and the specific gravity of pure sulfuric acid is 1.850. This means that when a mixture of 35% sulfuric acid and 65% water solution is used, and the battery is fully charged, the specific gravity should be 1.265; when the battery is discharged, the specific gravity should be close to 1.155.

State of charge The depth of discharge in a battery.

Step down transformer A transformer in which the amount of secondary voltage is smaller than the primary voltage.

Stepper motor A motor that is designed so that its shaft turns a specific number of degrees when voltage is applied. The stepper motor controller provides the voltage to the motor in steps to cause the shaft to turn a specific number of degrees.

Step up transformer   A transformer in which the amount of secondary voltage is larger than the primary voltage.

Stirling solar engine   Solar energy is used to heat gas and the energy in the gas is transmitted to a Stirling engine, whose pistons go up and down and cause a crankshaft to rotate and turn a generator.

Strings   Photovoltaic cells that are put together in multiple groups and are made into larger sheets before they are assembled into the solar panels. Each string has a specific voltage rating that the panel will produce when solar light is striking the cells. The strings are cut into sheets that are the same size as the tempered glass and they will fit into the solar panel frame.

Sub-array   A string of solar panels.

Substation   A small group of transformers located close to the location where solar panels create electrical power. The transformers at these substations step up the voltage to a level so it is large enough to transmit to the place where it will be used.

Summing junction   A point in a control algorithm or closed-loop control system where the set point is compared to the feedback signal (process-variable sensor). The difference between the set point and the feedback signal is called the *error*.

Surge protection   A protection circuit that is designed to take any voltage that is above a safe level and route it directly to a ground, where it can be dissipated harmlessly.

Switch   An electrical device with one or more sets of contacts. When the contacts are closed, current can flow through; when the contacts are open, the flow of current is interrupted.

Switch-mode power supplies (SMPS)   A power supply that uses electronic components to turn off and on at a high frequency. The switch mode power supply is more efficient than traditional power supplies.

Technical data sheet   The data sheet that is packed with each solar panel and shows all of the detailed parameters about it. The module parameters include the peak power, the rated voltage and rated current, a short-circuit current, an open-circuit voltage, and a nominal voltage, which are all established based on a Standard Test Condition of 1,000 watts per square meter (w/m^2) or irradiance (light shining on the panel), with an air mass of 1.5 and a cell temperature of 25°C. This test standard is used for every model of solar photovoltaic panel that is manufactured.

Tempered glass   The plate glass on the front side of a solar panel that is usually made of low-iron glass. Tempered glass is much stronger than standard glass and provides substantial protection to the solar cells in the solar panel.

Three-phase voltage   AC voltage that is supplied in three distinct circuits that are 120° apart. Each voltage source is called a *phase*.

Transformer   An electrical device with two coils (primary winding and secondary winding) that are located in close proximity. Voltage is applied to the primary winding and voltage is taken off the secondary winding.

Transistor   A three-terminal electronic device that has an emitter, collector, and base.

Transmission companies (Transco's)   Individual companies that manage the hardware and own the actual transmission lines, towers, transformers, and switch gear on the grid.

Trickle charge   Charging a battery when a small amount of current is allowed to flow during the recharge process.

Troubleshooting   Discovering and eliminating the cause of trouble in equipment. A process of finding faults in an electrical solar energy system. The troubleshooting process works best when you are able to identify everything in the system that is working correctly and then check only the remaining part of the circuit that is not working correctly.

Troubleshooting matrix   A table that shows the sequence of operations in troubleshooting.

True power   The voltage and current that flow only through resistive loads. Also called *real power*. True power is measured in watts and does not have any current from capacitive or inductive components.

Turn key project   A project that is completed to the extent it can be turned over to the owner and operated without any further work.

Turns ratios   The ratio of the number of turns in the primary winding as compared to the number of turns in the secondary winding of a transformer.

Two-axis solar tracking system   An electromechanical system that moves the face of a solar panel on two axes so that it tracks the sun and the solar panel receives the maximum amount of sunlight.

Two-stage battery charging   Battery charging that consists of a high rate of charge when the battery is discharged, and a trickle charge when the battery is nearly fully charged. The trickle charge continually keeps the battery at near full charge.

Underground transmission lines   Underground lines that carry electric energy from one point to another in an electric power system.

Underwriters Laboratory (UL)   An independent product safety certification organization that tests products and writes standards for safety.

Uninterruptible power supply (UPS)   A power supply that is backed by one or more batteries and an inverter that converts DC voltage to AC voltage. An uninterruptible power supply takes in AC voltage and converts it to DC voltage, which goes into the bank of batteries. When needed the electricity is then converted back to AC electricity that has a frequency of 60 Hz.

Valence electron   An electron in the outermost shell of an atom.

Valence shell   The outermost shell in an atom.

Valve-regulated lead-acid battery (VRLA)   A battery that is designed to limit the amount of electrolyte that can leak from the battery if it is inverted. This feature makes the battery require less maintenance and it does not require the addition of water to the cells as often. Absorbed glass mat battery and a gel battery (gel cell) are examples of this type of battery.

Voltage   The electrical force that causes electrons to move as current flow. It is also called electrical potential or electromotive force (EMF).

Voltage drop   The voltage that occurs when current flows through a resistor or load in an electrical circuit. In a series circuit, all of the voltage drops across each resistor can be added, and their total will equal the supply voltage to the circuit.

Voltage fading   A condition when voltage drops below a specified level. Typically occurs when large amounts of electrical power are consumed.

Volt-ampere (VA)   A rating for a transformer found by multiplying the voltage times the current (amperes). This rating does not take into consideration any impedance or reactance losses, so it is an apparent power rating.

Volts   A unit of electrical force given to electrons in an electrical circuit.

Volts AC (VAC)   Electrical force in an AC circuit.

Volts DC (VDC)   Electrical force in a DC circuit.

Watt hour   A unit of energy equal to one watt consumed over one hour. The electric utilities use this unit of measure to record the amount of energy used by a consumer and applies a rate to determine the electric bill. Typically the electric bill is measured in 1,000 watt hours (a kilowatt).

Wet cell battery   A battery that has a liquid electrolyte that covers the cells.

Wiring diagram   An electrical diagram that shows the location of all the electrical components in a circuit and the wires that connect to them.

Wound Rotor   A rotor in a motor that is made of coils of wire instead of laminated plates.

Wye-connected three-phase transformer   A three-phase transformer that has one end of each of its three coils connected at a point so that the overall shape of the windings looks like the letter "Y."

ACRONYMS

AC	Alternating Current	ITC	Investment Tax Credit
AEP	American Electric Power	KSC	Kennedy Space Center
AGM	Absorbed Glass Mat Battery	kW	Kilowatt
ANSI	American National Standards Institute	KWp	Kilowatt peak power
AP	Apparent Power	LED	Light-Emitting Diode
BIPV	Building Integrated Photovoltaics	LRA	Locked-Rotor Amperage
BP	British Petroleum	LVRT	Low-Voltage Ride Through
CA	Cranking Amps	MAAC	Mid-Atlantic Area Council
CAD	Computer-Aided Design or Computer Aided Drafting	MACRS	Modified Accelerated Cost Recovery System
CCA	Cold Cranking Amperage	MCA	Marine Cranking Amps
CdTe	Cadmium Telluride	MAIN	Mid-America Interconnected Network
CEMF	Counter Electromotive Force	MAPP	Mid-Continent Area Power Pool
CIGS	Copper Indium Gallium Selenide	MOSFET	Metal Oxide Semiconductor Field Effect Transistor
CIS	Copper Indium Selenium	MPPT	Maximum Power Point Tracking
CNC	Computer Numerical Control	NABCEP	North American Board Certified Energy Practitioner
CPV	Concentrated Photovoltaic		
CREBS	Clean Renewable Energy Bonds	NASA	National Aeronautical And Space Administration
CSI	Current-Source Input	NC	Normally closed
CSP	Concentrated solar power	NCPV	National Center For Photovoltaics
CT	Current Transformer	NEC	National Electrical Code
DC	Direct Current	NEMA	National Electrical Manufacturers Association
DER	Distributed Energy Sources	NFPA	National Fire Protection Association
DG	Distributed Generation	NiCad	Nickel Cadmium
DOD	Depth of Discharge	NO	Normally open
DOE	Department of Energy	NPCC	Northeast Power Coordinating Council
DPDT	Double-Pole Double-Throw	NREL	National Renewable Energy Laboratory
DPST	Double-Pole Single-Throw	PET	Polyethylene Terephthalate
DSIRE	Database of State Incentive for Renewables and Efficiency	PF	Power Factor
DSSC	Dye Sensitive Solar Cells	PLC	Programmable Logic Controller
ECAR	East Central Area of Reliability Coordination Agreement	PLL	Phase Locked Loop
		PSEG	Public Service Enterprise Group
EIA	Energy Information Administration	PTC	Production Tax Credit
EMF	Electromotive force	PURPA	Public Utility Regulatory Policies Act
EMI	Electromagnetic Interface	PV	Photovoltaic
EPS	Electric Power Systems	PV	Photovoltaic Cell
ERCOT	Electric Reliability Council of Texas	PV	Process Variable
EVA	Ethylene Vinyl Acetate	PVB	Polyvinyl-butiral
FEMP	Federal Energy Management Program	PWM	Pulse-Width Modulation
FEPA	Federal Energy Policy Act	RC	Reserve Capacity
FRCC	Florida Reliability Coordinating Council	REAP	Rural Energy for America Program
GaAs	Gallium Indium	REPI	Rural Energy Production Incentive
GaInAs	Gallium, Indium, and Arsenide	ROI	Return on Investment
GaInP	Gallium, Indium, and Phosphorous	SCADA	Supervisory Control and Data Acquisition
GPS	Global Positioning System	SCR	Silicon Controlled Rectifier
HCA	Hot Cranking Amps	SEGS	Solar Energy Generating Systems
GWh	Gigawatt-Hours	SEIA	Solar Energy Industries Association
HCPV	High Concentrated Photovoltaic	SERC	Southeastern Electric Reliability Council
HEV	Hybrid Electrical Vehicle	SLA	Sealed Lead-Acid Battery
HPF	High-Powered Flooded Battery	SLI	Starting, Lighting, and Ignition Battery
HVAC	Heating Ventilating and Air-Conditioning	SMPS	Switch-Mode Power Supply
IC	Integrated Circuit	SNL	Sandia National Lab
IEC	International Electrotechnical Committee	SP	Set Point
IEEE	Institute of Electrical and Electronics Engineers	SPDT	Single-Pole Double-Throw
IGBT	Insulated Gate Bipolar Transistor	SPP	Southwest Power Pool
InAlN	Indium, Aluminum Nitride	SPST	Single-Pole Single-Throw
IREC	Interstate Renewable Energy Council	SRV	Sealed Regulated Valve Battery
		TB	Terminal Board

TP	True Power
TRE	Texas Regional Entity
UL	Underwriter Laboratories
UPS	Uninterruptible Power Supply
USPS	United States Postal Service
UV	Ultraviolet

VA	Volt Ampere
VAR	Volt Ampere Resistance
VOM	Volt Ohm Milliameter
VRLA	Valve Regulated Lead-Acid Battery
VVI	Variable-Voltage Input
WSCC	Western Systems Coordinating Council

INDEX